AF356532

Modelling and Simulation of Sheet Metal Forming Processes

Modelling and Simulation of Sheet Metal Forming Processes

Special Issue Editors

Marta C. Oliveira
José Valdemar Fernandes

MDPI • Basel • Beijing • Wuhan • Barcelona • Belgrade • Manchester • Tokyo • Cluj • Tianjin

Special Issue Editors

Marta C. Oliveira José Valdemar Fernandes
University Coimbra University Coimbra
Portugal Portugal

Editorial Office
MDPI
St. Alban-Anlage 66
4052 Basel, Switzerland

This is a reprint of articles from the Special Issue published online in the open access journal *Metals* (ISSN 2075-4701) (available at: https://www.mdpi.com/journal/metals/special_issues/modelling_simulation_metal).

For citation purposes, cite each article independently as indicated on the article page online and as indicated below:

LastName, A.A.; LastName, B.B.; LastName, C.C. Article Title. *Journal Name* **Year**, *Article Number*, Page Range.

ISBN 978-3-03928-556-3 (Pbk)
ISBN 978-3-03928-557-0 (PDF)

© 2020 by the authors. Articles in this book are Open Access and distributed under the Creative Commons Attribution (CC BY) license, which allows users to download, copy and build upon published articles, as long as the author and publisher are properly credited, which ensures maximum dissemination and a wider impact of our publications.

The book as a whole is distributed by MDPI under the terms and conditions of the Creative Commons license CC BY-NC-ND.

Contents

About the Special Issue Editors

Marta C. Oliveira has been a Professor in the Department of Mechanical Engineering of the University of Coimbra since 2005. Her research interests mainly focus on applied and computational mechanics, in particular the numerical simulation of sheet metal forming processes, including the modelling of the metallic sheets constitutive behavior, algorithms for the treatment of contact with friction problems, and optimization techniques. She has been the coordinator of the Doctoral Program in Mechanical Engineering of the University of Coimbra since 2015. She has been involved in several research projects in the previously mentioned areas, three of them as coordinator, and is co-author of approximately 130 publications in international journals.

José Valdemar Fernandes is Full Professor at the University of Coimbra. His main research activities in recent years concern: (i) large plastic deformations: fundamental aspects, modelling, inverse analysis and applications to metal forming; and (ii) carbon nanotubes and their composites: production and evaluation and modelling of the mechanical properties. He was at the Foundation of the Centre for Mechanical Engineering, Materials and Processes (CEMMPRE) of the University of Coimbra (formerly Centre for Mechanical Engineering of the University of Coimbra (CEMUC)) where he was director for 15 years. He was President of the Portuguese Society for Microscopy. He has coordinated a dozen externally funded projects and participated in more than two dozen all in the areas mentioned above, and is co-author of around 100 publications in international journal.

Preface to "Modelling and Simulation of Sheet Metal Forming Processes"

The numerical simulation of sheet metal forming processes has become an indispensable tool for the design of components and their forming processes. This role was attained due to the huge impact in reducing time to market and the cost of developing new components in industries ranging from automotive to packing, as well as enabling an improved understanding of the deformation mechanisms and their interaction with process parameters. Despite being a consolidated tool, its potential for application continues to be discovered with the continuous need to simulate more complex processes, including the integration of the various processes involved in the production of a sheet metal component and the analysis of in-service behavior. The quest for more robust and sustainable processes has also changed its deterministic character into stochastic to be able to consider the scatter in mechanical properties induced by previous manufacturing processes. Faced with these challenges, this Special Issue presents scientific advances in the development of numerical tools that improve the prediction results for conventional forming process, enable the development of new forming processes, or contribute to the integration of several manufacturing processes, highlighting the growing multidisciplinary characteristic of this field.

Marta C. Oliveira, José Valdemar Fernandes
Special Issue Editors

Editorial

Modelling and Simulation of Sheet Metal Forming Processes

Marta C. Oliveira * and José V. Fernandes *

CEMMPRE, Department of Mechanical Engineering, University of Coimbra, Rua Luís Reis Santos, Pinhal de Marrocos, 3030-788 Coimbra, Portugal
* Correspondence: marta.oliveira@dem.uc.pt (M.C.O.); valdemar.fernandes@dem.uc.pt (J.V.F.)

Received: 12 December 2019; Accepted: 15 December 2019; Published: 17 December 2019

1. Introduction and Scope

Numerical simulation of sheet metal forming processes has become an indispensable tool for the design of components and their forming process, in industries ranging from the automotive, to the aeronautics, packing and household appliances. The strong contribution of virtual try-out to reduce the time-to-market and the cost of developing new components has been the main promoter for its extended application, along with the increasing computational power. The Finite Element Method (FEM) is the main numerical tool used in this context.

Nowadays, the automotive industry continues to drive the development of numerical simulation of sheet metal forming processes due to the strong environmental and safety standards that have led to the development of materials with better strength-to-weight ratio or new forming processes. Newly introduced materials allow to produce components from thinner sections, while maintaining satisfactory strength and stiffness, which ultimately results in a reduction of the overall structure mass, a crucial step to meet the ever-stringent standards of passenger safety and gas emissions [1]. However, as is well known, the increased mechanical strength of metallic materials is usually accompanied by a reduction of their ductility. This poses new challenges for predicting forming defects, leading to alternative strategies to the Forming Limit Diagram (FLD) concept. Moreover, the research effort has also been focused on the development of numerical models that enable the virtual try-out of forming processes, involving non-isothermal temperature environments or high strain rates, to try to explore the best formability of the material under these conditions. This requires an enhanced modelling of the material behaviour as well as process conditions, which demands for an improved integration between experimental and numerical analysis. The aim is to be able to integrate the several processes involved in the production of a sheet metal component in the virtual analysis of the forming processes and in-service behaviour. The analysis of the post-forming behaviour should also take into account the scatter in mechanical properties induced by the manufacturing processes. These research trends are reflected in the papers published in this issue, as analysed in the following section.

2. Contributions to the Special Issue

Researchers were invited to submit innovative research papers on modelling and numerical simulation of sheet metal forming processes. Thirteen research papers were published in this Special Issue of *Metals*, entitled "Modelling and Simulation of Sheet Metal Forming Processes", which highlight some of the research trends in the field [2–14].

The FEM virtual try-out for cold forming components is commonly supported by the concept of forming limit diagrams (FLDs), introduced to characterize the ductility of metal sheets [15,16]. The experimental determination of FLDs involves performing various mechanical tests on metal sheets with different samples geometries to reproduce a certain range of monotonic loading paths. However, this experimental approach is both expensive and time consuming, particularly at high temperatures.

For these reasons, different theoretical approaches have been develop for the prediction of FLDs, among which the Marciniak–Kuczynski (M–K) theory [17] is the most widely applied. In their paper, Fan et al. [10] used the M-K theory to predict the FLD for a TA32 alloy, for a temperature range between 700 °C and 800 °C, taking into account strain rate effects. The parameters of the thermomechanical hardening law considered in the M-K theory were fitted using experimental results from hot tensile tests. The results from these tests are also used to identify the normal anisotropy coefficient and select the Logan–Hosford yield criterion that allows a better correlation between the theoretical and experimental FLD, determined using the Nakazima test. Also, the theoretical FLD is used to study the initial blank shape of a component produced by Hot Press Forming (HPF), showing that the FLD concept can also be extended to this type of process.

The HPF technology is widely used to prevent formability problems and reduce springback. In addition to the parameters of conventional cold press forming, the blank temperature, the strain rate and the quenching methods also affect the formability and complicate the analysis of hot forming processes. Seo et al. [5] performed direct and indirect hot press forming of a ultra-high-strength steel (UHSS) boron steel, 22MnB5, considering different initial blank temperatures and blank-holding forces, in order to evaluate the formability but also the mechanical properties of the material after the forming process. The knowledge of such properties is very important to predict the failure in UHSS sheet metals during a car crash. Bayat et al. [7] propose the use of the M-K theory to predict the FLD for a 22MnB5, taking into account the scatter in material properties at different regions of formed components. Therefore, the hardening behaviour is characterized by performing uniaxial tensile tests of specimens extracted from structural components of a car, enabling the definition of a range for the hardening law parameters. In this context, Bayat et al. [7] suggest replacing the single FLD by a band of forming limits, using statistical approaches to calibrate its bounds.

It is well established that ductile failure in metals occurs due to the presence of defects such as voids and micro-cracks [18]. On the macroscopic scale, damage is observed as the degradation of material properties, e.g., the elastic stiffness, the yield stress or other measurable material properties. This is the approach adopted in continuous damage mechanics, which introduces an internal damage variable to be able to predict the ductile fracture. Cherouat et al. [11] coupled the damage potential, introduced by Lemaitre [19], with an elasto-visco-plastic material model in order to predict the onset of ductile damage for different forming processes. The occurrence of large inelastic deformations commonly implies a severe distortion of the computational domain, whose boundary is also altered by the elimination of the fully damaged elements. In this context, the authors propose a 3D adaptive remeshing scheme, for linear tetrahedral finite element, to enable tracking the evolution of large plastic deformations. The proposed model is used to analyse different processes, including blanking, multi-point and incremental forming and deep drawing of a front door panel. The results highlight the importance of the coupling between elastoplastic and damage behaviour on the damage evolution at large plastic deformations, but also of using remeshing techniques to assure the computational efficiency as well as to avoid convergence problems. Yue et al. [9] also used a coupled damage model to analyse the influence of kinematic hardening and ductile damage on springback prediction. In order to study the influence of the kinematic hardening, experimental three-point bending tests were performed for the AA7055 aluminium alloy, with specimens submitted to uniaxial tension until different pre-strain levels. The results show that both kinematic hardening and ductile damage influence the springback prediction, particularly for non-proportional strain paths.

The accuracy of the numerical results of sheet metal forming processes depends of the constitutive model selected for describing the material behaviour. In general, FEM analysis of complex forming processes is performed with phenomenological models. This was the approach adopted by Thuillier et al. [6] to describe the bake hardening effect, which is a thermal induced phenomenon that is widely explored, in particular by the automotive industry. Thuillier et al. [6] performed an experimental characterization of this effect for a low carbon steel (E220BH) and proposed a phenomenological model for its description. The specimens used in the experimental approach are pre-strained using a hydraulic bulge test device and a dedicated equipment was designed to characterize the dent resistance. The phenomenological model was validated through the numerical simulation of this multi-step process, i.e., bulge followed by dent at the pole by a vertical movement of a hemispherical punch.

The use of phenomenological models implicitly includes the strategy for identifying the model parameters, which is generally seen as an optimization problem [6,9,11,13]. Forming processes involving non-isothermal temperature conditions and/or high strain rates require constitutive models with more parameters. In this context, the Johnson–Cook hardening law was used to study the gas detonation forming [2], which is a high-speed forming process, with the potential to form complex geometries, including sharp angles and undercuts. The authors neglected the thermal softening, but included damage evolution [2]. Cherouat et al. [11] adopted the same hardening law only to take into account the strain rate effect. These authors used an inverse approach to identify the constitutive parameters, suggesting the use of the stage up to the maximum load of uniaxial tensile tests for the identification of the plastic parameters, while the stage after the maximum load is used only to identify the damage parameters. The results highlight that the damage evolution is sensitive to the element size, which can be mitigated by the adoption of the proposed remeshing technique. In fact, the identification of coupled damage models parameters still poses many challenges, as also mentioned in the work of Yue et al. [9].

Electromagnetic forming (EMF) is another widely used high-speed forming process, in which the deformation is promoted by the application of a magnetic force. Cui et al. [12] developed a three-dimensional (3D) sequential coupling method to analyse the electromagnetic uniaxial tensile test using a runway coil. As the magnetic force field depends on the specimen deformation, the mechanical and the electromagnetic problems must be coupled whenever the process involves large deformations. Sequential coupling allows the analysis of the influence of process parameters, such as tools conductivity, relative coil position and discharge voltages, for the strain paths observed in the specimen. In this context, numerical simulation is used to support the development of an experimental procedure and improve knowledge concerning the analysis of results.

Over the years, several benchmarks have been proposed to analyse the influence of the constitutive model and numerical strategies on formability and/or springback predictions, in particular within the NUMISHEET conference series. The "Benchmark 2 - Springback of a Jaguar Land Rover Aluminium" [20] is consider in the work by Mulidrán et al. [3], to analyse the influence of the yield criteria on springback prediction. The results show that the use of more advanced yield functions may improve the results accuracy. Nevertheless, the identification of the anisotropy parameters of these type of yield criteria requires experimental data covering a wide range of stress/strain paths. In the collected works, the characterization of the mechanical behaviour of the metallic sheets was mainly performed with uniaxial tensile tests [3,4,6,7,10,13], although some researchers have also resorted to shear tests [3,9,11]. Simões et al. [8] presented a numerical study that contributes for the understanding of the mechanical phenomena that occur in the material under Knoop indention, enhancing and simplifying the analysis of the results obtained in Depth-sensing indentation tests. This hardness test is particularly attractive for the determination of the near-surface properties, the characterization of brittle materials and post-forming properties; also, it is sensitive to the indenter orientation, making it a useful tool to analyse the materials anisotropy.

Besides the unquestionable advantages of applying numerical simulation in forming process design, numerical models also allow for better understanding of the influence of process parameters. In this context, the work by Neto el al. [13] focused on the effects of the geometry and dimensions of the forming tools on the formability and final thickness distribution of metallic thin stamped bipolar plates (BPPs) for fuel cells. Rufini et al. [4] analysed the influence of several geometrical characteristics and process parameters on the thinning prediction in the bending process of stainless steel pipes. They concluded that the ratio between the curvature radius and the pipe diameter dictates the failure or success of the operation, which is in agreement with empirical knowledge and the experimental results.

Despite the increasing computational power, the numerical analysis of complex geometries, involving features with small curvature still poses challenges. In this context, the adoption of simplifying assumptions can help to understand specific details of the process. This approach is commonly adopted, for instance, in the BPPs formability analysis, where most of the numerical studies of the stamping process reported in the literature consider plane strain conditions. This formability analysis can be complemented by studying specific zones of the BPPs plates, in particular the area including a U-bend channel section [13]. Process conditions can also be simplified, without compromising the numerical results accuracy. This approach was adopted by Patil et al. [2] to simulate the gas detonation forming of a cylindrical cup, by directly applying the detonation pressure as a load in the finite element (FE) model. This requires the proper experimental acquisition of the averaged pressure evolution during the process, close to the blank. The numerical model enables the analysis of the influence of the peak pressure on the damage prediction. However, the fracture prediction requires the modelling of the complete blank, to account for improper alignment of the blank in experiments. Also, in the case of the bending process of stainless steel pipes, the authors report that some discrepancies between experimental and numerical results maybe related with the presence in the experimental tests of an additional support element on the machinery, which was not contemplated in the simulation model [4]. These results highlight the importance of an accurate analysis of the experimental process conditions, to enable a proper analysis of the numerical simulation results. This requires the validation of the model using experimental results, which can be difficult in the early design stages for large size components. Tomáš et al. [14] propose the use of the similitude theory to help engineers to select the blank material using a scaled model. In their study, this theory was applied to study the deep drawing process of a bathtub made from cold rolled low carbon aluminium-killed steel, using both a numerical and a physical model. The comparison between numerical and experimental thickness variations, along some predefined sections, is used to select the constitutive model that enables a better description of the material mechanical behaviour.

3. Conclusions and Outlook

The Special Issue "Modelling and Simulation of Sheet Metal Forming Processes" presents a collection of research articles covering the relevant topics in the field in innovative ways. The guest editors are aware of the quality of the contributions and hope that this collection of works may be useful to researchers working in the field, promoting more research studies, debates, and discussions that will continue to bridge the gap between physical and virtual reality.

Acknowledgments: The guest editors would like to thank all who have contributed directly and indirectly for the successful development of this Special Issue. Thanks to all the authors who submitted their manuscripts and were willing to share results of their research activities in this Special issue. Special acknowledgements are due to reviewers who agreed to revise the articles and provide feedback to improve the quality of the manuscripts. Credits should also be given to the editors and to the Assistant Editor Kinsee Guo as well as to all the staff of the Metals Editorial Office for their contribution and support in the publication process of this issue.

Conflicts of Interest: The authors declare no conflicts of interest.

References

1. Badreddine, H.; Labergère, C.; Saanouni, K. Ductile damage prediction in sheet and bulk metal forming. *Comptes Rendus Mec.* **2016**, *344*, 296–318. [CrossRef]
2. Patil, S.P.; Prajapati, K.G.; Jenkouk, V.; Olivier, H.; Markert, B. Experimental and numerical studies of sheet metal forming with damage using gas detonation process. *Metals* **2017**, *7*, 556. [CrossRef]
3. Mulidrán, P.; Šiser, M.; Slota, J.; Spišák, E.; Sleziak, T. Numerical prediction of forming car body parts with emphasis on springback. *Metals* **2018**, *8*, 435. [CrossRef]
4. Rufini, R.; Di Pietro, O.; Di Schino, A. Predictive simulation of plastic processing of welded stainless steel pipes. *Metals* **2018**, *8*, 519. [CrossRef]
5. Seo, H.Y.; Jin, C.K.; Kang, C.G. Effect on blank holding force on blank deformation at direct and indirect hot deep drawings of boron steel sheets. *Metals* **2018**, *8*, 574. [CrossRef]

6.   Thuillier, S.; Zang, S.L.; Troufflard, J.; Manach, P.Y.; Jegat, A. Modeling bake hardening effects in steel sheets—Application to dent resistance. *Metals* **2018**, *8*, 594. [CrossRef]

7.   Bayat, H.R.; Sarkar, S.; Anantharamaiah, B.; Italiano, F.; Bach, A.; Tharani, S.; Wulfinghoff, S.; Reese, S. Modeling of forming limit bands for strain-based failure-analysis of ultra-high-strength steels. *Metals* **2018**, *8*, 631. [CrossRef]

8.   Simões, M.I.; Antunes, J.M.; Fernandes, J.V.; Sakharova, N.A. Numerical simulation of the depth-sensing indentation test with knoop indenter. *Metals* **2018**, *8*, 885. [CrossRef]

9.   Yue, Z.; Qi, J.; Zhao, X.; Badreddine, H.; Gao, J.; Chu, X. Springback prediction of aluminum alloy sheet under changing loading paths with consideration of the influence of kinematic hardening and ductile damage†. *Metals* **2018**, *8*, 950. [CrossRef]

10.   Fan, R.; Chen, M.; Wu, Y.; Xie, L. Prediction and experiment of fracture behavior in hot press forming of a TA32 titanium alloy rolled sheet. *Metals* **2018**, *8*, 985. [CrossRef]

11.   Cherouat, A.; Borouchaki, H.; Jie, Z. Simulation of sheet metal forming processes using a fully rheological-damage constitutive model coupling and a specific 3D remeshing method. *Metals* **2018**, *8*, 991. [CrossRef]

12.   Cui, X.; Zhang, Z.; Yu, H.; Cheng, Y.; Xiao, X. Dynamic uniform deformation for electromagnetic uniaxial tension. *Metals* **2019**, *9*, 425. [CrossRef]

13.   Neto, D.M.; Oliveira, M.C.; Alves, J.L.; Menezes, L.F. Numerical study on the formability of metallic bipolar plates for proton exchange membrane (PEM) fuel cells. *Metals* **2019**, *9*, 810. [CrossRef]

14.   Tomáš, M.; Evin, E.; Kepič, J.; Hudák, J. Physical modelling and numerical simulation of the deep drawing process of a box-shaped product focused on material limits determination. *Metals* **2019**, *9*, 1058. [CrossRef]

15.   Keeler, S.P.; Backofen, W.A. Plastic instability and fracture in sheets stretched over rigid punches. *Trans. Am. Soc. Met.* **1963**, *56*, 25–48.

16.   Goodwin, G.M. Application of strain analysis to sheet metal forming problems in the press shop. *SAE Tech. Pap.* **1968**, *77*, 380–387.

17.   Marciniak, Z.; Kuczyński, K. Limit strains in the processes of stretch-forming sheet metal. *Int. J. Mech. Sci.* **1967**, *9*, 609–620. [CrossRef]

18.   McClintock, F.A. A Criterion for Ductile Fracture by the Growth of Holes. *J. Appl. Mech.* **1968**, *35*, 363. [CrossRef]

19.   Lemaitre, J. A continuous damage mechanics model for ductile fracture. *J. Eng. Mater. Technol. Trans. ASME* **1985**, *107*, 83–89. [CrossRef]

20.   Allen, M.; Oliveira, M.; Hazra, S.; Adetoro, O.; Das, A.; Cardoso, R. Benchmark 2-Springback of a Jaguar Land Rover Aluminium. *J. Phys. Conf. Ser.* **2016**, *734*, 022002. [CrossRef]

© 2019 by the authors. Licensee MDPI, Basel, Switzerland. This article is an open access article distributed under the terms and conditions of the Creative Commons Attribution (CC BY) license (http://creativecommons.org/licenses/by/4.0/).

Article

Prediction and Experiment of Fracture Behavior in Hot Press Forming of a TA32 Titanium Alloy Rolled Sheet

Ronglei Fan, Minghe Chen, Yong Wu * and Lansheng Xie

College of Mechanical and Electrical Engineering, Nanjing University of Aeronautics and Astronautics, Nanjing 210016, China; fanronglei_nuaa@163.com (R.F.); meemhchen@nuaa.edu.cn (M.C.); meelsxie@nuaa.edu.cn (L.X.)
* Correspondence: wuyong@nuaa.edu.cn; Tel.: +86-178-2602-6738

Received: 29 October 2018; Accepted: 20 November 2018; Published: 23 November 2018

Abstract: In aerospace and automotive industries, hot press forming (HPF) technology is widely used for rapid and precise deformation of the complex sheet metal component, where the fracture behavior has always been a focused problem. In this study, the hot tensile test and the Nakazima test were carried out, in order to establish the Misiolek constitutive equation and determine the forming limit strain points at an elevated temperature, respectively. The microstructure evolution during the tensile test was also investigated by optical microscope. In addition, the Marciniak–Kuczynski (M–K) model, considering the Mises, Hill48, and Logan–Hosford yield criteria, was utilized to calculate the theoretical forming limit curve (FLC). Furthermore, the fracture behavior of the TA32 alloy sheet during the HPF process was accurately predicted by inserting the predicted FLC into finite element simulation, and the qualified complex component was obtained by optimizing the shape of the sheet.

Keywords: TA32 titanium alloy; fracture behavior; forming limit curve; M-K theory; finite element simulation

1. Introduction

Nowadays, titanium and titanium alloys are extensively used in the aerospace, marine, automotive, and medical industries. This is due to their superior high temperature performance, high specific strength, low density, corrosion resistance, good creep resistance, and excellent biocompatibility [1]. The TA32 alloy is a new type of near-α high temperature titanium alloy with good comprehensive performance. The alloy's long-term working temperature can reach 550 °C, and it has wide application prospects in the cylinder of the advanced aeroengine afterburner and the structure of the cruise missile [2]. However, there are plenty of difficulties, such as large forming forces, low formability, and the occurrence of springback during the cold forming. As one of the more advanced manufacturing technologies, the hot press forming (HPF) process has been actively developed, which can reduce forming time and improve dimensional precision [3]. Metal additive manufacturing, which can directly produce structural components without a mold or additional machining, has also received much attention in recent years. However, HPF technology is preferred in sheet metal forming, due to its advantages of higher production efficiency and lower manufacturing cost compared with the metal additive manufacturing [4]. Therefore, accurately predicting the fracture behavior of TA32 alloy in HPF has important significance for engineering applications.

The forming limit is an important performance index for the fracture behavior of sheet metal forming, and reflects the maximum amount of deformation that can be reached before the plastic deformation of the material is unstable during the forming process. Keeler and Backofen [5] first proposed the concept of forming limit diagrams (FLD), and they obtained the right-hand side of FLD.

Goodwin [6] obtained the left-hand side of FLD, by changing the width and thickness of the sheet, and since then FLD has been broadly used in sheet forming industry. However, it is time-consuming and expensive to experimentally determine FLD, especially at high temperatures; thus, many researchers use numerical models to predict FLDs, of which the Marciniak–Kuczynski (M–K) theory is the most widely applied [7–10].

In recently years, an increasing number of researchers have used finite element method (FEM) to simulate the process of HPF. Nedoushan [11] simulated the hot forming process of AA5083 aluminum alloy by combining a constitutive model, which considers inter-granular deformation and grain boundary mechanisms, and finite element software, and the simulation datas were consistent with the experimental results. Odenberger et al. [12] analyzed the hot forming process parameters of two Ti-6Al-4V prototype components by using FEM. Zhao et al. [13] applied a three-dimensional FEM model to simulate the forming process of atitanium fan blade, and discussed the influence of key factors like descending velocity and frictional coefficient on the forming force. The above studies indicate that FEM is an effective way to accurately predict the process of HPF.

To better predict the fracture behavior of a TA32 alloy sheet during the HPF process, it is necessary to perform a comprehensive study on the tensile properties, microstructural evolution, and forming limit of the TA32 alloy. In the present work, the hot tensile test and the Nakazima test were conducted to evaluate the alloy's mechanical properties and its forming limit strain points at an elevated temperature, respectively. The Misiolek constitutive equation was used to characterize the flow stress of TA32 alloy, and the microstructure evolution during the tensile test was also investigated. Then, the M–K model considering three different yield criteria was used to theoretically predict the forming limit curves (FLCs) of TA32 alloy. The initial inhomogeneity factor f_0 at different temperatures was adjusted by minimizing the average distances between the necking points of the hot tensile test and the theoretical FLCs under different strain rates of a certain temperature, and the accuracy of the theoretical FLC was evaluated by the Nakazima test results. Finally, ABAQUS software, which was developed by Dassault company in America, combined with the theoretical FLC predicted the fracture behavior of a TA32 alloy sheet during the HPF process, and the reliability of the simulation results were discussed by actual hot forming experiments.

2. Materials and Experimental Procedure

2.1. Materials

In this paper, the thickness of the as-received sheet was 1.5 mm. The nominal chemical composition of the sheet was Ti–5.5Al–3.5Sn–3.0Zr–0.7Mo–0.3Si–0.4Nb–0.4Ta (wt %). The β transformation temperature (at which $\alpha+\beta/\beta$) of the TA32 alloy was 1000 °C [2]. The microstructure of the as-received TA32 alloy is shown in Figure 1. It can be seen that there are a certain amount of intergranular β phase grains in the equiaxed α phase matrix, and there are some fine and highly dispersed rare earth phases in the matrix [2]. The contents of the α and β phases, measured by using Image Pro Plus software (Version 6.0, Media Cybernetics, Inc., Rockville, MD, USA), were about 84% and 8%, respectively, and the rest were the rare earth phase.

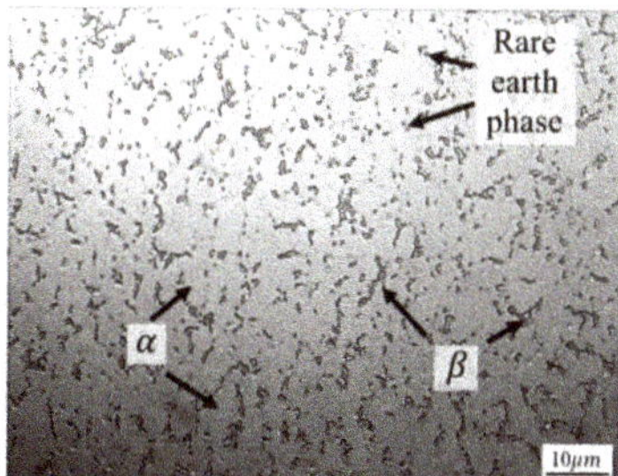

Figure 1. Microstructure of as-received TA32 alloy.

2.2. Hot Tensile Test

The tensile specimen size is shown in Figure 2a, and the length direction of the specimen is the rolling direction. The hot tensile tests were carried out on the UTM 5504X electronic universal testing machine which produced by SUNS company in China, and the test process is shown in Figure 2b. The furnace was heated to the test temperature at a heating rate of 10 °C/s, and the temperature was maintained for 30 min so that the temperature of the stretching chuck and the furnace chamber could be sufficiently exchanged. Then we opened the furnace door and quickly placed the sample, held it for 10 min, and performed tensile deformation at a predetermined tensile rate until the sample was broken. The specimen was taken out quickly and water quenched to retain the high temperature microstructure.

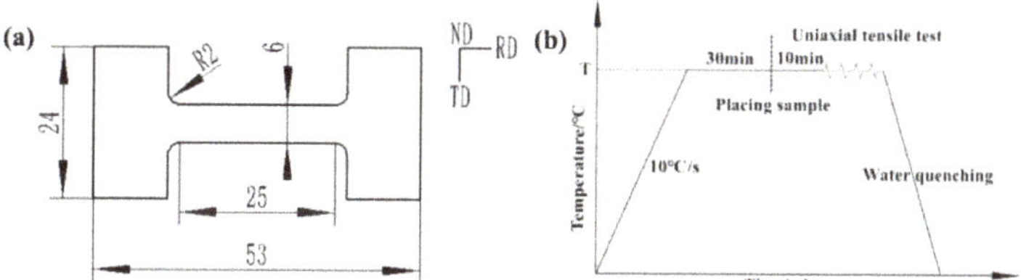

Figure 2. (**a**) Tensile specimen size and (**b**) the hot tensile test process.

Figure 3 shows the specimens before and after the hot tensile test. It can be seen that the elongation and the section shrinkage of the specimens increases significantly with decreasing strain rates and increasing deformation temperature. The maximum elongation of the TA32 alloy at 700 °C, 750 °C, and 800 °C was 92%, 142%, and 204%, respectively.

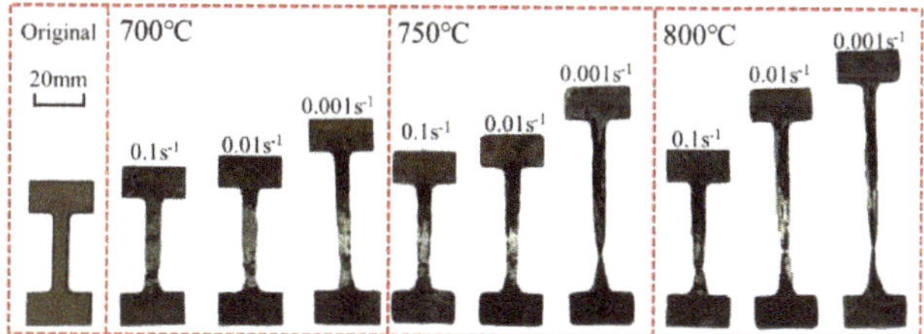

Figure 3. The specimens before and after the hot tensile test.

2.3. Metallography Procedure

After the hot tensile test, the metallographic specimens with size of 6 mm × 8 mm were cut off at a distance of 10 mm from the fracture. After being polished by SiC sandpaper, the surfaces of the specimens were polished by mechanical polishing, until no visible scratches were observed. The polished specimens were etched with Kroll reagent at a volume ratio of 3:5:100 (HF: HNO$_3$: H$_2$O) for 3 seconds, then quickly rinsed with clean water and blown dry. In this paper, the microstructure of TA32 alloy was observed by an MR 5000 optical microscope (Jiangnan Novel Optics CO., Ltd, Nanjing, China).

2.4. Nakazima Test

The FLC of the TA32 alloy sheet was obtained by the Nakazima test method, which is a hemispherical rigid punch bulging test [14]. The schematic diagram of the Nakazima test is shown in Figure 4a. The edge of the sheet was pressed by the blank holder and the concave die, and a certain part of the sheet was locally necked or broken by punch bulging, which reached its forming limit. The test apparatus is an improved thermoforming platform, as shown in Figure 4b.

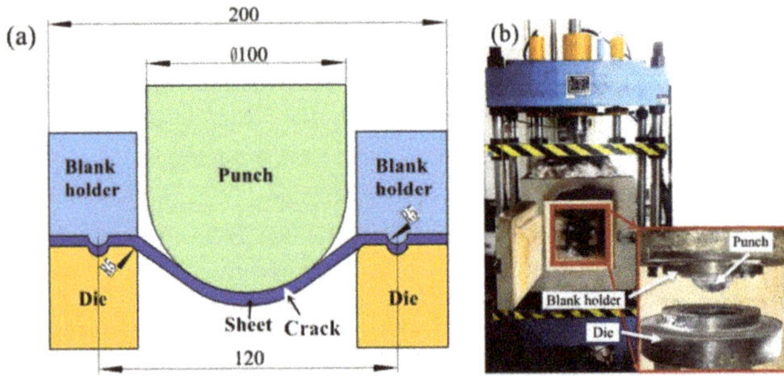

Figure 4. (**a**) Schematic diagram of the Nakazima test. (**b**) The forming limit curve (FLC) experimental platform.

The Nakazima test method requires different specimen geometries to produce all possible strain and stress states. One set of FLC specimens, with the length of 160 mm and different widths of 20 mm, 40 mm, 60 mm, 80 mm, 100 mm, 120 mm, and 140 mm, were obtained by wire cutting, in which each specimen represents one strain path on the FLD. The size of the FLC specimens is shown in Figure 5a. The square grid with the grid size of 5 mm was marked on the surface of specimens for limit strain analysis.

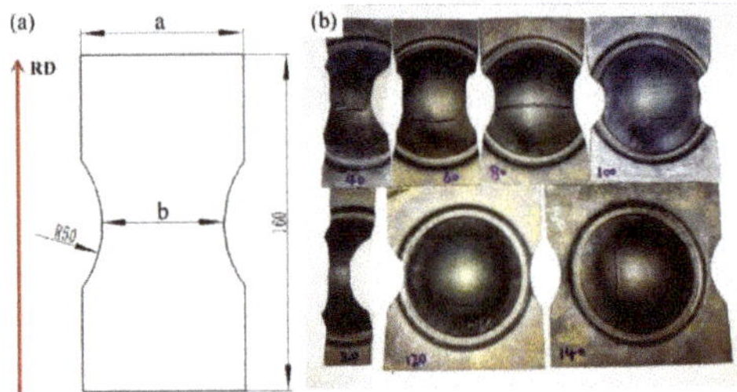

Figure 5. (**a**) The sizes of FLC specimens (b = 20406080100120140; a = b + 20. (**b**) The deformed FLC specimens.

In the experiment, the heating furnace and the FLC experimental device were heated to 750 °C, respectively, and both were kept for 30 min. High temperature lubricant was applied to the both sides of the FLC specimens. After waiting for the temperature in the furnace to be uniform, the specimen with the grid was placed in the heating furnace first, and kept it for 10 min to make the temperature of the sheet uniform. Then the specimen was removed and quickly put into the FLC experimental device for the forming experiment. The speed of the punch pressing was 50 mm/min, and the experiment stopped until the first crack was generated on the specimen. This process should be carried out as quickly as possible, so as to avoid the specimen losing too much heat and affecting the experimental results. The deformed FLC specimens are shown in Figure 5b.

2.5. Hot Press Forming Test

The schematic diagram of the HPF test is shown in Figure 6a. There are twelve independent resistive heating zones placed in the upper and lower workbench of the machine, and the forming temperature is controlled by the Proportion Integration Differentiation (PID) controller which produced by ENVADA company in China. The movement speed of the upper mold can be adjusted by controlling the flow rate of hydraulic oil. All the activity signals of the machine tool were sent and accepted by electronic computer. The appearance of the test platform is shown in Figure 6b.

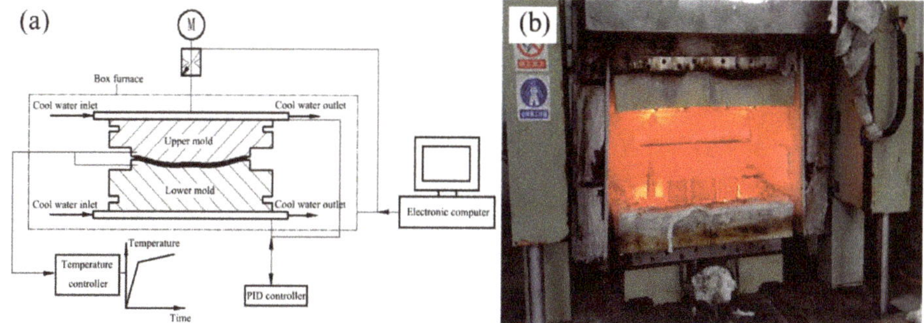

Figure 6. (**a**) Schematic diagram of hot forming test. (**b**) The appearance of the test platform.

In the HPF test, the box furnace was heated to 750 °C and held for 30 min to make the temperature inside the furnace uniform. The TA32 sheet with a high temperature lubricant sprayed on both sides was placed in the middle of the upper and lower molds. The temperature was kept for 10 min to make the temperature of the sheet uniform. Then the upper mold was moved downward at a speed of 50 mm/min, by controlling the computer until it contacted with the lower mold, and kept the pressure at 4 MPa for 15 minutes to make the sheet fully deform. After the hot forming, the component was taken out and air cooled to room temperature.

2.6. FEM Simulation Model

ABAQUS (version 6.14) software was used to simulate the HPF process of the TA32 alloy sheet. The assembly diagram of the model is shown in Figure 7. In the simulation, the sheet was modeled using a shell element with four integration points (S4R), and about 40,000 elements were divided in total. The molds, which were obtained by extracting the cavity surfaces of the upper and lower molds, were regarded as the rigid body. The process of HPF was simulated by way of fixing the lower mold and controlling the displacement of the upper mold. The whole simulation was calculated by the dynamic explicit method.

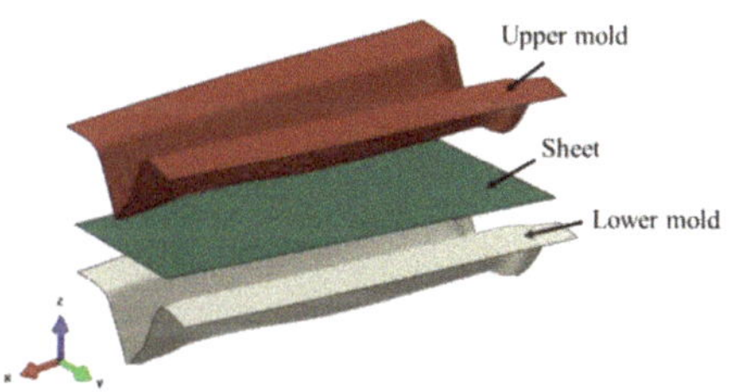

Figure 7. The assembly diagram of the model.

3. Results and Discussion

3.1. Hot Tensile Behavior and Microstructure Evolution

Considering both hardening and softening effects during hot plastic deformation in TA32, the Misiolek constitutive equation [15] was used to describe the stress–strain relationship:

$$\sigma = K\varepsilon^{n} \exp(n_1\varepsilon) \tag{1}$$

where σ and ε represent the true stress and true strain, respectively, and n and n_1 are the hardening index and the softening coefficient, respectively. The parameter K is the equation coefficient. Taking the natural logarithm of both sides, the Misiolek equation can be expressed as

$$ln\sigma = lnK + nln\varepsilon + n_1\varepsilon, \tag{2}$$

Then $n = dln\sigma/dln\dot{\varepsilon}$. In order to reduce the error, at the strain rates of 0.1, 0.01, and 0.001 s^{-1}, the flow stresses with true strains of 0.04–0.15 were extracted for linear fitting, and the n-values of the different strain rates were averaged. Therefore, the values of n obtained at 700, 750, and 800 °C were 0.091, 0.080, and 0.069, respectively. According to Equation (2), using a method similar to the solution of the n-value, the values of n_1 at 700, 750, and 800 °C were -0.398, -0.553 and -0.697, respectively. As shown in Figure 3, the flow stress curves are influenced by temperature and strain rate, wherein the effect of strain rates are reflected in the K-values, which are calculated from Equation (2) based on the above results. Therefore, the flow stress equation of the TA32 alloy at temperature of 700, 750, and 800 °C, with a strain rate of 0.1, 0.01, 0.001 s^{-1}, respectively, can be denoted as

$$\sigma = (3419.03337 + 73.10624ln\dot{\varepsilon} - 2.56667T)\varepsilon^{0.30506-2.2\times10^{-4}T}exp((2.519 - 0.003T)\varepsilon) \tag{3}$$

The comparison between the experimental and fitted curve under different strain rates at 700–800 °C is shown in Figure 8. It can be seen that the flow curve could be divided into three stages in the tensile test. At the first stage, the flow stress increases rapidly with the increase of strain. At the second stage, the flow stress tends to be stable during the deformation process. Since the stacking fault energy of titanium alloy is relatively low, the softening behavior in this stage could be attributed to adiabatic deformation heating or the generation of dynamic recrystallization [16]. At the third stage, the flow stress gradually decreases, and the localized necking occurs at the specimens until fracture. The fitting curve accurately reveals this flow behavior. Therefore, the Misiolek equation can represent the flow stress of the TA32 alloy at the temperature range of 700–800 °C with the strain rate of 0.1–0.001 s^{-1}, and could be used for the theoretical computation of forming limits.

The microstructure of the TA32 alloy under different temperatures, with strain rate of 0.001 s^{-1}, is shown in Figure 9. When the strain rate is constant, the content of the primary α phase decreases with the increasing of the deformation temperature, and the volume fraction and grain size of the β phase continue to increase. When the deformation temperature is 700 °C, the grain shape of the α phase is obviously elongated along the tensile direction, and the grain size of the β phase is slightly larger than that of the as-received microstructure. When the temperature is raised to 750 °C, the grains are uniformly distributed, and finely equiaxed recrystallized grains appear near the grain boundaries, indicating that dynamic recrystallization occurs at this temperature [17]. As the temperature is further increased to 800 °C, the fine recrystallized grains grow significantly.

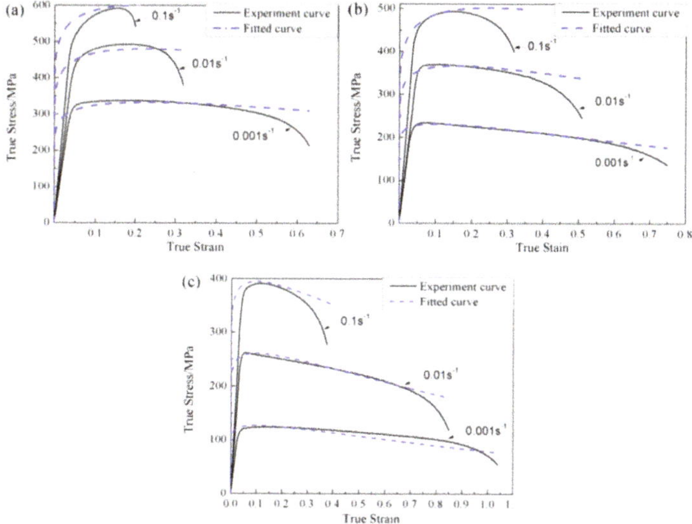

Figure 8. Comparison between the experimental and fitted curve under different strain rates at (**a**) 700 °C, (**b**) 750 °C, and (**c**) 800 °C.

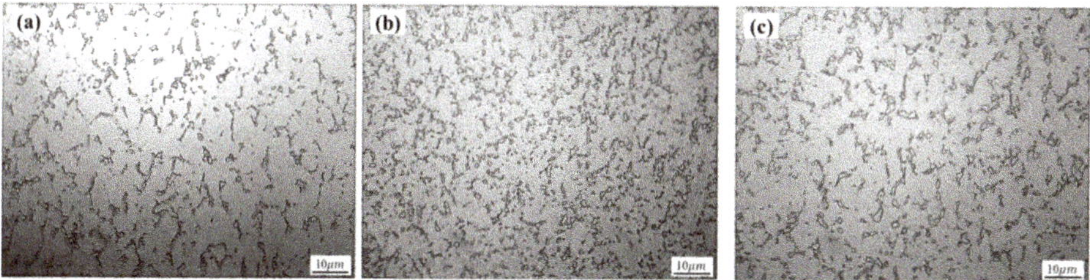

Figure 9. Microstructure of the TA32 alloy under different temperatures, with a strain rate of 0.001 s^{-1}: (**a**) 700 °C, (**b**) 750 °C, and (**c**) 800 °C.

The microstructure of the TA32 alloy under different strain rates at 750 °C is shown in Figure 10. When the deformation temperature is constant, with the change of strain rate, the change of α phase is mainly reflected in the grain shape—the phase content has no obvious change. Moreover, because the α phase grains are in a hexagonal, close-packed structure (HCP), and the β phase grains are in a body-centered cubic structure (BCC), the latter lattice structure has better plasticity, due to it possessing more slip systems. Therefore, the shape of the β grains are elongated more obviously. When the strain rate is 0.1 s^{-1}, the distortion activation energy can meet the energy requirement of recrystallization, but the deformation time is too short to allow the atoms to fully diffuse, so the recrystallization phenomenon is not obvious [17]. When the strain rate is reduced to 0.001 s^{-1}, the dynamic recrystallization process has enough time to further refine the microstructure, so that the plasticity of the material is enhanced and the elongation is continuously increased, which reasonably reflects the macroscopic mechanical properties of the TA32 alloy.

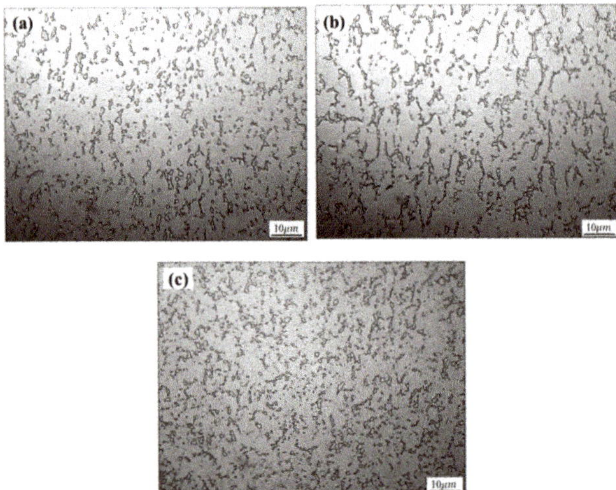

Figure 10. Microstructure of the TA32 alloy under different strain rates at 750 °C: (**a**) $\dot{\varepsilon} = 0.1$ s^{-1}, (**b**) $\dot{\varepsilon} = 0.01$ s^{-1}, (**c**) $\dot{\varepsilon} = 0.001$ s^{-1}

3.2. The Forming Limit Curve at an Elevated Temperature

After the Nakazima test, the grids near the fracture region of the deformed specimen were selected to measure the forming limit strain point. In the limit strain evaluation process, the fracture region should be located near the centerline of the deformed specimen. Furthermore, the distance between the measured grid and the crack cannot exceed the size of one grid, and the strain points of the selected grid are measured three times for determining the average value. In this paper, the grid–triangle nodes method proposed by Vogel and Lee [18] were used to calculate the limit strain. Figure 11 shows the measured forming limit strain points from each deformed FLC specimen. It can be seen that as the specimen width increases, the position of the limit strain point in the strain space moves from left to right, and the major limit strain of the TA32 alloy sheet at the plane strain state was about 0.31.

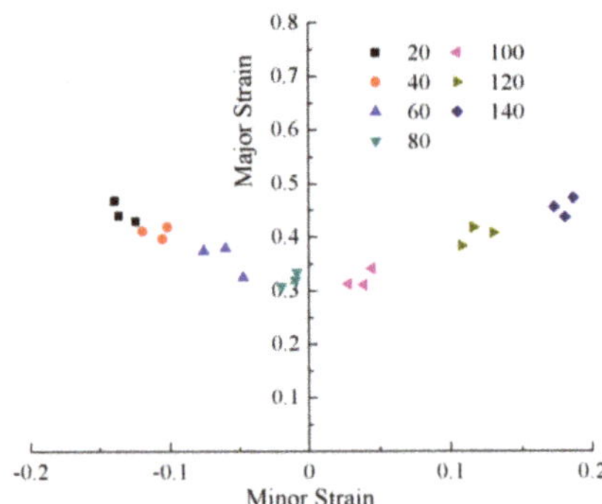

Figure 11. The measured forming limit strain points from each deformed FLC specimen.

In the M–K model, it is assumed that there is a shallow groove on the sheet surface which causes the localized necking, as shown in Figure 12. The safe region is called "A" and the groove region is named "B". The safe region is subjected to proportional strains, and it is assumed that strains at the groove direction are equal in the two regions. The initial inhomogeneity factor of the groove f_0 is defined as the thickness ratio $f_0 = t_{B0}/t_{A0}$, where t represents the thickness and subscript "0" represents the initial state. This initial inhomogeneity grows continuously with plastic straining to form a localized neck eventually [19]. During the deformation process, the strain ratio $\rho(\rho = \varepsilon_2/\varepsilon_1)$ outside the groove is constant, but decreases in the groove region.

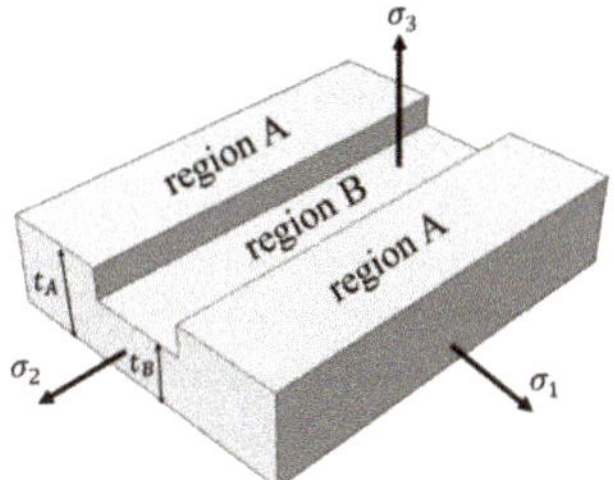

Figure 12. Schematic of the Marciniak–Kuczynski (M–K) model.

To start the analysis, a small certain value for $d\bar{\varepsilon}_A$ (0.001) was applied, with which we could calculate $d\varepsilon_{A1}$ and $d\varepsilon_{A2}$, and got value of $d\varepsilon_{A3}$ with the volume condition $d\varepsilon_1 + d\varepsilon_2 + d\varepsilon_3 = 0$. The equivalent strain $\bar{\varepsilon}_A$ would be obtained by $\bar{\varepsilon}_{new} = \bar{\varepsilon}_{old} + d\bar{\varepsilon}$. For each strain increment $d\bar{\varepsilon}_A$ in the safe region, there was a corresponding strain increment $d\bar{\varepsilon}_B$ in the groove region. Finding the value of $d\bar{\varepsilon}_B$ involved an iterative procedure. A force equilibrium equation ($\sigma_{A1}t_A = \sigma_{B1}t_B$) and compatibility condition ($d\varepsilon_{A2} = d\varepsilon_{B2}$) was used to link region A and region B with the following equation:

$$\frac{K_A}{\varphi_A}(\bar{\varepsilon}_A + d\bar{\varepsilon}_A)^n \exp(n_1(\bar{\varepsilon}_A + d\bar{\varepsilon}_A)) \exp(\varepsilon_{A3})$$
$$= \frac{K_B}{\varphi_B} f_0 (\bar{\varepsilon}_B + d\bar{\varepsilon}_B)^n \exp(n_1(\bar{\varepsilon}_B + d\bar{\varepsilon}_B)) \exp(\varepsilon_{B3}) \tag{4}$$

where K, n, and n_1 are the material coefficients of the Misiolek equation, and $\varphi = \bar{\sigma}/\sigma_1$ can be determined from the associated yield criterion.

After $d\bar{\varepsilon}_B$ was obtained in each step, it was compared to $d\bar{\varepsilon}_A$, and if $d\bar{\varepsilon}_B/d\bar{\varepsilon}_A > 10$ the necking had begun and ε_{A1} and ε_{A2} were saved; otherwise, a greater value for $d\bar{\varepsilon}_A$ was assumed and the process repeated. This procedure was done for different values of strain ratios until the whole diagram was computed. The calculation process was programmed by MATLAB software.

According to Equation (4), the determination of φ and $\bar{\varepsilon}$ are dependent on the employed yield criterion, the utilization of different yield criteria results in different critical strains, and correspondingly different FLCs. Three yield criteria that have been used extensively to study forming limit of sheet metals, namely Von Mises, Hill48, and Logan–Hosford, were used to predict the FLCs for the TA32 alloy.

The Von Mises yield criterion [20] states that under certain deformation conditions, when the second invariant of the stress deflection tensor at a point in the stressed object reaches a certain value, the point begins to enter a plastic state. Its yield function is represented by principal stress as

$$2\bar{\sigma}_{YF}^2 = (\sigma_1 - \sigma_2)^2 + (\sigma_2 - \sigma_3)^2 + (\sigma_3 - \sigma_1)^2, \tag{5}$$

The Hill48 yield criterion [21] considers that the material was supposed to have an anisotropy property with three orthogonal symmetric planes. The yield criterion can be written as a function of principal stresses:

$$(1+r)\bar{\sigma}_{YF}^2 = R(\sigma_1 - \sigma_2)^2 + (\sigma_2 - \sigma_3)^2 + (\sigma_3 - \sigma_1)^2, \tag{6}$$

where R is an average anisotropic parameter, determined from hot tensile texts at $0°$, $45°$, and $90°$ to the rolling direction ($R = (R_0 + 2R_{45} + R_{90})/4$). The measured R values of the TA32 alloy at the temperatures of 700 °C, 750 °C, and 800 °C were 0.836, 0.839, and 0.726, respectively.

Independently of Hill, Hosford proposed a yield criterion in the form [22]

$$(1+R)\bar{\sigma}_{YF}^a = R|\sigma_1 - \sigma_2|^a + |\sigma_1|^a + |\sigma_2|^a, \tag{7}$$

where the exponent parameter a is an integer greater than two. Hosford associated a to the crystallographic structure of the material and concluded that the best approximation was given by $a = 6$ for BCC materials and $a = 8$ for FCC materials. However, in this paper, the temperature of the Nakazima test was 750 °C, which is different from the transformation point of TA32 alloy; the lattice structure was mainly the close-packed hexagonal, and the value of a was not certain. Therefore, the values of the exponential parameter a were considered to be 4, 6, and 8, respectively, in order to calculate the theoretical FLC.

By measured the width strain ε_b and the thickness strain ε_t at the necking position of the specimens in the hot tensile test, the strain ε_l in the tensile direction was obtained according to the volume invariance principle, and the necking points $(\varepsilon_l, \varepsilon_b)$ under different temperature and strain rates were obtained. The M–K model based on the Misiolek constitutive equation and Mises yield criterion was selected to predict the FLCs under the corresponding conditions. In addition, the effect of the initial inhomogeneity factor to the prediction of forming limits cannot be ignored. By comparing the distances between the measured necking points and the theoretical FLCs under different strain rates of a certain temperature, the value of f_0 is adjusted to minimize the average distance. Using this method, the values of f_0 at the temperature of 700, 750, and 800 °C were determined to be 0.98, 0.993, and 0.996, respectively. Comparisons between the necking points of the hot tensile test and theoretical FLCs under different strain rates at 700, 750, and 800 °C is shown in Figure 13. It can be seen that the theoretical FLCs can agree well with every necking point.

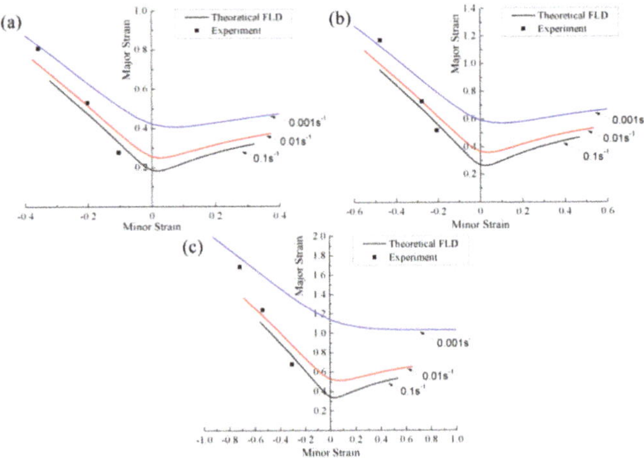

Figure 13. Comparisons between the necking points of hot tensile test and theoretical FLCs under different strain rates at (**a**) 700 °C, (**b**) 750 °C, and (**c**) 800 °C.

Figure 14 shows the comparison between theoretical FLCs with different yield criteria and experimental forming limit strain points at 750 °C. For theoretical analysis, yield surfaces were described by Mises, Hill48, and Logan–Hosford yield functions, and the hardening model was expressed by the Misiolek equation. In Figure 14, on the left-hand side of the FLD, the application of different yield criteria has little effect on the theoretical calculation results, while the differences are obvious on the right-hand side of the FLD. The results of two predicted FLCs based on the Mises and Hill48 yield criterion are very close. This is because the anisotropic parameter R of the sheet is 0.839 at 750 °C, which is close to 1, and the yield function of the two yield criteria is the same when $R = 1$. In addition, it can be seen that when the Logan–Hosford yield criterion is used for theoretical prediction, the change of the exponential parameter a has little effect on the prediction result of the hot tensile zone—while in the biaxial tensile zone, the theoretical FLCs increase with a decreasing a-value. The theoretical prediction result is the most consistent with the experimental data when $a = 4$.

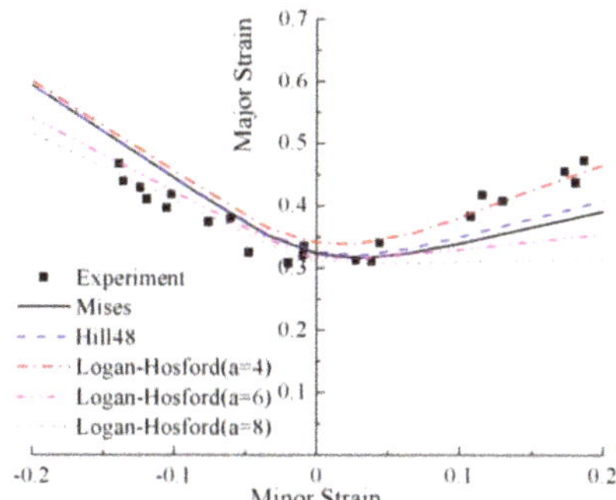

Figure 14. Comparisons between theoretical FLCs with different yield criteria and experimental forming limit strain points at 750 °C.

3.3. Prediction of Fracture Behavior in Hot Press Forming

This paper adopted the FEM to predict fracture behavior in HPF. The theoretical FLC, based on the Logan–Hosford yield criterion with $a = 4$, was imported into the FLD Damage module in the ABAQUS software, and the calculated forming limit diagram damage initiation criterion (FLDCRT) chart is shown in Figure 15a. The FLDCRT chart can visually reflect the fracture position of the component. When FLDCRT = 1, it indicates that the strain state just reaches the limit strain, and FLDCRT < 1 indicates that the sheet is safe—otherwise, fracture occurs. Therefore, the fracture zone will be generated at the sharp corner position of the sheet. Figure 15b shows the comparison between the strain states of the grids of the fracture zone in the model and the theoretical FLC. Obviously, there are many fracture points above the FLC. Figure 15c shows the actual deformed component, and it can be found that fracture indeed occurred at the sharp corner of the component, which is consistent with the simulation result.

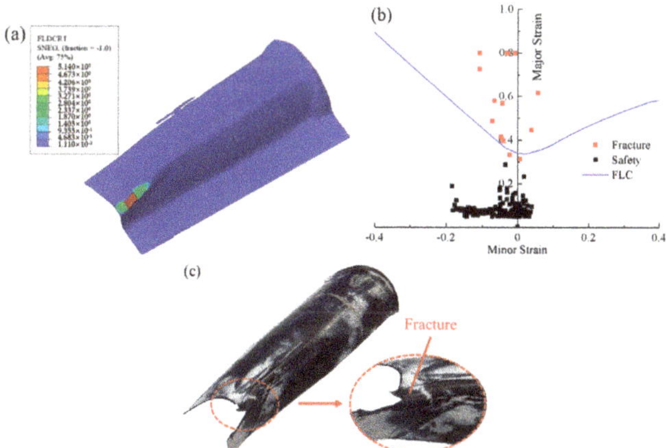

Figure 15. (**a**) FLDCRT chart, (**b**) forming limit diagram (FLD) result, and (**c**) the deformed component.

The reason for this phenomenon was that the material mobility at the sharp corner position of the component was restricted by the molds. After optimizing the shape of the sheet, the FLDCRT chart simulated by ABAQUS software is shown in Figure 16a. It can be seen that the maximum value of FLDCRT at the sharp corner of sheet is 0.86, which is less than 1, indicating that the complex component can be properly formed. Figure 16b shows that the strain state of the grids of the fracture zone in the model are all at the safety point below the predicted FLC. Figure 16c shows the actual deformed component after optimizing the sheet. It can be seen that the sharp corner of sheet was well formed without cracking, which is again consistent with the simulation result.

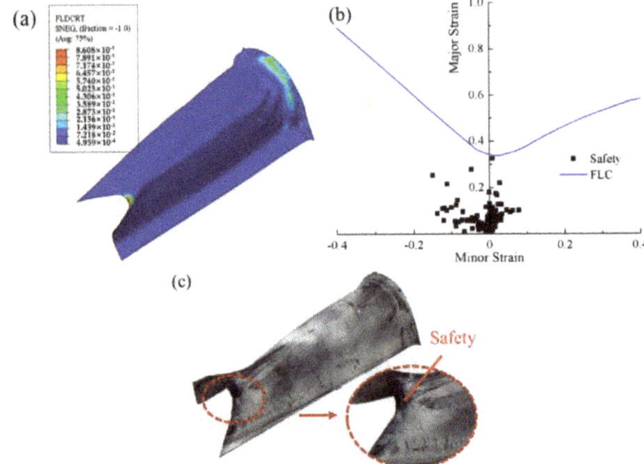

Figure 16. (**a**) FLDCRT chart of optimized sheet, (**b**) FLD result, and (**c**) the deformed component after optimizing the sheet.

In this study, the theoretical FLC calculated by the M–K theory was in conjunction with the finite element simulation to accurately predict the fracture behavior of TA32 alloy sheet during the HPF process, which can effectively optimize the shape of the sheet and process parameters. This method is efficient and reliable for the industrial applications of a TA32 titanium alloy rolled sheet.

4. Conclusions

This work focused on the fracture behavior of a TA32 alloy rolled sheet in hot press forming. The hot tensile behavior, microstructure evolution, and forming limit of a TA32 alloy rolled sheet were studied in this paper. Furthermore, FEM was used to accurately predict the fracture behavior of the TA32 alloy sheet during the HPF process, and the qualified complex component was obtained by optimizing the shape of the sheet. Some conclusions were summarized as follows:

(1) The flow stresses of TA32 alloy at the temperature range of 700–800 °C, with the strain rate of 0.1–0.001 s^{-1}, are accurately characterized by the Misiolek constitutive equation, which is expressed as $\sigma = \left(3419.03337 + 73.10624 ln\dot{\varepsilon} - 2.56667T\right)\varepsilon^{0.30506-2.2\times10^{-4}T}exp\left((2.519 - 0.003T)\varepsilon\right)$; this equation was used for the calculation of a theoretical FLC. The microstructure evolution of the TA32 alloy is related to the temperature and the strain rate. The dynamic recrystallization temperature at the strain rate of 0.001 s^{-1} is 750 °C. When the temperature is constant, the lower strain rate provides sufficient time for the dynamic recrystallization process to further refine the microstructure.

(2) The forming limit of a TA32 alloy at the temperature of 750 °C was measured and predicted by the Nakazima test and the M-K theory, respectively. The predicted FLC calculated by the Logan–Hosford yield criterion with the exponential parameter $a = 4$ is the optimal result for predicting the forming limit strain points of the Nakazima test.

(3) The fracture behavior of the TA32 alloy sheet during the HPF process was accurately predicted by combining the predicted FLC and ABAQUS software, and the qualified complex component was obtained by optimizing the shape of the sheet. This method can be used to optimize the initial configuration of a metal sheet in HPF, and provides guidance for the further application of TA32 alloy in engineering practice.

Author Contributions: Conceptualization, R.F., M.C. and Y.W.; Methodology, R.F.; Software, R.F.; Validation, R.F. and Y.W.; Formal Analysis, R.F.; Investigation, R.F.; Resources, M.C. and L.X.; Data Curation, R.F.; Writing-Original Draft Preparation, R.F.; Writing-Review & Editing, R.F., M.C. and Y.W.; Visualization, R.F.; Supervision, M.C. and L.X.; Project Administration, M.C. and Y.W.; Funding Acquisition, Y.W.

Funding: The authors gratefully acknowledge the financial support from the National Natural Science Foundation of China under Grant No.51805256, and appreciate the helpful comments from the reviewers.

Conflicts of Interest: The authors declare no conflicts of interest.

References

1. Banerjeea, D. Perspectives on Titanium Science and Technology. *Acta Mater.* **2013**, *61*, 844–879. [CrossRef]
2. Wang, Q.J.; Liu, J.R.; Yang, R. High Temperature Titanium Alloys: Status and Perspective. *J. Aeronaut. Mater.* **2014**, *34*, 1–26.
3. Lee, M.G.; Kim, S.J.; Han, H.N. Finite element investigations for the role of transformation plasticity on springback in hot press forming process. *Comp. Mater. Sci.* **2010**, *47*, 556–567. [CrossRef]
4. Saboori, A.; Gallo, D.; Biamino, S.; Fino, P.; Lombardi, M. An Overview of Additive Manufacturing of Titanium Components by Directed Energy Deposition: Microstructure and Mechanical Properties. *Appl. Sci.* **2017**, *7*, 883. [CrossRef]
5. Keeler, S.P.; Backofen, W. Plastic instability and fracture in sheets stretched over rigid punches. *ASM Trans.* **1963**, *56*, 25–48.
6. Goodwin, G. *Application of Strain Analysis to Sheet Metal Forming Problems in the Press Shop*; SAE International: Warrendale, PA, USA, 1968.
7. Marciniak, Z.; Kuczynski, K.; Pokora, T. Influence of the plastic properties of a material on the forming limit diagram for sheet metal in tension. *Int. J. Mech. Sci.* **1973**, *15*, 789–800. [CrossRef]
8. Ahmadi, S.; Eivani, A.R.; Akbarzadeh, A. An experimental and theoretical study on the prediction of forming limit diagrams using new BBC yield criteria and M–K analysis. *Comput. Mater. Sci.* **2009**, *44*, 1272–1280. [CrossRef]
9. Li, H.; Wu, X.; Li, G. Prediction of Forming Limit Diagrams for 22MnB5 in Hot Stamping Process. *J. Mater. Eng. Perform.* **2013**, *22*, 2131–2140. [CrossRef]
10. Kotkunde, N.; Srinivasan, S.; Krishna, G.; Gupta, A.K.; Singh, S.K. Influence of material models on theoretical forming limit diagram prediction for Ti–6Al–4V alloy under warm condition. *Trans. Nonferr. Met. Soc.* **2016**, *26*, 736–746. [CrossRef]
11. Nedoushan, R.J.; Farzin, M.; Banabic, D. Simulation of hot forming processes: Using cost effective micro-structural constitutive models. *Int. J. Mech. Sci.* **2014**, *85*, 196–204. [CrossRef]
12. Odenberger, E.L.; Oldenburg, M.; Thilderkvist, P.; Stoehr, T.; Lechler, J.; Merklein, M. Tool development based on modelling and simulation of hot sheet metal forming of Ti–6Al–4V titanium alloy. *J. Mater. Process. Technol.* **2011**, *211*, 1324–1335. [CrossRef]
13. Zhao, B.; Li, Z.; Hou, H.; Liao, J.; Bai, B. Three dimensional FEM simulation of titanium hollow blade forming process. *Rare Met. Mater. Eng.* **2010**, *39*, 963–968.
14. Nakazima, K.; Kikuma, T.; Hasuka, K. Study on the formability of steel sheets. *Yawata Tech. Rep.* **1968**, *284*, 140–141.
15. Gronostajski, Z. The constitutive equations for FEM analysis. *J. Mater. Process. Technol.* **2000**, *106*, 40–44. [CrossRef]
16. Saboori, A.; Dadkhah, M.; Pavese, M.; Manfredi, D.; Biamino, S. Hot deformation behavior of Zr-1%Nb alloy: Flow curve analysis and microstructure observations. *Mater. Sci. Eng. A* **2017**, *696*, 366–373. [CrossRef]
17. Humphreys, F.J.; Hatherly, M. *Recrystallization and Related Annealing Phenomena*, 2nd ed.; Elsevier: Amsterdam, The Netherlands, 2004.
18. Vogel, J.H.; Lee, D. An automated two-view method for determining strain distributions on deformed surfaces. *J. Mater. Shap. Technol.* **1988**, *6*, 205–216. [CrossRef]
19. Assempour, A.; Safikhani, A.R.; Hashemi, R. An improved strain gradient approach for determination of deformation localization and forming limit diagrams. *J. Mater. Process. Technol.* **2009**, *209*, 1758–1769. [CrossRef]
20. Mises, R.V. Mechanik der festen Körper im plastisch-deformablen Zustand. *Göttin Nachr. Math.Phys.* **1913**, *1*, 582–592.

21. Hill, R. A theory of the yielding and plastic flow of anisotropic metals. *Proc. Royal. Soc. A* **1948**, *193*, 281–297. [CrossRef]
22. Logan, R.W.; Hosford, W.F. Upper-bound anisotropic yield locus calculations assuming ⟨111⟩-pencil glide. *Int. J. Mech. Sci.* **1980**, *22*, 419–430. [CrossRef]

© 2018 by the authors. Licensee MDPI, Basel, Switzerland. This article is an open access article distributed under the terms and conditions of the Creative Commons Attribution (CC BY) license (http://creativecommons.org/licenses/by/4.0/).

Article

Effect on Blank Holding Force on Blank Deformation at Direct and Indirect Hot Deep Drawings of Boron Steel Sheets

Hyung Yoon Seo [1], Chul Kyu Jin [2,*] and Chung Gil Kang [3]

[1] Department of Computer Software Engineering, Changshin University, 262 Paryong-ro, Masanhoiwon-gu, Changwon-si, Gyeongsangnam-do 51352, Korea; hyseo@cs.ac.kr
[2] School of Mechanical Engineering, Kyungnam University, 7 Kyungnamdaehak-ro, Masanhappo-gu, Changwon-si, Gyeongsangnam-do 51767, Korea
[3] School of Mechanical Engineering, Pusan National University, San 30 Chang Jun-dong, Geum Jung-Gu, Busan 46241, Korea; cgkang@pusan.ac.kr
* Correspondence: cool3243@kyungnam.ac.kr; Tel.: +82-55-249-2346; Fax: +82-505-999-2160

Received: 6 July 2018; Accepted: 24 July 2018; Published: 25 July 2018

Abstract: This study involves performing direct and indirect hot press forming on ultra-high-strength steel (UHSS) boron steel sheets to determine formability. The indirect hot press process is performed as a cold deep drawing process, while the direct hot press process is performed as a hot deep drawing process. The initial blank temperature and the blank holding force are set as parameters to evaluate the performance of the direct and indirect deep drawing processes. The values of punch load and forming depth curve were obtained in the experiment. In addition, the hardness and microstructure of the boron steel sheets are examined to evaluate the mechanical properties of the material. The forming depth, maximum punch load, thickness, and thinning rate according to blank holding force were examined. The result shows that a larger blank holding force has a more significant effect on the variation of the thickness and thinning rate of the samples during the drawing process. Furthermore, the thinning rate of the deep drawing part in with and without fracture boundary was respectively examined.

Keywords: hot deep drawing; cold deep drawing; boron steel; deformation characteristics; direct forming; indirect forming

1. Introduction

Boron steel currently represents the ultra-high-strength steel (UHSS) applied in the automotive industry because of the demand for higher passive safety and weight reduction. However, ultra standard high-strength steels, like boron steel, are difficult to manufacture with cold forming because of disadvantages such as large forming forces, the difficulty of forming complex components, and the occurrence of serious springback at room temperature [1,2]. Therefore, requirements regarding complexity and accuracy increase.

Hot press forming is used widely in the automotive industry. Hot forming can vastly improve the tensile strength of the components. Nowadays, hot press forming at an elevated temperature makes it possible to produce high strength. Hot stamping is not only an innovative technique that is used to produce UHSS components like side impact and bumper beams, but it also reduces the springback under high-temperature forming and achieve good formability. The low springback attributed to in-die cooling gives boron steel an unparalleled edge in dimension control and subsequent assembly process [3,4]. Naderi et al. presented hot stamping as a non-isothermal, high-temperature forming process, in which complex ultra-high-strength parts are produced, with the goal of no springback [5].

Altan studied the formability of boron-alloyed steels at high temperatures of 650 °C to 850 °C [6]. The results showed that the material has excellent formability and can be formed into a complex shape in a single stroke. He also studied the tensile strength and microstructure change during hot stamping. Xing et al. set up a material model under the hot stamping condition of quenchable steel, based on the experimental data for the mechanical and physical properties [7].

In addition to the parameters of conventional cold press forming, such as the blank holding force, punch velocity, punch and die radii, and friction coefficient, the blank temperature and quenching methods also affect the formability and complicate the hot forming process. So et al. provided information on cold and warm blanks of the quenchable boron steel 22MnB5 [8]. From their experimental research, it can be concluded that higher quality and more economical production can be achieved by adjusting the blanking process parameters for the commonly used ultra-high-strength steel sheet 22MnB5. Nakagawa et al. examined the springback and the deformation behavior in hot stamping of a steel sheet with 0.6 mm thickness [9]. Löbbe and Tekkaya investigated the mechanisms influencing the geometrical and mechanical properties of the products after heat-assisted sheet forming processes [10].

Borsetto et al. investigated the influence of the thermal process parameters on the chemical behavior of the Al-Si layer coating, in terms of the heating temperature, holding time, and cooling rate [11]. Naderi et al. described the hot stamping facilities and methods used in experiments [12]. They studied the effect of a hot stamping process on the microstructural and mechanical properties of boron-alloyed and non-boron-alloyed steels and presented an innovative method to carry out a metallographic analysis by the application of lateral and surface hardness maps. Ryan et al. reported on the hot forming die by which various mechanical properties can be partially obtained through the control of the cooling rate in hot forming [13]. Ouyang et al. proposed a friction coefficient and heat transfer coefficient between die and blank with cold and hot deep drawing processes [14]. The sheet material is boron sheet with Al-Si coating layer. During the hot press forming, coating layer is fractured by bending and heating during hot deep drawing. Moon et al. and Seok et al. proposed the deformation behavior of the coating layer on boron sheet [15,16].

As shown in reported research recently, the comparison study of formability and deformation behavior considering blank holding force in hot deep drawing (direct hot press forming) and cold deep drawing (indirect hot press forming) is limited. Therefore, in this study, the relationship between the forming depth and forming load has been investigated with both deep drawing experimental data for variation of blank holding force.

This study, using a boron steel as the blank material, investigated the formability of direct and indirect hot deep drawing under different blank holding forces. Experiments were carried out for direct and indirect hot deep drawing. In direct hot deep drawing, the drawing process carried out with different initial blank temperatures ranged from 850–950 °C, while in indirect hot deep drawing, the blanks were firstly pre-formed in room temperature, and then the cold pre-formed parts were heated to 900 °C and quenched in water. After forming, the forming depth and maximum punch loads for direct and indirect hot deep drawing were examined. Moreover, the thinning rate, microstructure, and hardness at different positions of the drawn part were examined for direct and indirect hot deep drawings.

2. Experimental Methods

2.1. Equipment

The deep drawing equipment used in this study consisted of a 50 ton hydraulic cylinder for moving the forming punch, a 30 ton load cell that was capable of measuring the forming load, a 20 ton hydraulic cylinder that clamped the upper die, and another 5 ton load cell for measuring the clamping force. Additionally, the equipment had a computer which has a linear variable differential transformer (LVDT). When the steel sheet was formed in this equipment, the LVDT measured the punch load according to punch stroke. For the heating, a heat chamber was used that was capable of maintaining

1200 °C of atmosphere temperature. 22MnB5 material with a thickness of 0.6 mm was used in these experiments. The chemical composition of 22MnB5 boron steel is shown in Table 1. Figure 1 shows the engineering stress and engineering strain curve of 22MnB5 boron steel. Yield strength, tensile strength and elongation are 470 MPa, 650 MPa and 27%, respectively.

Table 1. Chemical composition of 22MnB5 boron steel (wt %).

C	Si	Mn	Cr	B
0.2123	0.0806	1.4840	0.4063	0.0016

Figure 1. Engineering stress and engineering strain curve of 22MnB5 boron steel.

Figure 2 shows a schematic diagram of the deep drawing die. As shown in Figure 2, there were several cartridge heaters in the upper die, lower die, and punch. Hereby, a constant tool temperature for the dies and punch can be obtained. The blank sheet was Ø75 mm in diameter (D) and 0.6 mm in thickness (t_0). The clearance (C_L) between the punch and the die was 1.2 mm. Diameter (d) and radius (R_p) of punch was Ø37.6 mm and 4 mm, respectively. Die radius (R_d) was 5 mm. The material and hardness of the die is AISI H13 and 54 HRC, respectively.

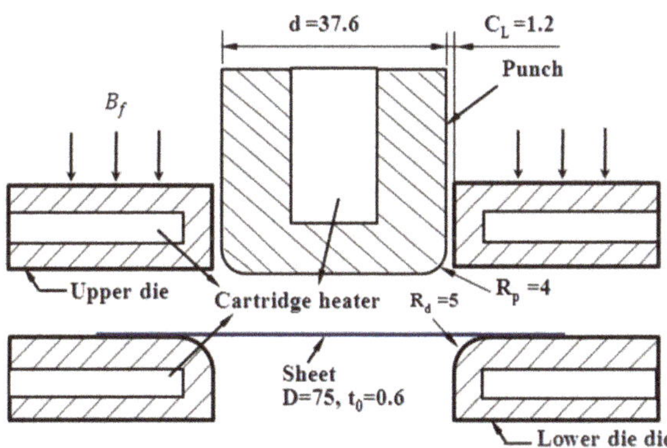

Figure 2. Die construction and dimension for deep drawing (unit: mm, B_f: Blank holding force, C_L: clearance between punch and die).

2.2. Process

The hot forming process currently exists in two different main processes: the direct hot forming method and the indirect hot forming method. In the direct hot forming process, as shown in Figure 3a, the blanks are austenitized at temperatures between 850 and 950 °C for about 5 min inside a furnace and subsequently transferred to the die set via a transfer unit. Afterwards, the blank is subsequently formed and quenched in the closed tool. Unlike the direct process, indirect hot forming provides for a part to be drawn, unheated, to about 90% to 95% of its final shape in a conventional die, sometimes followed by a partial trimming operation, depending on edge tolerance. The cold pre-formed part is then heated to austenitization temperature in a continuous furnace, sometimes followed by a calibration process (also called 2nd forming process), depending on the complexity and accuracy of the part. The formed part is quenched in the die or water, as shown in Figure 3b. The reason for the additional step is to extend the forming limits for highly complex shapes by heat-treating the cold pre-formed parts. Full martensite transformation in the material causes an increase in the tensile strength of up to 1500 MPa.

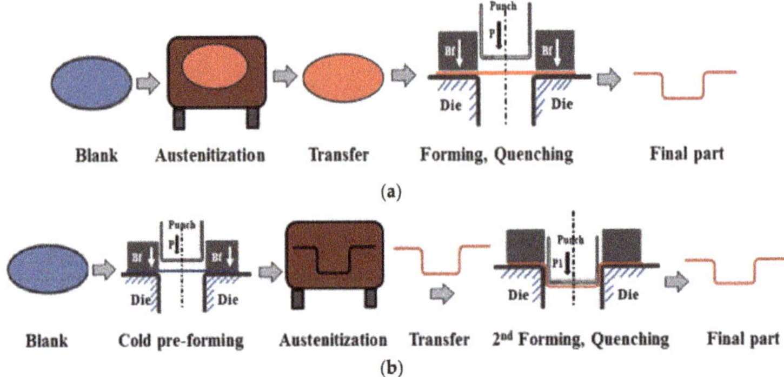

Figure 3. Procedure of hot press forming and cold press forming: (**a**) direct hot press forming (hot deep drawing process); (**b**) indirect hot press forming (cold deep drawing process).

Table 2 lists the experiment conditions for direct and indirect hot deep drawing, where T_s represents the initial blank temperature, and T_d and T_p represent the die temperature and punch temperature, respectively. In the direct hot deep drawing, samples were separately heated to 850, 900, and 950 °C for 5 min in the furnace to give the sheet material an austenitic microstructure over the entire blank. They were then transferred manually to the heated tooling device used for the drawing process, which were kept at 300 °C. After direct hot drawing, the parts were quenched in the closed die. The total time elapsed before the actual forming process—which included the transfer time and blank holder moving time was approximately 5 s. On the other hand, in indirect hot deep drawing, the blanks were firstly pre-formed in room temperature, which is called the cold pre-forming step, and then the cold pre-formed parts were heated to 900 °C and quenched in water. Both in direct and indirect hot deep drawings, the punch velocity (V_P) was fixed at 6 mm/s. Blank holding forces (B_f) were 5 kN, 15 kN, 30 kN, and 90 kN.

After drawing, the properties of drawn cups, such as thinning rate, microstructure, and hardness, were examined. Positions for the measurement of the thinning rate, microstructure and hardness are shown in Figure 4. Position ③ is the middle point between position ② and position ④. For the hardness measurement, Vickers hardness was measured. The load was set to 500 g and measurements were taken five times in total for each specimen. The results were averaged. In order to investigate thinning, the cross section of formed specimens was observed using a digital microscope. The thinning rate shows the amount of change in the thickness after forming compared to the initial thickness. The thinning rate was calculated by Equation (1).

$$\text{thinning rate} = \frac{t_0 - t_1}{t_0} \times 100 \tag{1}$$

where, t_0 is thickness of the blank and t_1 is thickness of formed specimens.

Table 2. Experiment conditions for direct and indirect hot deep drawing process.

Parameters	Direct	Indirect
Blank temperature, T_s (°C)	850, 900, 950	Room temperature (RT)
Tool (die and punch) temperature, T_d, T_p (°C)	300	RT
Punch speed (mm/s)	6	6
Blank holding force, B_f (kN)	5, 15, 30, 90	5, 15, 30, 90

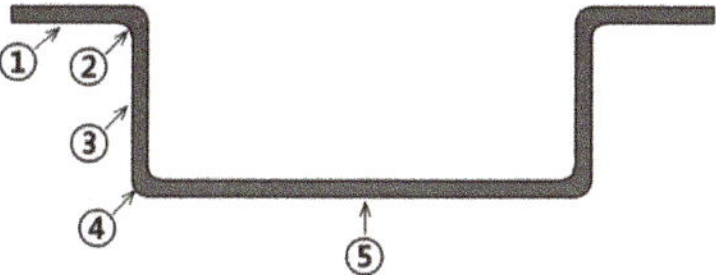

Figure 4. Position for the measurement of the thinning rate, microstructures and Vickers hardness.

3. Results

Experiments for hot deep drawing with different blank holding forces and initial blank temperatures were carried out. It was observed from the experiments that fracture started to occur at the time that the punch load reached close to the maximum value. Figure 5 shows the relationship between the punch load and the forming depth for cold pre-deep drawing and direct hot deep drawing at different values of blank holding forces. As shown in Figure 5a, when blank holding force was 5 kN, the forming depth was 18 mm in cold pre-deep drawing. In direct hot deep drawing, the forming depth was 7 mm when blank temperatures were 950 °C, decreasing to 5 mm as blank temperatures decreased to 850 °C. The maximum punch load in cold pre-deep drawing, which was found to be 42 kN, was much higher than that in direct hot deep drawing, which was 26 kN. As can be seen from Figure 5, the maximum punch load and forming depth in cold pre-deep drawing were larger than in direct hot deep drawing. Under the same blank holding forces, as blank temperatures increased from 850 °C to 950 °C, the forming depth increased and required a lower maximum punch load (less than 26 kN for 950 °C). The differences in the forming depth and punch load without any fracture according to blank holding forces are shown in the following figures.

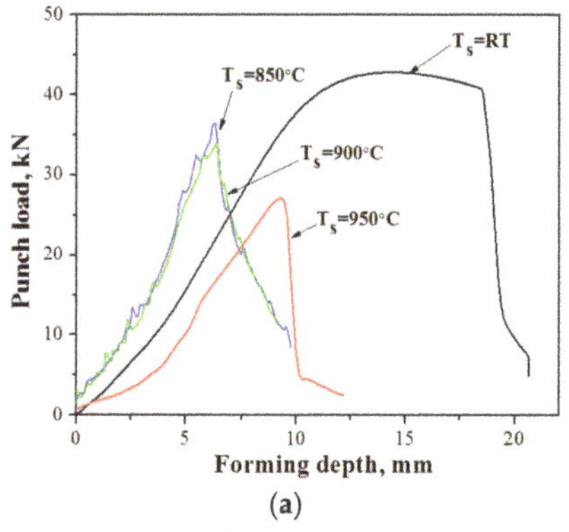

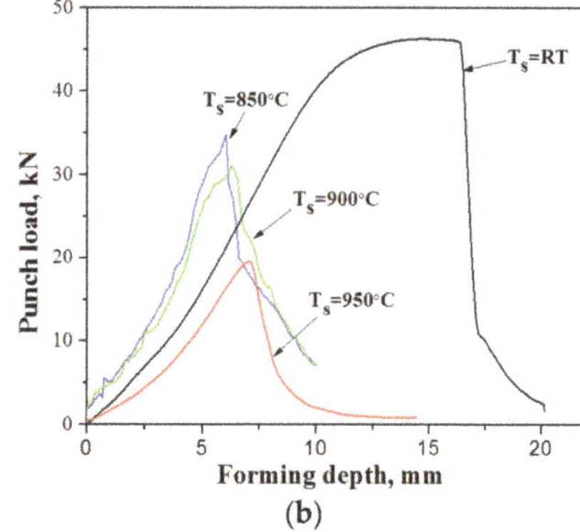

Figure 5. *Cont.*

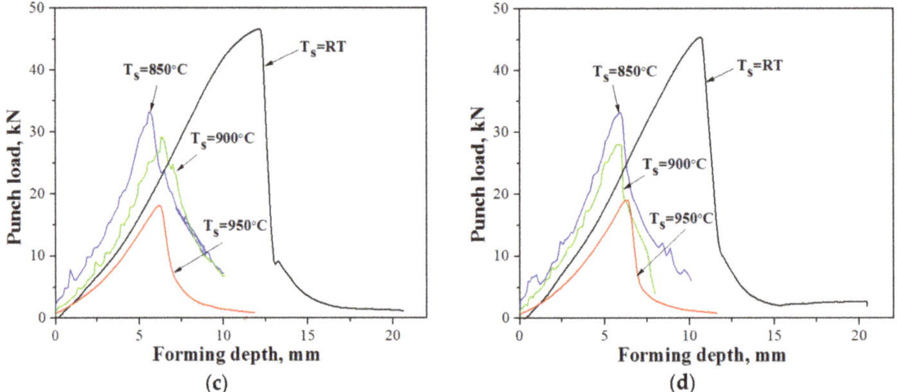

Figure 5. Punch load versus forming depth for indirect (T_s = RT) and direct (T_s = 850, 900 and 950 °C) deep drawing at different blank holding forces: (**a**) B_f = 5 kN; (**b**) B_f = 15 kN; (**c**) B_f = 30 kN; (**d**) B_f = 90 kN.

Figure 6 shows the shapes of the drawn cups with fracture and without any fracture at different blank holding forces. As shown in the figures, in both cold pre-deep drawing and in direct hot deep drawing, the fracture started to occur at the wall part of the drawn cups, which agrees with the analyzed results for the thinning rate. In the drawing process, the cup was under a tensile stress state, and the wall thickness decreased quickly. As blank holding forces increased, a fracture occurred more easily, and a lower forming depth was obtained. Further experiments need to be executed under different lubricant conditions and blank holding forces for cold pre-deep drawing and in direct hot deep drawing. This could also reduce the manufacturing cost by determining the optimal lubricant and improving the formability.

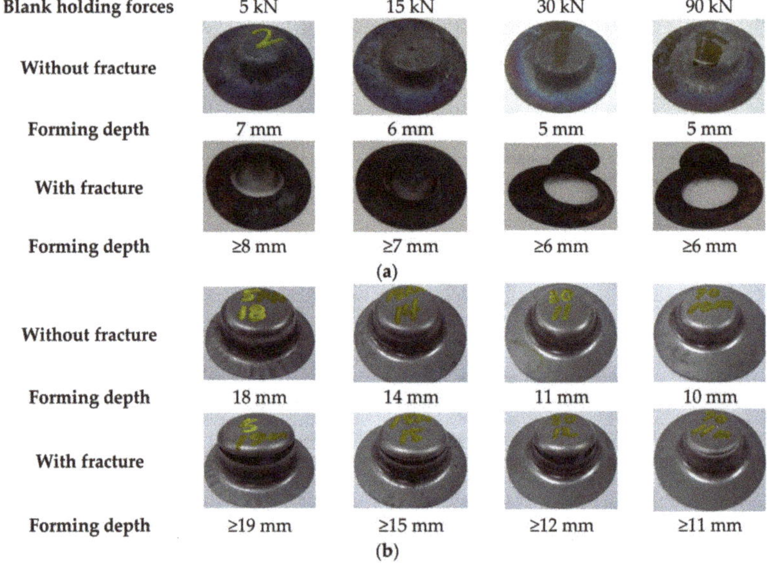

Figure 6. Shape of drawn cups without fracture and with fracture at different blank holding forces: (**a**) direct deep drawing (T_s = 950 °C); (**b**) indirect deep drawing (T_s = RT).

Figure 7 compares the forming depth versus blank holding forces for cold pre-deep drawing and direct hot deep drawing at different blank temperatures. The forming depth was much higher in cold pre-deep drawing than that in direct hot deep drawing. In cold pre-deep drawing (at room temperature), the forming depth values were 18 mm, 14 mm, 11 mm, and 10 mm when blank holding forces were 5 kN, 15 kN, 30 kN, and 90 kN, respectively. In direct hot deep drawing, the forming depth increased as blank temperatures increased from 850 °C to 950 °C. When blank temperatures were 950 °C, the forming depth values were 7 mm, 6 mm, 5 mm, and 5 mm at blank holding forces of 5 kN, 15 kN, 30 kN, and 90 kN, respectively. In addition, the forming depth was examined with T_s = 850 °C and 900 °C. Both in cold pre-deep drawing and direct hot deep drawing, the forming depth decreased with increasing blank holding forces. On the other hand, when blank holding force was larger than 30 kN, there was negligible difference in the forming depth regardless of blank holding forces. In deep drawing, fracture occurs at the wall part of a drawn cup because it lacks ductility under a tensile stress state, and the wall thickness of a drawn cup is affected by the blank holding force [17,18]. As blank holding forces increased, the wall thickness decreased quickly. Thus, fracture occurred more easily at a higher blank holding forces and a lower forming depth was obtained.

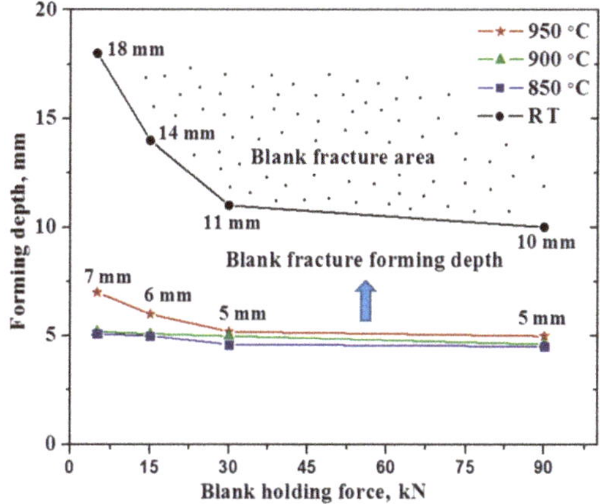

Figure 7. Relationship between forming depth and blank holding force at different blank heating temperatures.

Figure 8 compares the relationships between the maximum punch load without any fracture and the blank holding force for cold pre-deep drawing and direct hot deep drawing at different blank temperatures. The maximum punch load without any fracture in cold pre-deep drawing was much higher than that in direct hot deep drawing. In direct hot deep drawing, the maximum punch load decreased when blank temperatures increased from 850 °C to 950 °C. In cold pre-deep drawing, the maximum punch load without any fracture was approximately 45 kN and did not change much with increasing blank holding force. When T_s = 950 °C, the maximum punch loads were 26 kN, 19 kN, 18 kN, and 18 kN, when the blank holding forces were 5 kN, 15 kN, 30 kN, and 90 kN, respectively. In direct hot deep drawing, the maximum punch load decreased with increasing blank holding force. The maximum punch load without any fracture was much higher when blank holding force was 5 kN compared to a higher blank holding force, and the maximum punch load without any fracture was almost the same under a high blank holding force. A lower blank temperature in the drawing process led to a higher forming force, which could be expressed as a higher punch load. The cold pre-deep drawing experiment was executed at room temperature, and the temperature remained almost the same during the drawing process. On the other hand, in the direct hot deep drawing, the blank temperature decreased during the drawing process because of the temperature difference between the blank and the tools. When the blank holding force was 5 kN, the forming depth was deeper than when blank holding force was 15 kN to 90 kN, which caused a longer drawing time, leading to a lower temperature. Consequently, the lower temperature of the blank required a higher punch load.

As stated above, a deeper forming depth can be achieved in cold pre-deep drawing than direct hot deep drawing. Although cold pre-deep drawing required a higher punch load without any fracture (43–46 kN), it was lower than the normal values in other drawing processes [19]. Thus, when the designed components were complicated and could not be finished with one stroke, indirect hot deep drawing was used widely in the automotive industry. On the other hand, in direct hot deep drawing, when blank temperatures were 950 °C, although a deeper forming depth was achieved, a lower punch load was required than when blank temperatures was 850 °C or 900 °C. This implies that T_s = 950 °C was a better condition than T_s = 900 °C or 850 °C under the same blank holding force by considering the forming depth and maximum punch load. In addition, in both cold pre-deep drawing and direct hot deep drawing, a deeper forming depth could be obtained when blank holding force was 5 kN compared with a higher blank holding force.

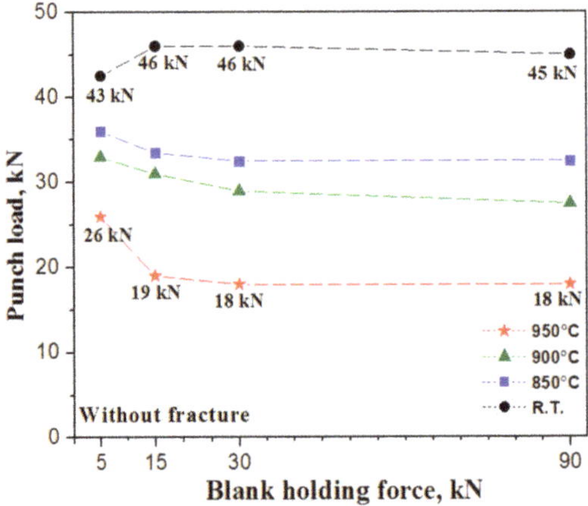

Figure 8. Relationship between maximum punch load without fracture and blank holding force at different blank heating temperatures.

Figure 9 compares the thinning rate at each position for samples without any fracture and with fracture for direct hot deep drawing (T_s = 950 °C) and indirect hot deep drawing (RT) at B_f = 5 kN. As shown in Figure 9, the thinning rate in position③was much higher than the rates at the other positions. This indicated that in both direct hot deep drawing and in cold pre-deep drawing, the fracture started to occur at the wall part of the drawn cups. In the drawing process, the cup was under a tensile stress state, and the wall thickness decreased quickly. In addition, the thinning rate at each position obtained in cold pre-deep drawing was higher than that in direct hot deep drawing. In the samples without any fracture, at position ③, the thinning rate was 17% and the thickness of the blank sheet was 0.5 mm in the case of direct hot deep drawing, whereas the thinning rate was 20% and the thickness of the blank sheet was 0.48 mm in the case of cold pre-deep drawing. This means that fracture occurred more easily in cold pre-deep drawing than in direct hot deep drawing considering the thinning rate. In addition, because of the material gathering in the drawing process, the thickness at position②was greater than at positions①,④, and⑤.

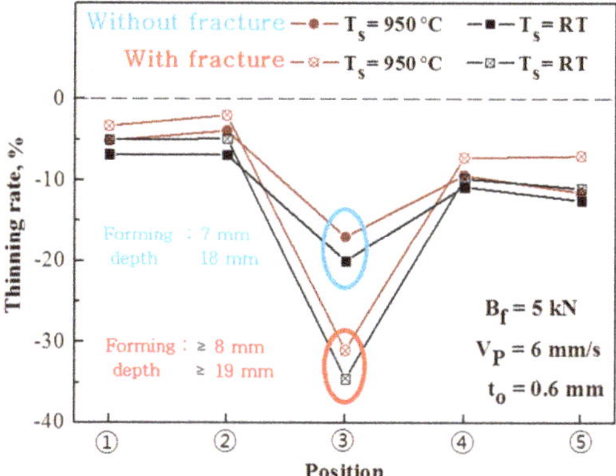

Figure 9. Thinning rate at each position for deep drawing samples with and without fracture.

The microstructure and hardness at different positions were checked for direct and indirect hot deep drawing. Figure 10 shows the microstructure at different positions of the drawn parts for direct hot deep drawing (T_s = 950 °C) and indirect hot deep drawing when B_f = 5 kN. Temperature plays an important role in the microstructure and hardness. The drawn part was quenched more rapidly in water quenching than in-die quenching. This fast quenching speed ensured the transformation of austenite to martensite and promoted the generation of the nucleuses, with smaller nucleuses leading to higher strength and hardness values. Martensite microstructure was not found in direct hot deep drawing because both T_p and T_d were kept at 300 °C in direct hot deep drawing, which led to a low quenching speed. Thus, a better microstructure was obtained in indirect hot deep drawing.

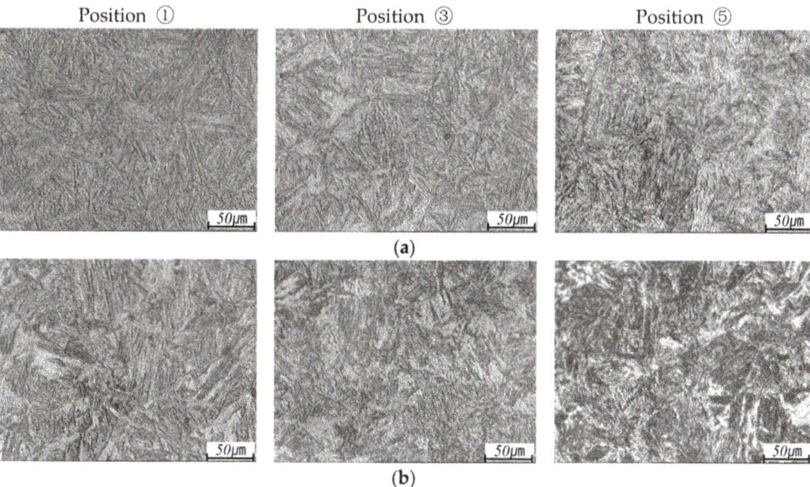

Figure 10. Microstructures at①,③, and⑤of positions for hot and cold deep drawn samples: (**a**) hot deep drawing (T_s = 950 °C); (**b**) cold deep drawing and water quenching after heating temperature to 900 °C.

Figure 11 shows the hardness values at different positions of the drawn part for direct hot deep drawing (T_s = 950 °C) and indirect hot deep drawing when B_f = 5 kN. As can be seen, the hardness after water quenching in indirect hot deep drawing was much higher than the die quenching in direct hot deep drawing for the measured positions, which was a result of the smaller nucleuses gained from water quenching. Both in direct and indirect hot deep drawings, the highest hardness was measured at a position ④ near the punch round, which was caused by work-hardening during the drawing process. The strength and hardness of indirect hot deep drawing with water quenching were better than those of direct hot deep drawing with die quenching.

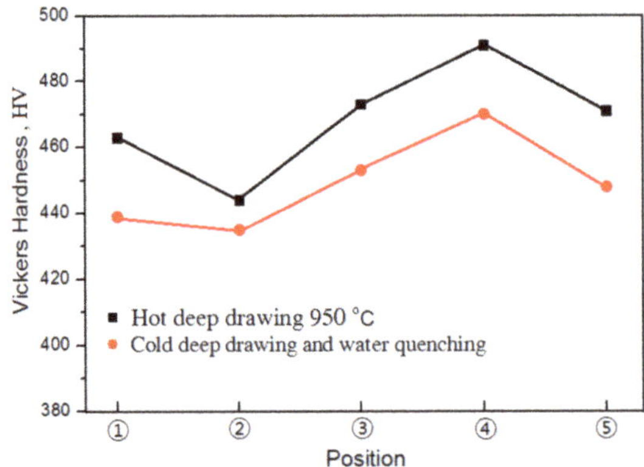

Figure 11. Vickers hardness at different positions of the drawn part for hot deep drawing (T_s = 950 °C) and cold deep drawing.

4. Conclusions

(1) The forming depth in cold deep drawing was noticeably deeper than that in hot deep drawing. A much greater maximum punch load was achieved in cold deep drawing than hot deep drawing. Cold deep drawing was found to be a better forming process.

(2) In hot deep drawing, for T_s = 950 °C, a deeper forming depth was achieved. T_s = 950 °C was required a lower maximum punch load than in the case of T_s = 850 °C or T_s = 900 °C.

(3) The thinning rate at each position obtained in cold deep drawing was higher than that in hot deep drawing.

(4) The microstructure and hardness in cold deep drawing, with water quenching after heating the drawn cups to 900 °C, were significantly better than those in hot forming at T_s = 950 °C.

Author Contributions: H.Y.S. and C.K.J. designed the experiment tools and performed the experiment. C.K.J. and C.G.K. analyzed the experimental results, whereas C.K.J. maintained and examined them. All authors have contributed to the discussions as well as revisions.

Acknowledgments: This work was supported by the National Research Foundation of Korea (NRF) grant funded by the Korea government (MSIT) through GCRC-SOP (No. 2011-0030013) and by the National Research Foundation of Korea (NRF) grant funded by the Korea government (MSIT) (No. 2017R1A2B4007884).

Conflicts of Interest: The authors declare no conflicts of interest.

References

1. Jarvinen, H.; Honkanen, M.; Jarvenpaa, M.; Peura, P. Effect of paint baking treatment on the properties of press hardened boron sheet. *J. Mater. Process. Technol.* **2018**, *252*, 90–104. [CrossRef]

2. Omer, K.; George, R.; Bardelcik, A.; Worswick, M.; Malcolm, S.; Detwiler, D. Development of a hot stamped channel section with axially tailored properties—Experiments and model. *Int. J. Mater. Form.* **2018**, *11*, 149–164. [CrossRef]

3. Taylor, T.; Clough, A. Critical review of automotive hot-stamped sheet steel from an industrial perspective. *Mater. Sci. Technol.* **2018**, *34*, 809–861. [CrossRef]

4. Li, Y.; Jeon, Y.P.; Kang, C.G. Experimental assessment of high temperature formability of boron steel sheet manufactured with a spring compound bending die. *J. Eng. Mater. Technol.* **2012**, *134*, 021019. [CrossRef]

5. Naderi, M.; Saeed, A.; Bleck, W. The effects of non-isothermal deformation on martensitic transformation in 22MnB5 steel. *Mater. Sci. Eng. A* **2008**, *487*, 445–455. [CrossRef]

6. Altan, T. Hot-stamping boron-alloyed steels for automotive parts. *Stamp. J.* **2007**, *19*, 10–13.

7. Xing, Z.W.; Bao, J.; Yang, Y.Y. Hot stamping processing experiments of quenchable boron steel. *Mater. Sci. Technol.* **2008**, *16*, 172–175.

8. So, H.; Faßmann, D.; Hoffmann, H.; Golle, R.; Schaper, M. An investigation of the blanking process of the quenchable boron alloyed steel 22MnB5 before and after hot stamping process. *J. Mater. Process. Technol.* **2012**, *212*, 437–449. [CrossRef]

9. Nakagawa, Y.; Mori, K.I.; Maeno, T. Springback-free mechanism in hot stamping of ultra-high-strength steel parts and deformation behavior and quenchability for thin sheet. *Int. J. Adv. Manuf. Technol.* **2018**, *95*, 459–467. [CrossRef]

10. Löbbe, C.; Tekkaya, A.E. Mechanisms for controlling springback and strength in heat-assisted sheet forming. *CIRP Ann.* **2018**, *67*, 273–276. [CrossRef]

11. Borsetto, F.; Ghiotti, A.; Bruschi, S. Investigation of the high strength steel Al–Si coating during hot stamping operations. *Key Eng. Mater.* **2009**, *410–411*, 289–296. [CrossRef]

12. Naderi, M. Hot Stamping of Ultra High Strength Steels. Ph.D. Thesis, RWTH Aachen University, Aachen, Germany, 2007.

13. Ryan, G.; Alexander, B.; Michael, W. Development of a hot forming die to produce parts with tailored mechanical properties-numerical study. In Proceedings of the Hot Sheet Metal Forming of High-Preformance Steel 2th International Conference, Lulea, Sweden, 15–17 June 2009; pp. 189–198.

14. Ouyang, Y.; Lee, M.S.; Moon, J.H.; Kang, C.G. The effect of the blank holding force on formability in hot deep drawing of boron steel considering heat transfer phenomena and friction coefficient by simulation and experimental investigation. *Proc. Inst. Mech. Eng. Part B J. Eng. Manuf.* **2012**, *226*, 1506–1518. [CrossRef]

15. Moon, J.H.; Lee, M.S.; Seo, P.K.; Kang, C.G. A study on mechanical properties of laser-welded blank of a boron sheet steel by laser ablation variable of Al-Si coating layer. *Int. J. Precis. Eng. Manuf.* **2013**, *14*, 283–288. [CrossRef]

16. Seok, H.H.; Mun, J.C.; Kang, C.G. Micro-crack in zinc coating layer on boron steel sheet in hot deep drawing process. *Int. J. Precis. Eng. Manuf.* **2015**, *16*, 919–927. [CrossRef]

17. Yoshihara, S.; Manabe, K.; Nishimura, H. Effect of blank holder force control in deep-drawing process of magnesium alloy sheet. *J. Mater. Process. Technol.* **2005**, *170*, 579–585. [CrossRef]

18. Wifi, A.; Mosallam, A. Some aspects of blank-holder force schemes in deep drawing process. *JAMME* **2007**, *24*, 315–323.

19. Kim, H.; Sung, J.H.; Sivakumar, R.; Altan, T. Evaluation of stamping lubricants using the deep drawing test. *Int. J. Mach. Tools Manuf.* **2007**, *47*, 2120–2132. [CrossRef]

© 2018 by the authors. Licensee MDPI, Basel, Switzerland. This article is an open access article distributed under the terms and conditions of the Creative Commons Attribution (CC BY) license (http://creativecommons.org/licenses/by/4.0/).

 *metals*

Article

Modeling of Forming Limit Bands for Strain-Based Failure-Analysis of Ultra-High-Strength Steels [†]

Hamid Reza Bayat [1],*, Sayantan Sarkar [1], Bharath Anantharamaiah [1], Francesco Italiano [2], Aleksandar Bach [2], Shashidharan Tharani [1], Stephan Wulfinghoff [1] and Stefanie Reese [1]

[1]   Institute of Applied Mechanics, RWTH Aachen University, Mies-van-der-Rohe-Str. 1, 52074 Aachen, Germany; sayantan.sarkar@hs-osnabrueck.de (S.S.); bharath.anantharamaiah@rwth-aachen.de (B.A.); shashidharan.tharani@rwth-aachen.de (S.T.); stephan.wulfinghoff@rwth-aachen.de (S.W.); stefanie.reese@rwth-aachen.de (S.R.)

[2]   Ford Werke GmbH, Research and Innovation Center Aachen, Suesterfeldstrasse 200, 52072 Aachen, Germany; fitalian@ford.com (F.I.); sabach3@ford.com (A.B.)

*   Correspondence: hamid.reza.bayat@rwth-aachen.de; Tel.:+49-(0)241-802-5004

†   This paper is an extended version of paper published in the 21st International ESAFORM Conference on Material Forming: ESAFORM 2018, Palermo, Italy, 23–25 April 2018.

Received: 9 July 2018; Accepted: 7 August 2018; Published: 10 August 2018

Abstract: Increased passenger safety and emission control are two of the main driving forces in the automotive industry for the development of light weight constructions. For increased strength to weight ratio, ultra-high-strength steels (UHSSs) are used in car body structures. Prediction of failure in such sheet metals is of high significance in the simulation of car crashes to avoid additional costs and fatalities. However, a disadvantage of this class of metals is a pronounced scatter in their material properties due to e.g., the manufacturing processes. In this work, a robust numerical model is developed in order to take the scatter into account in the prediction of the failure in manganese boron steel (22MnB5). To this end, the underlying material properties which determine the shapes of forming limit curves (FLCs) are obtained from experiments. A modified Marciniak–Kuczynski model is applied to determine the failure limits. By using a statistical approach, the material scatter is quantified in terms of two limiting hardening relations. Finally, the numerical solution obtained from simulations is verified experimentally. By generation of the so called forming limit bands (FLBs), the dispersion of limit strains is captured within the bounds of forming limits instead of a single FLC. In this way, the FLBs separate the whole region into safe, necking and failed zones.

Keywords: forming limit curve; inhomogeneity; boron steel; robustness evaluation

---

1. Introduction

The ability to assess with possibility of the forming limits of sheet metals is critical to avoid excessive thinning or localized necking. Forming limit curves (FLCs) are one of the highly recognized tools to foresee the failure in sheet metals. The concept of FLCs was first introduced by Keeler and Backofen for the tension-tension zone [1] and then further extended to the tension-compression zone by Goodwin [2]. Over the years, different experimental and numerical methods have been developed for the accurate determination of FLCs. The Nakazima out-of-plane test and the Marciniak in-plane test [3] are well-known approaches to experimentally generate FLCs. Besides experimental methods, numerical methods are used to investigate failure. Thereby, an important aspect is the strain localization during forming.

The available numerical methods can be divided into three main frameworks: the maximum force criteria, the Marciniak–Kuczynski (MK) models and the finite element (FE) methods. A basic necking criterion in a simple tension case was established by Considère [4] which was thereafter extended to biaxial stretching by Hill [5] and Swift [6]. Hill [5] expressed the localized necking as a discontinuity in the velocity and Swift [6] determined the instability condition in the plastic strain by expressing the yield stress as a function of the induced stress during the deformation of the diffused necking. Using Considère's criteria, Hora and Tong [7] introduced 'modified maximum force criteria' (MMFC) taking into account the strain path after diffuse necking. These criteria have been used for the FLC determination under non-linear strain paths by Tong et al. [8,9]. The enhanced modified maximum force criterion (eMMFC) was introduced by Hora et al. [10] in 2008 considering the sheet thickness and the curvatures of the parts. Another approach to determine the necking is the bifurcation analysis that was discussed by Hill [11].

To overcome the drawback in the maximum force model, Marciniak and Kuczynski [3] developed a new model considering a pre-existing inhomogeneity in sheet metals. The instability begins along the inhomogeneity due to a gradual strain concentration under biaxial stretching. Strain-rate sensitivity and plane anisotropy have been further studied in the later works as additional influencing parameters [12]. Hutchinson and Neale [13] analyzed an imperfection sensitive MK model with J_2 (von Mises) flow theory for principal strain states varying from uniaxial to equibiaxial tension. In the work of Chan et al. [14], localized necking is studied for the negative minor strain region (uniaxial tension to plane strain). Additionally, the inhomogeneity oriented in zero-extension direction is analyzed in [5]. The complete FLCs of anisotropic rate sensitive materials with orthotropic symmetry have been predicted for linear and complex strain paths in the work of Rocha et al. [15].

The shape of the FLC depends on the constitutive equations used in the MK model. Yield criteria as well as the hardening relation can alter the limit strains. Different parameters influencing the FLCs are studied in the literature. Lian et al. [16] studied the variation of sheet metal stretchability by varying the shape of the yield surface. A yield function that describes the behavior of orthotropic sheet metals under full plane stress state was introduced by Barlat and Lian [17]. Cao et al. [18] implemented the general anisotropic yield criterion for localized thinning in the sheet metals into the MK model. Additional failure modes in sheet metals, namely ductile and shear failure criteria were included in CrachFEM (MATFEM, Munich, Germany). Eyckens et al. [19] extended the MK model to predict localized necking considering through-thickness shear (TTS) during the forming operations.

In addition to MK and MMFC methods, the finite element (FE)-based approach, introduced by Burford and Wagoner [20], is another way to determine FLCs. Boudeau and Gelin [21] worked on the prediction of localized necking in sheet metals during the forming processes through FE simulations. Combining a ductile fracture criterion with finite element simulations, Takuda et al. [22] predicted the limit strains under biaxial stretching. Different finite element methods for the prediction of FLCs can be found in [23–26].

Due to the scatter in material properties at different regions of sheet metals, it becomes difficult to precisely predict the failure strains by a single FLC. Hence, a scatter of material properties has to be determined and a band of FLCs needs to be defined. Janssens et al. [27] introduced the concept of forming limit bands (FLBs) to have a reliable estimate of the uncertainty of FLCs. Strano and Colosimo [28] extended it further with logistic regression analysis to generate FLBs from experimental data. Chen and Zhou [29] applied percent regression analysis to the curve fitting of experimental data of limit strains. The above-mentioned studies analyze the experimental data statistically to obtain FLBs. On the other hand, Banabic et al. [30] was one of the first to incorporate the concept of FLBs in a theoretical approach to improve the robustness. The MK model with BBC2003 [31] plasticity criterion and Hollomon and Voce hardening laws were used in the derivation of the limit strains. Later, Comsa et al. [32] generated an FLB using Hora's MMFC model with the Hill'48 yield criterion and Swift's hardening law.

The aim of this work is to develop a methodology to predict the failure in UHSS sheet metals during a car crash. To this end, first, the limit strains of a car component are evaluated experimentally. Then, the scatter of the material properties is considered by defining a range of hardening relations using curve fitting of the experimental data. A modified MK model with a simplification related to the zero extension angle is applied to form the bounds of the limit strains numerically. In this way, the bounds can capture the material scatter via a parametric study.

The present paper is structured as follows. In Section 2, a modified MK model with an inclined groove is presented based on the work of Rocha et al. [15]. To solve the discrete equations, an implicit integration method is employed in Section 3. The obtained results are validated in Section 4 based on the theory of the method. Furthermore, the experimental data and their post-processing are described in Section 5 to determine the material scatter. Based on a statistical analysis of the material scatter, the forming limit bands are generated and discussed in Section 6. Eventually, the results are summarized and conclusions are drawn.

2. Theory of Forming Limit Curve

A sheet metal is subjected to biaxial stretching under a load given in terms of the principal stresses σ_1 and σ_2 as shown in Figure 1. The metal component has two regions with different thicknesses named as region A and region B. The groove, that is denoted as region B, is considered to make an angle ψ with respect to the minor principal stress axis as shown in Figure 1. Although the thickness variation is smooth in reality, a sharp variation is considered to simplify the calculations. Due to its smaller thickness, region B will represent the necking region during the stretching operation.

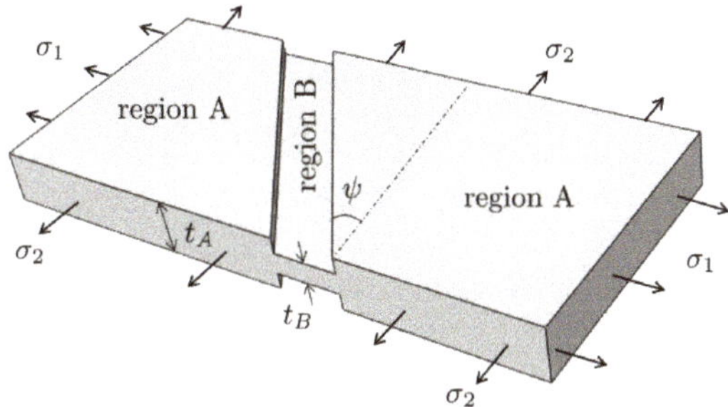

Figure 1. Modified MK model with initial in-homogeneity [33].

Marciniak [3] introduced an inhomogeneity parameter expressed in terms of the ratio of the initial thicknesses of region B (t_{0B}) and A (t_{0A}). In the present work, it is denoted by f_i and defined as

$$f_i = 1 - \frac{t_{0B}}{t_{0A}}, \tag{1}$$

which implies that, when f_i takes the value zero, the sheet is geometrically homogeneous. As stresses increase, the sheet metal undergoes different strain increments in non-necking and necking regions as indicated by subscripts A and B, respectively. The ratio of minor to major strain increments is assumed to remain constant within each load path. It is defined by

$$\gamma = \frac{d\varepsilon_{2A}}{d\varepsilon_{1A}}. \tag{2}$$

Strain and stress states of both regions (A and B) are monitored separately. In this way, one can define plastic instability as the increment of the equivalent plastic strain in region B becomes considerably greater than that of region A. The plastic instability indicator β is expressed as the ratio between the equivalent plastic strain increments in region A ($d\bar{\varepsilon}_A$) and region B ($d\bar{\varepsilon}_B$) [34]:

$$\beta = \frac{d\bar{\varepsilon}_A}{d\bar{\varepsilon}_B}. \tag{3}$$

In practice, the value of β is considered as 0.1 to indicate the loss of stability. The model is derived for an anisotropic material with an orthotropic symmetry. The classical Lankford coefficients R_0, R_{45} and R_{90} are considered as measures of anisotropy with respect to different directions of in-plane loading (angles $0°$, $45°$ and $90°$ with respect to the rolling direction). Additionally, planar anisotropy is assumed and characterized by a time-independent average R-value defined by

$$\bar{R} = \frac{R_0 + 2R_{45} + R_{90}}{4}. \tag{4}$$

In the case of isotropic material, the value of $\bar{R}$ becomes equal to unity. We define

$$\alpha = \sigma_2/\sigma_1 \tag{5}$$

as the ratio of stresses in the principal directions, namely σ_1 and σ_2 [15]. The Hill'48 yield criterion is used in combination with an associative flow rule and Swift hardening relation. Hill's criterion for in plane stress conditions is expressed as

$$2f = \bar{\sigma}^2 = F\sigma_{yy}^2 + G\sigma_{xx}^2 + H(\sigma_{xx} - \sigma_{yy})^2 + 2P\sigma_{xy}^2, \tag{6}$$

where $\bar{\sigma}$ denotes the equivalent stress and F, G, H, and P are functions of the Lankford coefficients. Thus, they are material specific constants. The reader is referred to the Appendix A for more information about these constants. In the following, strain rate dependent stress relation similar to the classical MK model [12] is defined:

$$\bar{\sigma} = C_1 \left(\varepsilon_0 + \bar{\varepsilon}\right)^n \dot{\bar{\varepsilon}}^m. \tag{7}$$

In the latter relation, C_1 is a strength coefficient and ε_0 denotes the initial yield strain. In addition, $\bar{\varepsilon}$ represents the equivalent plastic strain with n as isotropic hardening exponent. Furthermore, m is the strain rate exponent. The associated flow rule is given by

$$d\varepsilon_{ij} = d\lambda' \frac{\partial f}{\partial \sigma_{ij}}, \tag{8}$$

where $d\lambda'$ is the hardening parameter being expressed as

$$d\lambda' = d\bar{\varepsilon}/\bar{\sigma}$$

and ε_{ij} are logarithmic strain components. Similar to the classical MK model, the incompressibility condition is given by

$$d\varepsilon_{1A} + d\varepsilon_{2A} + d\varepsilon_{3A} = 0. \tag{9}$$

Considering the strain in thickness direction, the sheet thickness can be written as

$$t_A = t_{0A}\, e^{\varepsilon_{3A}}, \; t_B = t_{0B}\, e^{\varepsilon_{3B}}. \tag{10}$$

The force equilibrium conditions are given by

$$\sigma_A^{nn}\, t_A = \sigma_B^{nn}\, t_B, \tag{11}$$

$$\sigma_A^{nt}\, t_A = \sigma_B^{nt}\, t_B, \tag{12}$$

where n and t (in Equations (11) and (12)) denote the normal and tangential direction with respect to the inhomogeneity, respectively. The compatibility condition across the discontinuity is defined by

$$d\varepsilon_A^{tt} = d\varepsilon_B^{tt}. \tag{13}$$

First, the force equilibrium in normal direction (Equation (11)) is divided by $\bar{\sigma}$. Next, by using Equations (1) and (7) in (11) and upon simplification, the following relation is derived

$$\frac{\cos^2\psi + \alpha\,\sin^2\psi}{\sqrt{1 + (F+H)\,\alpha^2 - 2H\,\alpha}} = (1 - f_i)\,exp(\varepsilon_{3B} - \varepsilon_{3A}) \left(\frac{\varepsilon_0 + \bar{\varepsilon}_B}{\varepsilon_0 + \bar{\varepsilon}_A}\right)^n \left(\frac{d\bar{\varepsilon}_B}{d\bar{\varepsilon}_A}\right)^m \sqrt{\frac{B_1\,(d\bar{\varepsilon}_A/d\bar{\varepsilon}_B)^2 + B_2}{B_3}}. \tag{14}$$

Equation (14) represents the residual in the algorithm described in Section 3. Strain increments in the third direction for regions A and B are derived by the application of compatibility and incompressiblity conditions

$$d\varepsilon_{3A} = -\frac{G + F\,\alpha}{\sqrt{1 + (F+H)\,\alpha^2 - 2H\,\alpha}}\,d\bar{\varepsilon}_A, \tag{15}$$

$$d\varepsilon_{3B} = -\left[H_6\,\sqrt{\frac{B_1\,(d\bar{\varepsilon}_A/d\bar{\varepsilon}_B)^2 + B_2}{B_3}} + H_7\left(\frac{d\bar{\varepsilon}_A}{d\bar{\varepsilon}_B}\right)\right]d\bar{\varepsilon}_B, \tag{16}$$

where B_1, B_2, B_3, H_6 and H_7 are functions of R_0, R_{45}, R_{90}, ψ and α. For a detailed derivation the reader is referred to the Appendix A and to Rocha et al. [15].

The above mentioned mathematical model is capable of evaluating the limit strain for both positive and negative minor strains ε_2. There are four different fundamental strain paths, i.e., uniaxial tension, plane strain, biaxial and equibiaxial tension. The relation between the stress ratio α and the strain ratio γ for an isotropic material is given by [14,34]

$$\alpha = \frac{2\gamma + 1}{2 + \gamma}. \tag{17}$$

Table 1 shows the corresponding values of α calculated from Equation (17).

Table 1. Incremental strain and stress ratios for four different load paths.

-	Uniaxial	Plane Strain	Biaxial	Equibiaxial
γ	−0.5	0	0.5	1
α	0	0.5	0.8	1

In the tension-compression quarter of the FLC, the limit strains depend on the (initial) orientation of the groove. Therefore, an arbitrary angle ψ between the direction of the imperfection and the direction of the principal minor stresses is considered as described in Figure 1. As FLCs represent the maximum allowable strains, the minimum limit strains of all the possible orientations of the groove are to be identified. Upon implementing the rotation of angle from $0°$ to $45°$ to evaluate the minimum strain, the computational cost of the algorithm is increased. This can be simplified by using the concept of zero extension direction provided by Hill's theory [5]. As predicted by this theory, if the imperfection is oriented along the zero extension direction, there will be a substantial difference in the limit strain compared to that of $\psi = 0°$ [14].

According to the theory of Hill [5], localized necking occurs along the direction of zero extension, which is determined by [35]

$$\psi^* = tan^{-1}\sqrt{-\gamma}. \tag{18}$$

The zero extension angle in case of uniaxial tension for planar isotropic material is found to be $\psi^* = 35.3°$. From Chan et al. [14], the calculated limit strains using the zero extension direction are within 2% of the limit strains calculated based on the aforementioned rotation of the imperfection. Thus, for the purpose of simplification and faster computation the angular orientation of the discontinuity is set to the zero extension angle, which is investigated later in this work.

3. Numerical Solution and Algorithm

The numerical solution is performed by means of an implicit integration method. To this end, the algorithm is implemented using two loops in FORTRAN similar to that of Werner [34]. Here, the value of β is unknown and evaluated as a function of the equivalent plastic strain in the necking region. The computation continues until β reaches a critical value of 0.1. The iterative technique used to solve for β is shown in Figure 2. In the outer loop (n loop), $\bar{\varepsilon}_B$ is increased by a constant increment $d\bar{\varepsilon}_B$. In this work, $d\bar{\varepsilon}_B$ is set to 0.001 (refer to Figure 3). This is given by

$$\bar{\varepsilon}_B|_{n+1} = \bar{\varepsilon}_B|_n + d\bar{\varepsilon}_B. \tag{19}$$

The inner loop (k loop) is running for a maximum number of 100 times by varying β from 1 to 0 in steps of 0.01. This loop terminates when the sign of the residual represented in Equation (14) changes. The plastic strain $d\bar{\varepsilon}_A$ in the non-necking region is calculated by arithmetic averaging of two consecutive β values within the n loop (refer to Equation (22)). To start the algorithm $\bar{\varepsilon}_A$, $\bar{\varepsilon}_B$, ε_{3A}, ε_{3B} as well as $d\bar{\varepsilon}_A/d\bar{\varepsilon}_B = \beta$ must be initialized. Here, the initial conditions as set as follows:

$$\bar{\varepsilon}_A|_0 = \bar{\varepsilon}_B|_0 = 0,$$
$$\varepsilon_{3A}|_0 = \varepsilon_{3B}|_0 = 0 \text{ and} \tag{20}$$
$$\beta_0 = 1.$$

The inhomogeneity parameter f_i is manually input. In every k loop, first $d\varepsilon_{3B}$ is calculated using Equation (16) since the values of $d\bar{\varepsilon}_A$, $d\bar{\varepsilon}_B$, α and $\bar{R}$ are already known. In the next step, ε_{3B} is computed using the equation

$$\varepsilon_{3B}|_k = \varepsilon_{3B}|_n + d\varepsilon_{3B}|_k, \tag{21}$$

where the variable $\varepsilon_{3B}|_n$ is already initialized. Next, $d\bar{\varepsilon}_A|_k$ is calculated using the arithmetic averaging:

$$d\bar{\varepsilon}_A|_k = 0.5 \left(\beta_n + \beta_k \right) d\bar{\varepsilon}_B. \tag{22}$$

Having initialized β and $\bar{\varepsilon}_A|_n$, we determine $\bar{\varepsilon}_A|_k$ using the relation

$$\bar{\varepsilon}_A|_k = \bar{\varepsilon}_A|_n + d\bar{\varepsilon}_A|_k. \tag{23}$$

Next, ε_{3A} is calculated in analogy to the previous equation by using $d\varepsilon_{3A}$ from Equation (15)

$$\varepsilon_{3A}|_k = \varepsilon_{3A}|_n + d\varepsilon_{3A}|_k. \tag{24}$$

Hardening relations $\bar{\sigma}_A$ and $\bar{\sigma}_B$ are computed using $\bar{\varepsilon}_A|_k$, $\bar{\varepsilon}_B|_{n+1}$ and the expression

$$\dot{\bar{\varepsilon}}_B|_k = \dot{\bar{\varepsilon}}_A/\beta_k$$

which is obtained by

$$\frac{\dot{\bar{\varepsilon}}_B}{\dot{\bar{\varepsilon}}_A} = \frac{d\bar{\varepsilon}_B}{dt}\frac{dt}{d\bar{\varepsilon}_A}. \tag{25}$$

For different load cases, the value of α is considered in Table 1. Other load cases are set by defining the value of α between 0 to 1.

In this work, hardening relations are fitted to experiments and implemented in the code.

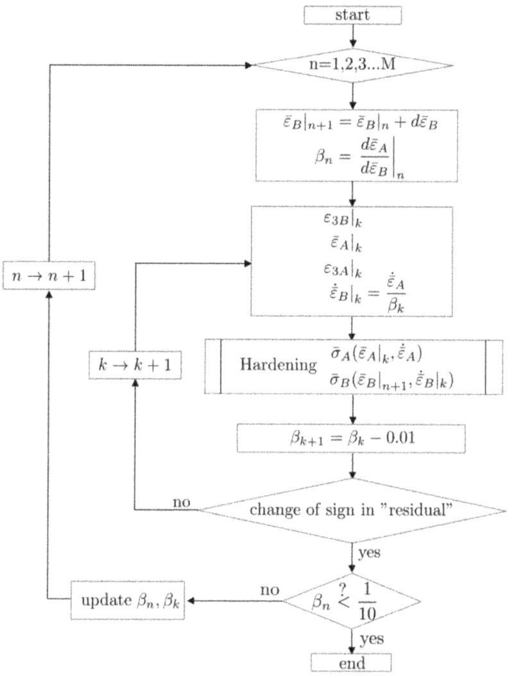

Figure 2. Implemented algorithm in FORTRAN.

4. Verification

To verify the convergence properties of the solution method, different step sizes of $d\bar{\varepsilon}_B$ are set ranging from 2×10^{-5} to 0.2. Without changing the other parameters such as hardening relations, anisotropy etc., the limit strains in region A are evaluated. It is clearly seen in Figure 3 that the limit strains ($\bar{\varepsilon}_A$) are close to convergence as $d\bar{\varepsilon}_B$ approaches 0.002. Therefore, the step size is set to 0.001. Henceforth, all strains and stress measures are normalized to the mean ultimate tensile strength (R_m) and the mean uniform strain of entire samples, respectively.

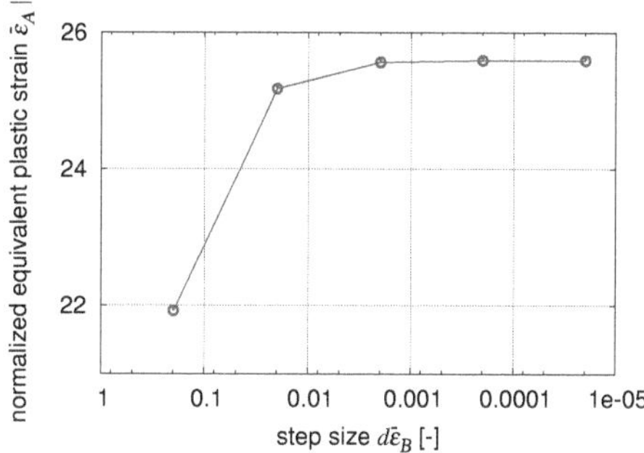

Figure 3. Convergence of the numerical method (strain normalized with respect to the mean uniform strain).

To verify the results, a hardening relation for a manganese boron steel (22MnB5) is applied. In addition, the material is assumed to be isotropic in this work. Nonetheless, due to the confidentiality of the project the parameters of the hardening relation (Equation (7)) are not explicitly revealed. In Figure 4, the evolution of the equivalent plastic strain ($\bar{\varepsilon}_A$) in the non-necking region is plotted against its counterpart ($\bar{\varepsilon}_B$) in the necking region. Initially, the slope of the curve is found to be approximately 1, since both regions are undergoing the same strain. At some point, the strain in region B starts to grow significantly faster than the one in region A. This is due to the localized necking in region B since it has a smaller cross-section. The difference between the strains in regions A and B is captured vividly for normalized strains in region B greater than 10. In addition, the stress-strain curve in region B is plotted in Figure 5 for sheet metals subjected to equibiaxial tension. This plot captures the implemented hardening relation and represents the material behavior in the non-linear regime.

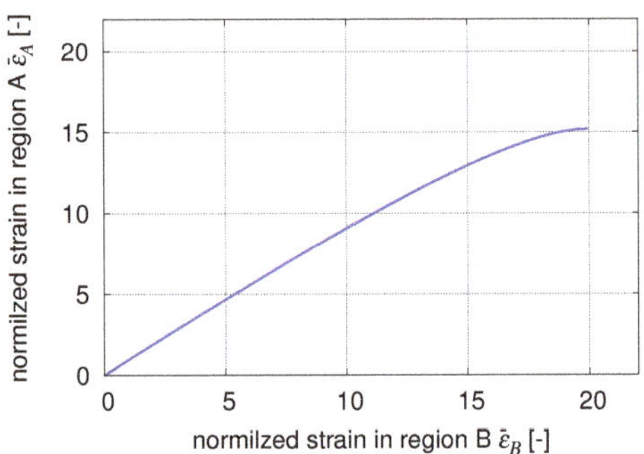

Figure 4. Normalized equivalent plastic strains in regions A and B (strain normalized with respect to the mean uniform strain).

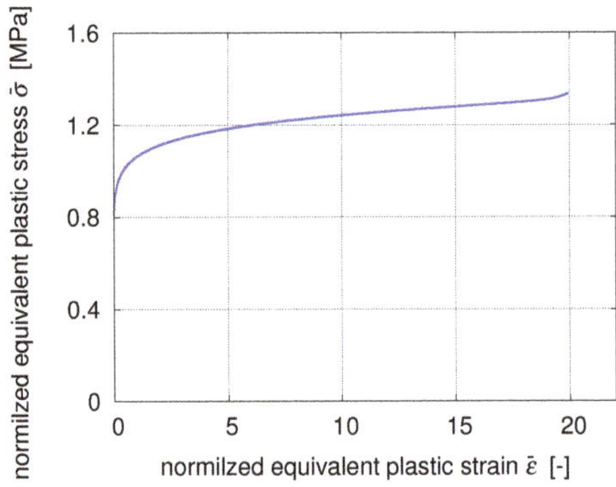

Figure 5. Normalized equivalent plastic stress as a function of normalized equivalent plastic strain under equibiaxial tension (stress and strain normalized with respect to the mean ultimate tensile strength and the mean uniform strain, respectively).

The maximum admissible strains that the material can withstand before the onset of necking are shown in Figure 6. Localized necking is evaluated by considering a constant predefined value of strain ratio γ during the deformation process. The entire curve is obtained by varying γ from -0.5 to 1 in steps of 0.5.

The limit strains can also be related to the equivalent plastic strain as shown in Figure 7. In this figure, the load path dependence of the limit strains is clearly evident. For instance, it is noticeable that sheet metals can deform to a higher extent under equibiaxial tension compared to uniaxial tension. Moreover, the choice of the inhomogeneity parameter f_i plays a crucial role in the position of an FLC.

The higher the inhomogeneity parameter, the lower is the entire FLC. In this section, it is set to 0.02 for all loading paths. A detailed investigation of the aforementioned parameter is discussed in Section 6.

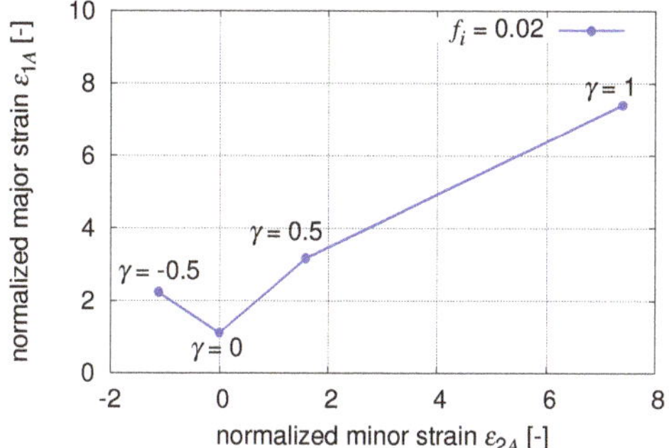

Figure 6. Forming limit curves (FLC)—major strain vs minor strain (strain normalized with respect to the mean uniform strain).

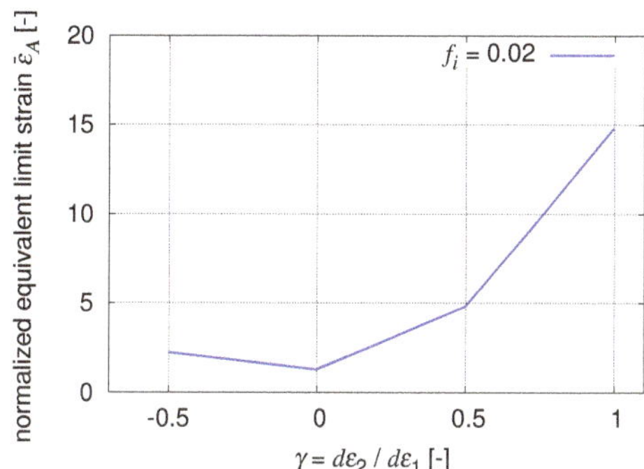

Figure 7. FLC—normalized equivalent limit strain for different loading paths (strain normalized with respect to the mean uniform strain).

The effect of the angular orientation of the inhomogeneity in the tension-compression regime is studied in Figures 8 and 9. It is evident that the choice of $\psi = 0°$ leads to an overestimation of the limit strains as depicted in Figure 9. This signifies the application of the angled groove for the tension-compression loading regime of the FLC. Furthermore, to compare the strains for any arbitrary orientation of the angle with that of the zero extension angle ($\bar{\varepsilon}_A^*$) from [35], an interval of 0°–50° is considered, due to the symmetric influence of the orientation of the groove. The minimum equivalent strain ($\bar{\varepsilon}_{A,min}$) is found at the angle 34°. On the other hand, the zero extension angle obtained here is equal to 35.3° which yields approximately the same limit strain $\bar{\varepsilon}_A^*$ as $\bar{\varepsilon}_{A,min}$. This is seen clearly in Figure 9, where both strains are mostly equal.

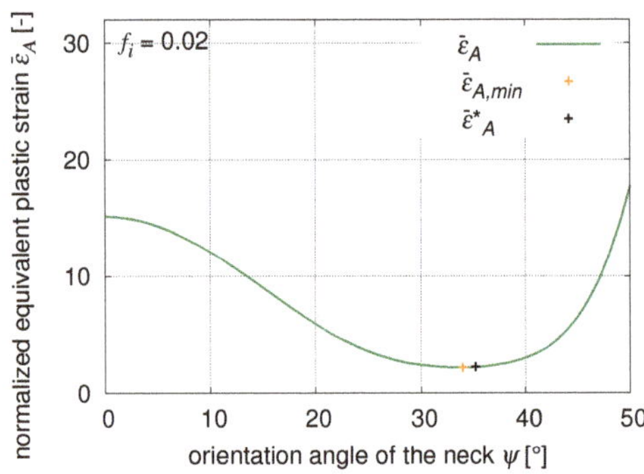

Figure 8. Variation of limit strains with respect to the angle of imperfection under uniaxial tension (strain normalized with respect to the mean uniform strain) [33].

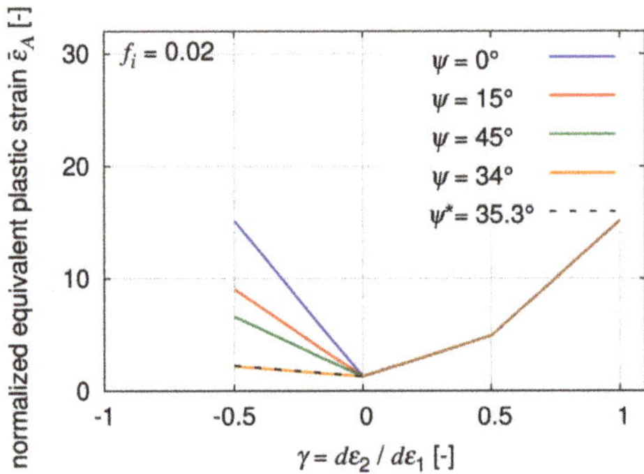

Figure 9. Effect of the groove orientation on the limit strain (strain normalized with respect to the mean uniform strain) [33].

5. Experiments

Once the numerical method is verified, it needs to be validated with the experimental data. To obtain a precise and accurate strain measurement, a digital image correlation (DIC) measurement system was employed in the experiments. This method is able to display full-field strain measurement in the localized necking region. Simultaneous evaluation of the major and minor strains at any point makes it suitable for FLC related applications. In this project, the non-contact and material independent measurement system ARAMIS (v6.3, GOM, Braunschweig, Germany) is used for strain measurements. The stress values are obtained from the universal testing machine Zwick Z100 (Zwick Roell, Ulm, Germany).

The experiments are performed by extracting samples from structural components of a car. These body parts are formed/stamped steels which are later heat treated to yield similar material properties like those of car components. To record the scattered material properties, samples from ten different batches are considered. Each batch consists of six components and each component delivers 12 test samples (extracted from six different regions). A total of 720 samples are extracted for two different test categories, namely notched and A50 specimens (see Figure 10).

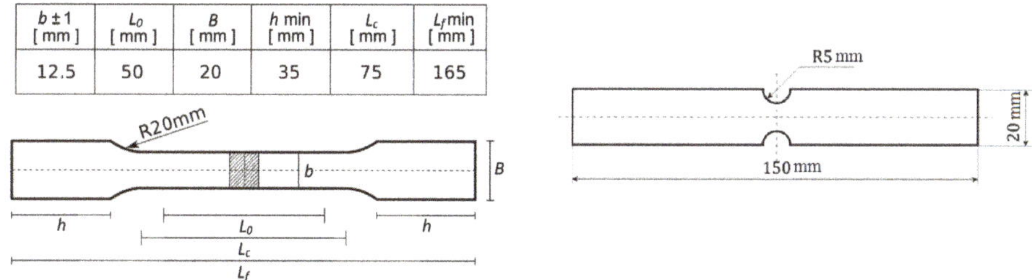

Figure 10. Geometry and dimension of the A50 (**left**) and notched (**right**) samples.

A selection of the notched samples is applied for the validation of the FLCs whereas the A50 samples are used in uniaxial tensile tests so that the material parameters can be obtained for both, the scatter determination and fitting purposes. These results are presented batch wise in terms of the normalized ultimate tensile strengths of the specimens. For the statistical analysis of the tensile test results, only the results of the A50 samples are considered.

The samples are selected in such a way that they can capture the largest scatter. The ultimate tensile strength of the material is selected as the decisive parameter since it is associated to the initiation of the necking. The stress distribution is represented in a box and whisker plot in Figure 11 for the A50 samples.

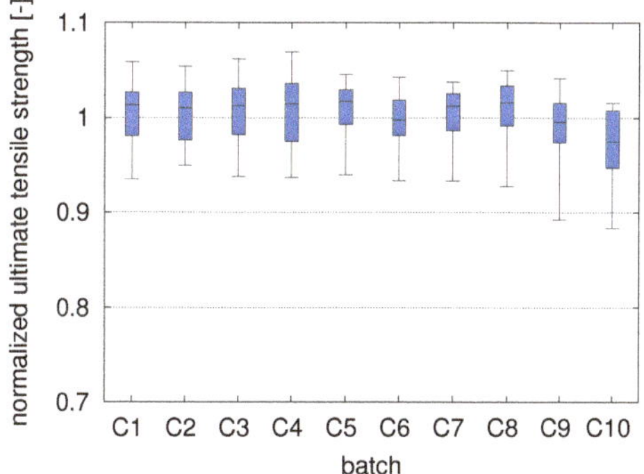

Figure 11. Distribution of the normalized stress with respect to the mean ultimate tensile strength (R_m) in A50 samples [33].

The onset of material instability can be observed from the exhibition of the shear bands as depicted in Figure 12. These bands correspond to an abrupt loss of homogeneity in deformation. Hereafter, the localized deformations rapidly intensify, leading to necking and rupture of the specimen. Figure 12a shows the non-uniform strain distribution along the section length in the pre-failure regime. The closer to the shear band, the greater the localized strain. Points 0, 1 and 2 in Figure 12b,c illustrate different major strains under uniaxial loading. Since point zero (P_0) lies on the shear band, it undergoes severe strains. In contrary, points one and two (P_1 and P_2) show much smaller strain values as they locate far from the shear bands where the localized necking occurs.

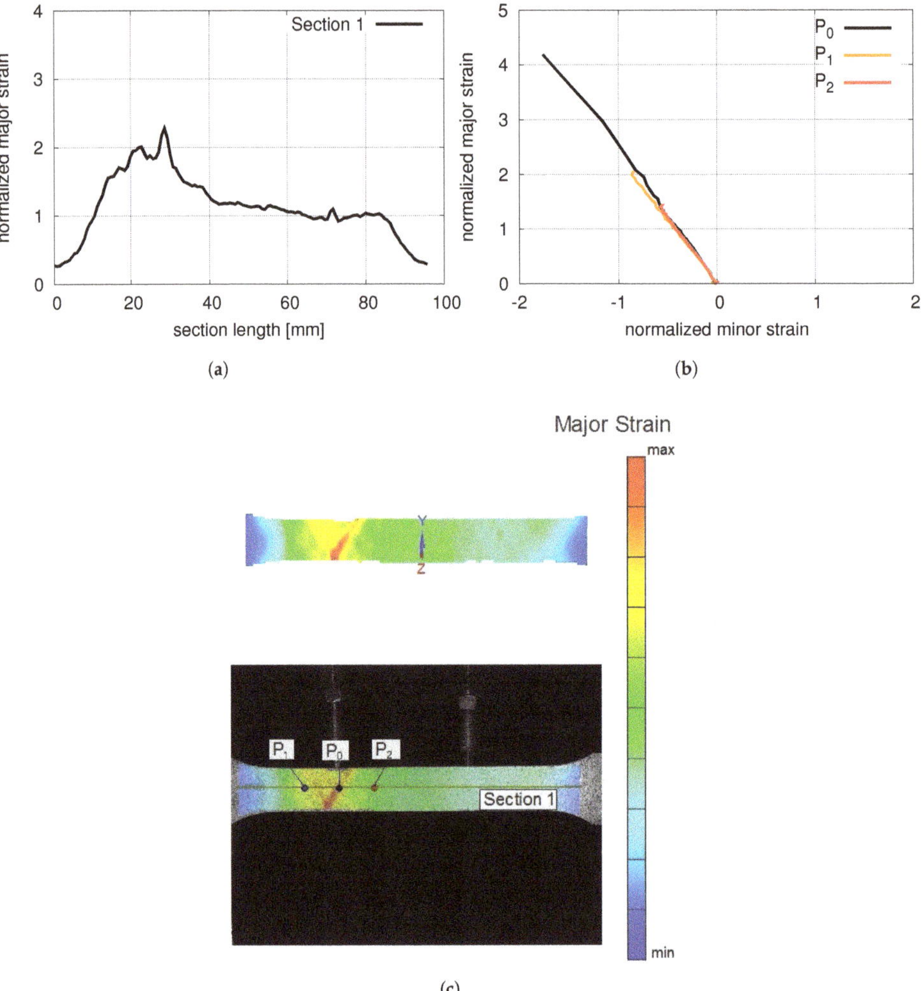

Figure 12. Strain distribution along the specimen length normalized with respect to the mean uniform strain (**a**), experimentally determined strain growth for different points lying on the shear bands and far from them with strains normalized with respect to the mean uniform strain (**b**) and the evolution of shear bands (**c**) in the pre-failure regime of an A50 sample [33].

According to Werner [34], the ratio β of the increment of the effective plastic strain in the non-necking region ($d\bar{\varepsilon}_{P_i}$, $i = 1, 2$) to that of the necking region ($d\bar{\varepsilon}_{P_0}$) indicates the plastic instability expressed as

$$\beta = \frac{d\bar{\varepsilon}_{P_i}}{d\bar{\varepsilon}_{P_0}}, \quad i = 1, 2,$$

where the points P_1 and P_2 belong to the non-necking region and the point P_0 belongs to the necking region (refer to Figure 12). Additionally, the growth of the major strains at the aforementioned points are plotted with respect to time in Figure 13.

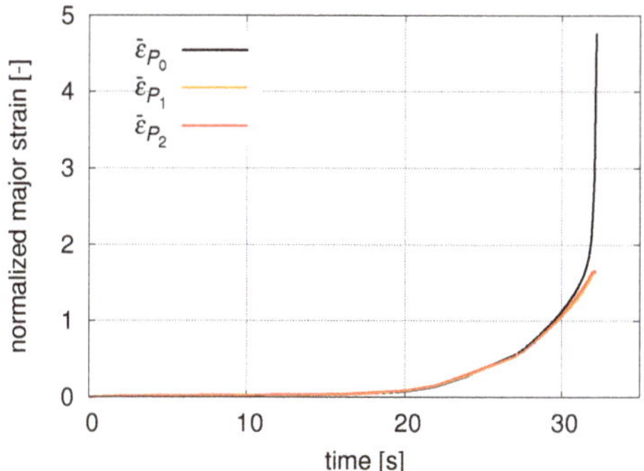

Figure 13. Growth of the major strains (normalized with respect to the mean uniform strain) in a sample in the necking and non-necking regime [33].

In order to guarantee the accuracy of the strains during the abrupt localization phenomenon, a frequency of 10 frames per second is applied in the tensile tests. At point P_0, exponential growth of strains is observed. This is not the case at other points (P_1 and P_2) far away from the necking region. However, this can be better illustrated in terms of the strain rates. Figure 14 shows the last two seconds before a sample ruptures. At this time, strain rates in the necking region (P_0) observe a sharp increment whereas the ones in the non-necking regions (P_1 and P_2) end up with a slight decrease. This is due to the fact that from second 31.5, all plastic strains flow though the shear band where the P_0 lies and the points P_1 and P_2 behave as though they are unloaded. Notice that the plastic strains remain in the latter points since they are permanent. Finally, Figure 15 illustrates the corresponding strains over the strain increment ratio $(1/\beta)$. From these results, the corresponding values of the major strains for a specific ratio (here $1/\beta = 10$) are extracted for each sample to prescribe the onset of the instability in the numerical solution.

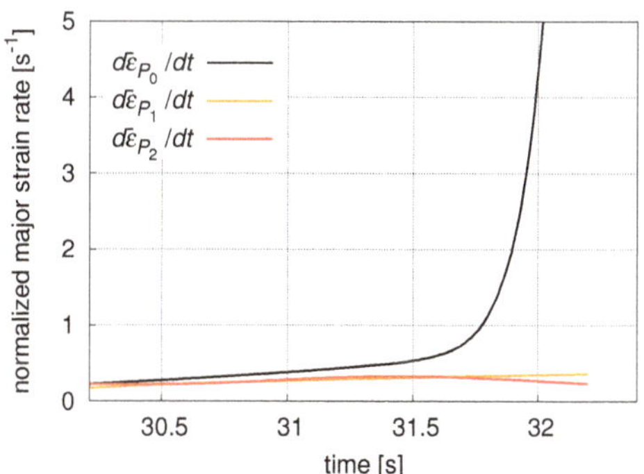

Figure 14. Major strain rates (normalized with respect to the mean uniform strain) during the onset of instability in the necking and non-necking regime.

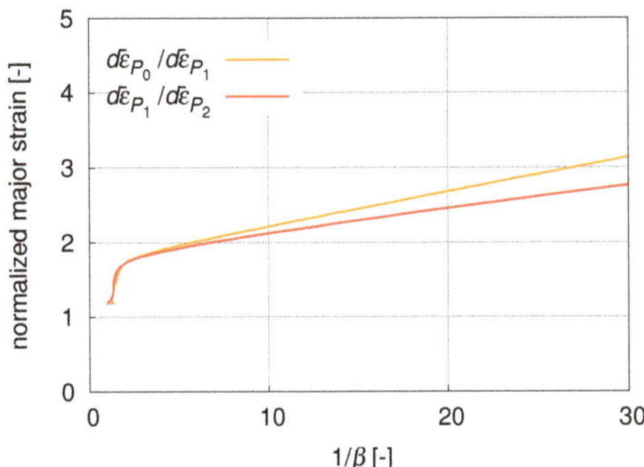

Figure 15. Corresponding major strains (normalized with respect to the mean uniform strain) for different strain increment ratios ($1/\beta$) for a sample.

The plane strain condition occurs in the middle of the notched samples as the dimension of the sample along one direction is much larger compared to the other two dimensions. Similar to the A50 samples, the strain field is analyzed across the shear bands for determining the onset of necking. The distribution of the normalized ultimate tensile strengths of the notched samples are depicted in Figure 16. Limit strains from notched samples are used as a reference in the forming limit curves for the plane strain loading path.

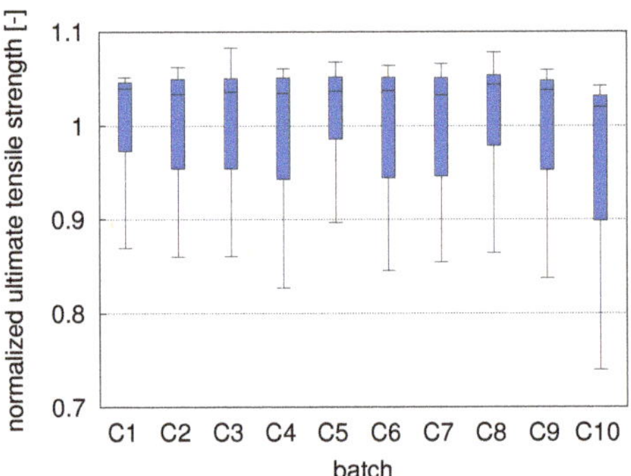

Figure 16. Stress distribution normalized with respect to the mean ultimate tensile strength (R_m) in notched samples [33].

Figure 17a shows the distribution of the normalized major strains along the width of the notch during necking. The stress state of a point in the middle of a notched sample of this form reflects a plane strain condition with a small deviation [36] (refer to Figure 17b). This is shown in the experimentally determined limit strain of the notched sample from ARAMIS in the necking regime (see Figure 17c) as well.

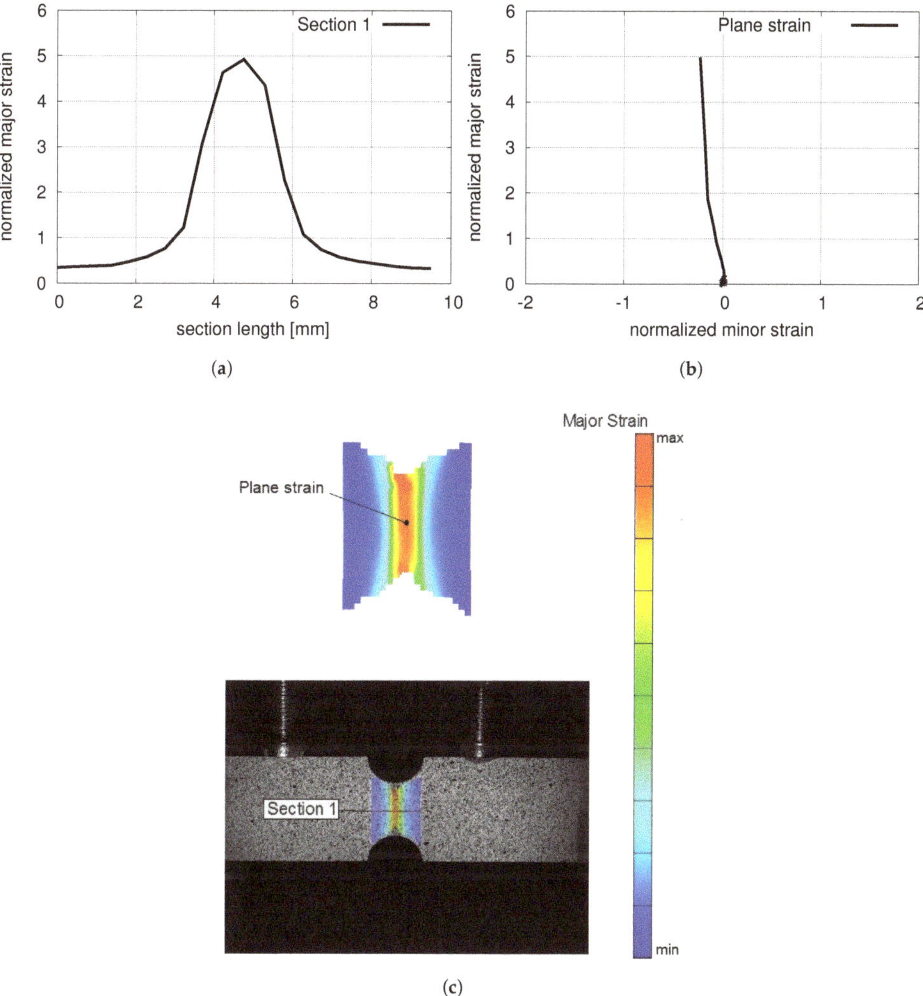

Figure 17. Strain distribution along the notch width normalized with respect to the mean uniform strain (**a**), experimentally determined limit strain with normalized strains with respect to the mean uniform strain (**b**) and the evolution of necking (**c**) in the pre-failure regime of a notched sample.

The flow diagram of the material is produced using the A50 samples to obtain the parameters related to the hardening relation (see Figure 18). As the initial cross-sectional area is considered for determining the stress values, the flow diagram typically represents the engineering stress-strain values. For sufficiently small deformations, the engineering stresses and strains are almost equivalent to the Cauchy stresses and logarithmic strains (in the sequel called "true" stresses and strains). Since the forming limit is characterized by large scale permanent plastic deformation, the hardening relations must be based on the true stress-strain values. Consequently, the engineering stress-strain diagram is first transformed to the corresponding true stress-strain diagram by the relations.

$$\sigma_{\text{true}} = \sigma_{\text{eng}} \left(1 + \varepsilon_{\text{eng}}\right),$$
$$\varepsilon_{\text{true}} = \ln \left(1 + \varepsilon_{\text{eng}}\right).$$

(26)

This transformation is illustrated also in the stress strain curve of a sample in Figure 18.

To capture the scatter of the material, 36 different hardening curves are generated from the flow diagrams of the corresponding samples.

In Figure 18, the necking of the specimen is observed around the normalized strain of 1.4 in the engineering flow curve ($\bar{\sigma}_{eng}$). However, by transferring the latter into the true stress-strain curve ($\bar{\sigma}_{true}$), only the material response until the ultimate tensile strength is considered. Due to the negligible contribution of the elastic regime (under 2 %) in this material, only the plastic response of the material ($\bar{\sigma}_{true-plast}$) is taken into account in fitting as well as in the numerical solution (see Figure 18) [34].

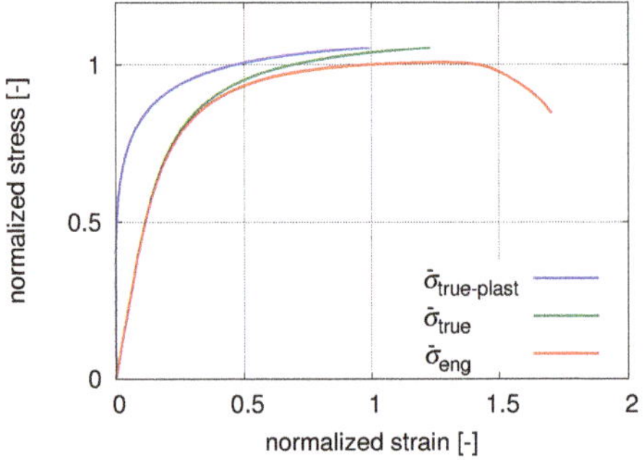

Figure 18. Engineering and true stress-strain normalized with respect to the mean ultimate tensile strength and mean uniform strain, respectively [33].

Ultimately, the curve is fitted by minimizing the square of the difference between the experimental values obtained from the plastic true stress-strain curve and the derived hardening relation from Equation (7). Figure 19 illustrates the experimental true stress-strain curve and the fitted Swift hardening relation for a specimen.

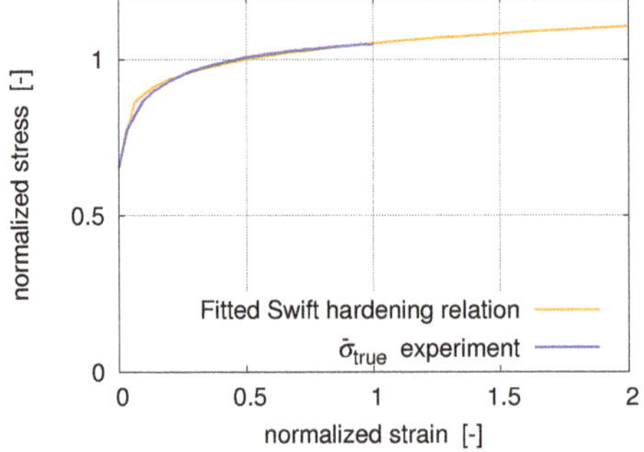

Figure 19. Fitted curve to experimental results, normalized with respect to the mean ultimate tensile strength and mean uniform strain, respectively.

By curve fitting, the set of coefficients is generated for all 36 samples. Since equivalent stresses at regions A and B are expressed as a ratio in the residual (Equation (14)) of the mathematical model, the influence of strength coefficient C_1 on the limit strains is nullified. Therefore, this coefficient is kept constant in all computations. On the other hand, the variation of n results in different hardening relations and consequently different FLCs. Since the material used here is assumed to be close to rate independent, the variation of the corresponding values are not considered later in the forming limit band generation. The strain rate $\dot{\varepsilon}$ is set manually to 0.0033 s^{-1} in the experiments (10 mm/min) whereas the strain rate exponent m is found to be 0.008 from experiments. The evaluation of the term $\dot{\varepsilon}^m$ leads to the value $\dot{\varepsilon} = 0.9553$, which confirms the assumption of rate-independence.

6. Forming Limit Bands

Due to the scatter in the material properties, the prediction of the failure limits by a single FLC will result in either under- or overestimation. Unlike FLCs, forming limit bands are statistical approaches towards a robust design methodology. By implementation of a band of forming limits instead of a single curve, it is possible to distinguish among safe, necking and failed zones in the forming limit diagrams.

In order to generate an FLB, first and foremost, the associated theoretical FLC model is determined. Various parameters that influence the behavior of the FLCs are identified. The material parameters are obtained from the experiments and the scatter in the mechanical properties is measured. Next, the relation between the mechanical properties and the process parameter (f_i) is derived. Finally, the range of the material parameters is defined by a statistical approach (here standard deviation ($\pm 2\sigma$)) and the FLBs are generated. The generated FLBs are furthermore experimentally validated.

Different parameters obtained here can be categorized as follows:

- mechanical parameters: rolling anisotropy (R_0, R_{45}, R_{90}); strength coefficient C_1, initial yield strain (ε_0), hardening exponent n, strain rate exponent m,
- process parameter: measure of inhomogeneity f_i,
- method parameter: plastic instability indicator β.

Since the material is assumed to be planar isotropic, the influence of Lankford's coefficients is eliminated. Moreover, as discussed in the previous chapter, the roll of the coefficient C_1 during the incorporation of stress ratio in the residual (Equation (14)) is omitted. Due to the assumption of rate-independence, the strain rate exponent m is applied only in the numerical calculations and therefore not varied in the forming limit curves. Since β is defined by the user, it is not considered as a material parameter. Finally, the two parameters f_i and n are needed to get the FLBs. This can be established by either fixing one parameter and varying the other or by varying both.

In spite of the fact that the inhomogeneity parameter f_i has a physical interpretation, it cannot be measured in reality and thus is considered as a process parameter. Within the numerical solution, a pre-defined inhomogeneity value is set as an input for the FLC computation. This is chosen in a way to fit the FLC and is not obtained from the fitting of the hardening relations. In contrast to f_i, the hardening exponent n is determined through the fitting of the hardening relation to the experiments. Ultimately, both parameters define the shape of the forming limit curves.

A parameter study is performed to identify the FLB and the influence of the parameters on it. The strain hardening exponent n is found to be within the range of 0.05645–0.14549 by fitting of the hardening relations to the 36 A50 samples. From Figure 20a, it is seen that the higher the value of the hardening exponent, the higher the limit strains are. In addition, experimental limit strains for uniaxial tension and plane strain states are plotted in Figure 20. The inhomogeneity parameter is kept constant ($f_i = 0.02$) while altering the hardening exponent. As it is evident, for a certain value of f_i, two different hardening relations can contain a big range of the material scatter. However, the upper bound of the FLB corresponding to $n = 0.14549$ overestimates the limit strains in the uniaxial tension regime ($\gamma = -0.5$).

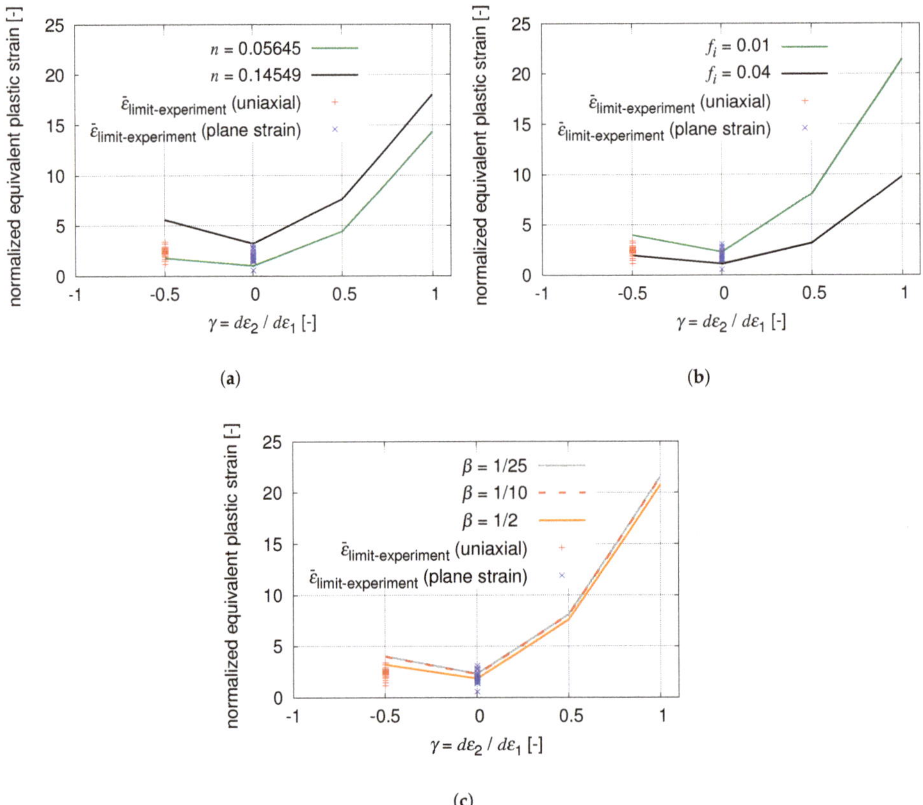

Figure 20. (**a**) Parameter study of strain hardening exponent n with constant $f_i = 0.02$, (**b**) Parameter study of inhomogeneity parameter f_i with constant n and (**c**) Influence of plastic instability indicator β with constant $f_i = 0.01$ and $n = 0.08579$ (all strains are normalized with respect to the mean uniform strain).

Similar to the study of the strain hardening exponent, the influence of inhomogeneity parameter f_i is studied in Figure 20b. In this study, the hardening exponent n is set to 0.08579 and f_i is varied from 0.01 to 0.04 so that the FLB can capture the largest range of the the experimental limit strains for both, uniaxial and plane strain states. Lower inhomogeneity values, i.e., smaller thickness variations, result in a higher potential to resist necking. Consequently, setting f_i to 0.01 results in substantially higher limit strains than the ones with $f_i = 0.04$ as shown in Figure 20b. Here, by fixing n and varying f_i, the FLB is not only overestimating the strains in the equibiaxial loading but also not covering the entire scatter from the experiment.

Limit strains are defined as the starting point of material instability. While generating the numerical FLC, the limiting value of the plastic instability indicator β is considered as 0.1. However, it is not possible to denote the onset of necking by a single value. In Figure 20c, normalized limit strains with three different instability criteria are shown. It is apparent that changes in β values do not have a strong influence on the limit strain values, provided that β is chosen small enough. Therefore, the change of β from 0.1 to 0.04 yields negligible changes in the limit strains. Nonetheless, the choice of a large value such as $\beta = 0.5$ will lead to an underestimation of the limit strains.

It is observed that fixing a parameter and varying another puts a big constraint either on the numerical solution or on the form of FLB so that the material scatter is not captured anymore (see Figure 20). Therefore, both remaining parameters, namely the hardening exponent n and the inhomogeneity parameter f_i must be set simultaneously.

To this end, the scatter of a data set is quantified using the standard deviation. As the material scatter is expressed in terms of the hardening exponent n, the standard deviation can provide a band where most of the n values will lie. Beforehand, the normal distribution of n is checked using a quantile-quantile plot. As shown in Figure 21, results approximately follow a straight line, which indicates the distribution to be normal.

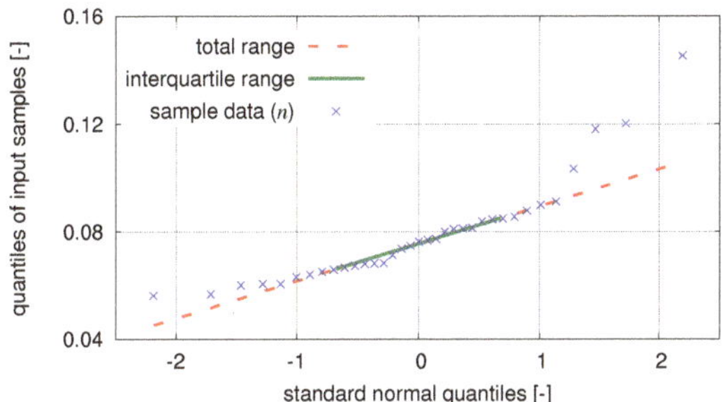

Figure 21. Quantile-quantile plot of sample data n.

With the help of standard deviation ($\pm 2\sigma$) of the mean (μ), a band of n can be defined, which statistically contains 95.45% of all values. The mean of the distribution of n is found to be 0.0791. Employing standard deviation, the interval of n is found within the interval of $[0.0410, 0.1171]$. Therefore, these limiting values of n measure the scatter in the material properties. Next, for the statistically obtained n bounds, the corresponding f_i values are set to 0.02 and 0.018 to cover the experimentally determined limit strains.

In Figure 22, upper and lower bounds of FLB are plotted in terms of normalized major-minor strains. Experimental results of 72 representative samples are found to be within the range of numerically generated bands. Evidently, by considering a range for n and f_i, the FLBs divide the region into safe, necking and failed zones. Here, the failed points are captured experimentally immediately before the rupture of the specimen whereas the safe points present the strains before the onset of the necking. Figure 23 depicts the numerically determined forming limit band in terms of equivalent plastic strains.

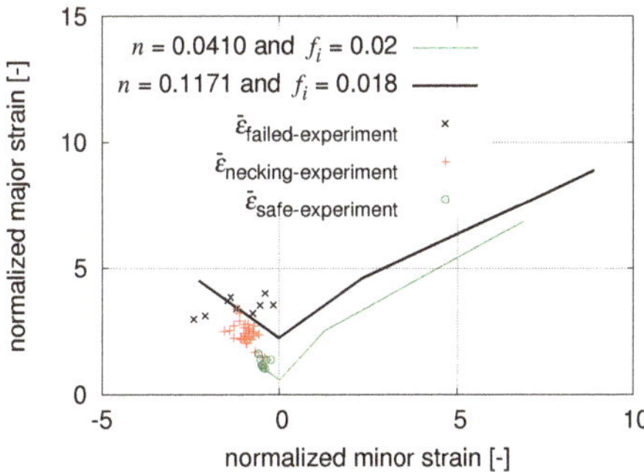

Figure 22. Forming limit band as a function of major-minor strains (normalized with respect to the mean uniform strain) with experimental strains from uniaxial loading [33].

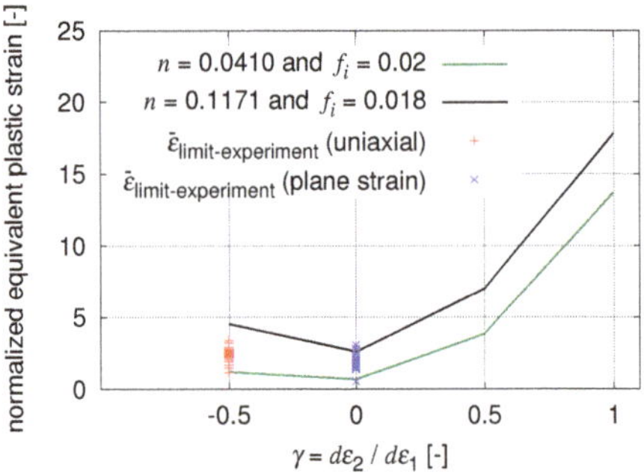

Figure 23. Forming limit band for different loading paths with strains normalized with respect to the mean uniform strain [33].

7. Conclusions

In the present work, a modified Marciniak–Kuczynski model with an inclined groove is implemented to generate forming limit bands. The model is derived for planar anisotropic rate dependent material. However, due to the application of manganese boron steel (22MnB5), it is simplified later to planar isotropy and rate independence. In this model, the concept of zero extension angle for the tension-compression quarter of the FLCs is applied. The form of the numerically determined FLCs is governed by different material and numerical input parameters. The material parameters are obtained by fitting the Swift hardening relation to the material response of the tensile tests. To this end, different samples are extracted from car body components and subjected to tensile loading. Since the material properties show a considerable scatter, statistical analysis is established to incorporate the scatter into the FLCs along with the numerical parameters. In order to capture the full field strains during the tests, digital image correlation is used in addition to conventional measuring systems. As expected, manganese boron-steel exhibits a considerable material scatter which cannot be captured by a single FLC. Hence, a band of FLCs, namely a forming limit band is generated by incorporating the effects of the material scatter (hardening exponent) as well as the numerical parameters, namely inhomogeneity and instability parameters. Furthermore, the material scatter is statistically analyzed to calibrate the bounds of the FLB. From the generated FLB, the limit strains of the material are segregated into various regimes, i.e., safe, necking and failure. In this way, the necking of a material during a car crash is not represented by a single curve, but by a band of curves.

Author Contributions: Conceptualization, H.B., S.R., S.W., A.B. and F.I.; Methodology, H.B., S.R. and S.S.; Software, H.B., S.S., B.A. and S.T.; Validation, H.B., S.R. and A.B.; Formal Analysis, S.R. and S.W.; Investigation, H.B., S.S. and S.T.; Resources, S.R., S.W., A.B. and F.I.; Writing—Original Draft Preparation, H.B., S.S. and B.A.; Writing—Review and Editing, H.B., S.W., S.R., A.B. and F.I.; Visualization, H.B., S.S. and S.T.; Supervision, S.R., S.W., A.B., F.I. and H.B.; Project Administration, S.R.; Funding Acquisition, S.R.

Funding: This research received no external funding.

Acknowledgments: Financial support for this work funded by Ford Research and Innovation Center Aachen a division of Ford Werke GmbH is gratefully acknowledged.

Conflicts of Interest: The authors declare no conflict of interest.

Appendix A

The mathematical formulation of forming limit curve is developed based on the modified Marciniak–Kuczynski model and written according to the paper of Rocha et al. [15]. To describe the plastic behavior of the material, the yield criterion, flow rule, and hardening relation are defined

Hill's criterion to define the yield surface of the material is given by

$$2f = F\,\sigma_{yy}^2 + G\,\sigma_{xx}^2 + H\,(\sigma_{xx} - \sigma_{yy})^2 + 2\,P\,\sigma_{xy}^2, \tag{A1}$$

where material constants are expressed as

$$
\begin{aligned}
F &= \frac{R_0}{R_{90}\,(1 + R_0)}, \\
G &= \frac{1}{1 + R_0}, \\
H &= \frac{R_0}{1 + R_0}, \\
P &= \frac{0.5\,(R_0 + R_{90})\,(2R_{45} + 1)}{R_{90}\,(1 + R_0)}.
\end{aligned}
\tag{A2}
$$

The Levy-von Mises flow rule is given by

$$d\varepsilon_{ij} = d\lambda'\,\frac{\partial f}{\partial \sigma_{ij}}, \tag{A3}$$

where $d\lambda'$ is the hardening parameter and expressed as

$$d\lambda' = d\bar{\varepsilon}/\bar{\sigma}. \tag{A4}$$

Finally the material hardening relation with the strain rate dependency is determined by

$$\bar{\sigma} = C_1\,(\varepsilon_0 + \bar{\varepsilon})^n\,\dot{\bar{\varepsilon}}^m. \tag{A5}$$

Apart from the above-mentioned material behavior, the force equilibrium, compatibility relations and incompressibility condition specific to the model are used to develop the mathematical model of FLC. The incompressibility condition is given as

$$d\varepsilon_{1A} + d\varepsilon_{2A} + d\varepsilon_{3A} = 0. \tag{A6}$$

The force equilibrium conditions are

$$
\begin{aligned}
\sigma_A^{nn}\,t_A &= \sigma_B^{nn}\,t_B, \\
\sigma_A^{nt}\,t_A &= \sigma_B^{nt}\,t_B,
\end{aligned}
\tag{A7}
$$

where n and t denote the normal and tangential direction with respect to the inhomogeneity, respectively. The compatibility condition across the discontinuity is defined by

$$d\varepsilon_A^{tt} = d\varepsilon_B^{tt}. \tag{A8}$$

Force equilibrium equation along the normal direction can be represented as

$$\bar{\sigma}_A\left(\frac{\sigma_A^{nn}}{\bar{\sigma}_A}\right) t_A = \bar{\sigma}_B\left(\frac{\sigma_B^{nn}}{\bar{\sigma}_B}\right) t_B. \tag{A9}$$

Incorporating the material inhomogeneity factor f_i and hardening relation (Equation (A5)) in the above equation, the residual can be derived as

$$\frac{\sigma_A^{nn}/\bar{\sigma}_A}{\sigma_B^{nn}/\bar{\sigma}_B} = (1 - f_i)\,exp(\varepsilon_{3B} - \varepsilon_{3A})\left(\frac{\varepsilon_0 + \bar{\varepsilon}_B}{\varepsilon_0 + \bar{\varepsilon}_A}\right)^n\left(\frac{d\bar{\varepsilon}_B}{d\bar{\varepsilon}_A}\right)^m. \tag{A10}$$

Since anisotropy directions coincide with the principal stress axes in the homogeneous region (A), the numerator of the left hand side of the above equation can be written as

$$\frac{\sigma_A^{nn}}{\bar{\sigma}_A} = \frac{A_1}{A_2}, \tag{A11}$$

with

$$A_1 = (cos\psi)^2 + \alpha \, (sin\psi)^2$$

and

$$A_2 = \sqrt{1 + (F + H)\,\alpha^2 - 2\,H\,\alpha}. \tag{A12}$$

To evaluate the denominator of left-hand side of residual (Equation (A10)), the stress tensor is transformed to the x-y reference axes, as anisotropy directions do not overlap with the principal stress axes in inhomogeneous region (B). At first, $\sigma_B^{nn}/\bar{\sigma}_B$ is expressed in terms of material anisotropy and groove orientation by means of simultaneous division of both force equilibrium equations, which reads as

$$\frac{\sigma_B^{nt}}{\sigma_B^{nn}} = \frac{\sigma_A^{nt}}{\sigma_A^{nn}} = K_0. \tag{A13}$$

After tensor transformation, the above equation becomes

$$\frac{\sigma_B^{xy}}{\sigma_B^{xx}} = K_1 \frac{\sigma_B^{yy}}{\sigma_B^{xx}} + K_2, \tag{A14}$$

with:

$$K_0 = \frac{(sin\psi \; cos\psi)(\alpha - 1)}{A_1},$$

$$K_1 = \frac{sin\psi \; cos\psi}{\alpha \, (sin\psi)^2 - (cos\psi)^2}, \tag{A15}$$

$$K_2 = -\alpha \, K_1.$$

Using the compatibility relation (Equation (A8)) and stress transformation, the following relation is established

$$\frac{A_3 \, d\bar{\varepsilon}_A/d\bar{\varepsilon}_B}{\sigma_B^{xx}/\bar{\sigma}_B} = sin^2\psi - H\,cos^2\psi + \frac{\sigma_B^{yy}}{\sigma_B^{xx}}\,[(F+H)\,cos^2\psi - H\,sin^2\psi] - 2\,P\,\frac{\sigma_B^{xy}}{\sigma_B^{xx}}\,sin\psi \; cos\psi, \tag{A16}$$

with

$$A_3 = \frac{(1 - H\,\alpha)\,sin^2\psi + \{(F+H)\,\alpha - H\}\,cos^2\psi}{A_2}. \tag{A17}$$

Finally, using the above equations, the residual is expressed as

$$\frac{cos^2\psi + \alpha \, sin^2\psi}{\sqrt{1 + (F+H)\,\alpha^2 - 2H\,\alpha}} = (1 - f_i)\,exp(\varepsilon_{3B} - \varepsilon_{3A})\,\left(\frac{\varepsilon_0 + \bar{\varepsilon}_B}{\varepsilon_0 + \bar{\varepsilon}_A}\right)^n \left(\frac{d\bar{\varepsilon}_B}{d\bar{\varepsilon}_A}\right)^m \sqrt{\frac{B_1\,(d\bar{\varepsilon}_A/d\bar{\varepsilon}_B)^2 + B_2}{B_3}}, \tag{A18}$$

with

$$
\begin{aligned}
B_1 &= B_4\,B_8\,A_3^2 - B_6\,B_5\,A_3^2, \\
B_2 &= B_5\,B_7^2 - B_4\,B_9^2, \\
B_3 &= B_8\,B_7^2 - B_6\,B_9^2, \\
B_4 &= \{(sin\psi)^2 + 2\,K_1\,sin\psi\,cos\psi\}^2, \\
B_5 &= \{(cos\psi)^2 + 2\,K_2\,sin\psi\,cos\psi\}^2, \\
B_6 &= F + H + 2\,P\,K_1^2, \\
B_7 &= (F + H)\,(cos\psi)^2 - H\,(sin\psi)^2 - 2\,P\,K_1\,sin\psi\,cos\psi, \\
B_8 &= 1 + 2\,P\,K_2^2, \\
B_9 &= (sin\psi)^2 - H\,(cos\psi)^2 + 2\,P\,K_2\,sin\psi\,cos\psi.
\end{aligned}
\tag{A19}
$$

The strain increments in thickness direction in homogeneous region (A) is calculated by using the flow rule and the incompressibility equation

$$
d\varepsilon_{3A} = A_4\,d\bar{\varepsilon}_A,
\tag{A20}
$$

where

$$
A_4 = -\frac{G + F\,\alpha}{A_2}.
\tag{A21}
$$

Due to the presence of angular groove, the calculation of the strain increment in inhomogeneous region is lengthy but still straight-forward. Using Levy-von Mises flow rule along with previously derived equations, thickness strain increments in inhomogeneous region are computed as:

$$
d\varepsilon_{3B} = -\left[H_6\,\sqrt{\frac{B_1\,(d\bar{\varepsilon}_A / d\bar{\varepsilon}_B)^2 + B_2}{B_3}} + H_7\left(\frac{d\bar{\varepsilon}_A}{d\bar{\varepsilon}_B}\right) \right] d\bar{\varepsilon}_B
\tag{A22}
$$

where

$$
\begin{aligned}
H_1 &= (sin\psi)^2 - H\,(cos\psi)^2, \\
H_2 &= (F + H)\,(cos\psi)^2 - H\,(sin\psi)^2, \\
H_3 &= -2\,P\,sin\psi\,cos\psi, \\
H_4 &= H_1\,(cos\psi)^2 + H_2\,(sin\psi)^2 + 2\,K_0\,sin\psi\,cos\psi\,(H_2 - H_1) + H_3\,(sin\psi\,cos\psi) - H_3\,K_0\,\{(sin\psi)^2 - (cos\psi)^2\}, \\
H_5 &= H_1\,(sin\psi)^2 + H_2\,(cos\psi)^2 - H_3\,sin\psi\,cos\psi, \\
H_6 &= G\,(cos\psi)^2 + F\,(sin\psi)^2 + 2\,K_0\,sin\psi\,cos\psi\,(F - G) - \frac{\{G\,(sin\psi)^2 + F\,(cos\psi)^2\}\,H_4}{H_5}, \\
H_7 &= \frac{A_3\{G\,(sin\psi)^2 + F\,(cos\psi)^2\}}{H_5}.
\end{aligned}
\tag{A23}
$$

References

1. Keeler, S.P.; Backofen, W. Plastic instability and fracture in sheets stretched over rigid punches. *ASM Trans. Q.* **1963**, *56*, 25–48.
2. Goodwin, G. *Application of Strain Analysis to Sheet Metal Forming Problems in the Press Shop*; SAE International: Warrendale, PA, USA, 1968. [CrossRef]
3. Marciniak, Z.; Kuczyński, K. Limit strains in the processes of stretch-forming sheet metal. *Int. J. Mech. Sci.* **1967**, *9*, 609–620. [CrossRef]
4. Considère, M. *Memoire sur L'emploi du fer et de L'acier dans les Constructions*; Dunod: Malakoff, France, 1885.
5. Hill, R. On discontinuous plastic states, with special reference to localized necking in thin sheets. *J. Mech. Phys. Solids* **1952**, *1*, 19–30. [CrossRef]

6.   Swift, H. Plastic instability under plane stress. *J. Mech. Phys. Solids* **1952**, *1*, 1–18. [CrossRef]

7.   Hora, P.; Tong, L.; Reissner, J. A prediction method for ductile sheet metal failure in FE-simulation. In Proceedings of the 3rd International Conference and Workshop on Numerical Simulation of 3D Sheet Metal Forming Processes, Dearborn, MI, USA, 29 September–3 October 1996; Voume 96, pp. 252–256.

8.   Tong, L.; Hora, P.; Reissner, J. Prediction of forming limit with nonlinear deformation paths using modified maximum force criterion. In Proceedings of the 5th International Conference and Workshop on Numerical Simulation of 3D Sheet Metal Forming Processes, Jeju Island, Korea, 21–25 October 2002.

9.   Nurcheshmeh, M.; Green, D.E. Prediction of forming limit curves for nonlinear loading paths using quadratic and non-quadratic yield criteria and variable imperfection factor. *Mater. Des.* **2016**, *91*, 248–255. [CrossRef]

10.   Hora, P.; Tong, L. Theoretical prediction of the influence of curvature and thickness on the FLC by the enhanced modified maximum force criterion. In Proceedings of the 7th International Conference and Workshop on Numerical Simulation of 3D Sheet Metal Forming Processes, Interlaken, Switzerland, 1–5 September 2008; pp. 205–210.

11.   Hill, R. A general theory of uniqueness and stability in elastic-plastic solids. *J. Mech. Phys. Solids* **1958**, *6*, 236–249. [CrossRef]

12.   Marciniak, Z.; Kuczyński, K.; Pokora, T. Influence of the plastic properties of a material on the forming limit diagram for sheet metal in tension. *Int. J. Mech. Sci.* **1973**, *15*, 789–800. [CrossRef]

13.   Hutchinson, J.; Neale, K. Sheet necking-II. Time-independent behavior. In *Mechanics of Sheet Metal Forming*; Springer: Boston, MA, USA, 1978; pp. 127–153.

14.   Chan, K.; Koss, D.; Ghosh, A. Localized necking of sheet at negative minor strains. *Metall. Mater. Trans. A* **1984**, *15*, 323–329. [CrossRef]

15.   Da Rocha, A.; Barlat, F.; Jalinier, J. Prediction of the forming limit diagrams of anisotropic sheets in linear and non-linear loading. *Mater. Sci. Eng.* **1985**, *68*, 151–164. [CrossRef]

16.   Lian, J.; Barlat, F.; Baudelet, B. Plastic behaviour and stretchability of sheet metals. Part II: Effect of yield surface shape on sheet forming limit. *Int. J. Plast.* **1989**, *5*, 131–147. [CrossRef]

17.   Barlat, F.; Lian, K. Plastic behavior and stretchability of sheet metals. Part I: A yield function for orthotropic sheets under plane stress conditions. *Int. J. Plast.* **1989**, *5*, 51–66. [CrossRef]

18.   Cao, J.; Yao, H.; Karafillis, A.; Boyce, M. Prediction of localized thinning in sheet metal using a general anisotropic yield criterion. *Int. J. Plast.* **2000**, *16*, 1105–1129. [CrossRef]

19.   Eyckens, P.; Van Bael, A.; Van Houtte, P. Marciniak—Kuczynski type modelling of the effect of through-thickness shear on the forming limits of sheet metal. *Int. J. Plast.* **2009**, *25*, 2249–2268. [CrossRef]

20.   Burford, D.; Wagoner, R. A more realistic method for predicting the forming limits of metal sheets. In *The Minerals, Metals & Materials Society, Forming Limit Diagrams: Concepts, Methods, and Applications*; Tms: Warrendale, PA, USA, 1989; pp. 167–182.

21.   Boudeau, N.; Gelin, J. Prediction of the localized necking in 3D sheet metal forming processes fr om FE simulations. *J. Mater. Proc. Technol.* **1994**, *45*, 229–235. [CrossRef]

22.   Takuda, H.; Mori, K.; Takakura, N.; Yamaguchi, K. Finite element analysis of limit strains in biaxial stretching of sheet metals allowing for ductile fracture. *Int. J. Mech. Sci.* **2000**, *42*, 785–798. [CrossRef]

23.   Kolasangiani, K.; Shariati, M.; Farhangdoost, K. Prediction of forming limit curves (FLD, MSFLD and FLSD) and necking time for SS304L sheet using finite element method and ductile fracture criteria. *J. Comput. Appl. Res. Mech. Eng.* **2015**, *4*, 121–132.

24.   Mahboubkhah, M. Determination of the forming limit diagrams by using the fem and experimental method. *J. Mech. Sci. Technol.* **2013**, *27*, 1437–1442. [CrossRef]

25.   Paraianu, L.; Comsa, D.; Jurco, P.; Cosovici, G.; Banabic, D. Finite Element Calculation of Forming Limit Curves. Available online: http://jamme.acmsse.h2.pl/papers_amme02/1190.pdf (accessed on 18 December 2002).

26.   Safari, M.; Hosseinipour, S.J.; Azodi, H.D. Experimental and numerical analysis of forming limit diagram (FLD) and forming limit stress diagram (FLSD). *Mater. Sci. Appl.* **2011**, *2*, 496–502. [CrossRef]

27.   Janssens, K.; Lambert, F.; Vanrostenberghe, S.; Vermeulen, M. Statistical evaluation of the uncertainty of experimentally characterised forming limits of sheet steel. *J. Mater. Proc. Technol.* **2001**, *112*, 174–184. [CrossRef]

28.   Strano, M.; Colosimo, B. Logistic regression analysis for experimental determination of forming limit diagrams. *Int. J. Mach. Tools Manuf.* **2006**, *46*, 673–682. [CrossRef]

29. Chen, J.; Zhou, X. A new curve fitting method for forming limit experimental data. *J. Mater. Sci. Technol.* **2005**, *21*, 521–525.
30. Banabic, D.; Vos, M.; Paraianu, L.; Jurco, P.; Cueto, E.; Chinesta, F. Theoretical prediction of the forming limit band. In Proceedings of the AIP Conference, Zaragoza, Spain, 18–20 April 2007; Volume 907, pp. 368–373.
31. Banabic, D.; Kuwabara, T.; Balan, T.; Comsa, D.; Julean, D. Non-quadratic yield criterion for orthotropic sheet metals under plane-stress conditions. *Int. J. Mech. Sci.* **2003**, *45*, 797–811. [CrossRef]
32. Sorin, C.; Dragos, G.; Paraianu, L.; Banabic, D.; Chinesta, F.; Chastel, Y.; El Mansori, M. Prediction of the forming limit band for steel sheets using a new formulation of Hora's criterion (MMFC). In Proceedings of the AIP Conference, Paris, France, 24–27 October 2010; Volume 1315, pp. 425–430.
33. Bayat, H.R.; Sarkar, S.; Italiano, F.; Bach, A.; Wulfinghoff, S.; Reese, S. Generation of forming limit bands for ultra-high-strength steels in car body structures. In *Proceedings of the AIP Conference, 23–25 April 2018*; AIP Publishing: Palermo, Italy, 2018; Volume 1960, p. 160001.
34. Werner, H. Influence of Hardening Relations on Forming Limit Curves Predicted by the Theory of Marciniak, Kuczyński, and Pokora. In *Predictive Modeling of Dynamic Processes*; Springer: Boston, MA, USA, 2009; pp. 43–65.
35. Lian, J.; Baudelet, B. Forming limit diagram of sheet metal in the negative minor strain region. *Mater. Sci. Eng.* **1987**, *86*, 137–144. [CrossRef]
36. Schwindt, C.D.; Stout, M.; Iurman, L.; Signorelli, J.W. Forming limit curve determination of a DP-780 steel sheet. *Procedia Mater. Sci.* **2015**, *8*, 978–985. [CrossRef]

© 2018 by the authors. Licensee MDPI, Basel, Switzerland. This article is an open access article distributed under the terms and conditions of the Creative Commons Attribution (CC BY) license (http://creativecommons.org/licenses/by/4.0/).

Article

Simulation of Sheet Metal Forming Processes Using a Fully Rheological-Damage Constitutive Model Coupling and a Specific 3D Remeshing Method

Abel Cherouat [1,*], Houman Borouchaki [1] and Jie Zhang [2]

[1] Department Recherche Opérationnelle, Statistiques Appliquées, University of Technology of Troyes-GAMMA3-Team INRIA, 12 rue Marie Curie, 10004 Troyes, France; houman.borouchaki@utt.fr
[2] Department of Materials Science and Engineering, Shanghai Jiao Tong University, 1954 Huashan Rd, Xuhui Qu, Shanghai 200000, China; zhangjie@126.com
* Correspondence: abel.cherouat@utt.fr; Tel.: +33-325-715-674

Received: 17 October 2018; Accepted: 20 November 2018; Published: 26 November 2018

Abstract: Automatic process modeling has become an effective tool in reducing the lead-time and the cost for designing forming processes. The numerical modeling process is performed on a fully coupled damage constitutive equations and the advanced 3D adaptive remeshing procedure. Based on continuum damage mechanics, an isotropic damage model coupled with the Johnson–Cook flow law is proposed to satisfy the thermodynamic and damage requirements in metals. The Lemaitre damage potential was chosen to control the damage evolution process and the effective configuration. These fully coupled constitutive equations have been implemented into a Dynamic Explicit finite element code Abaqus using user subroutine. On the other hand, an adaptive remeshing scheme in three dimensions is established to constantly update the deformed mesh to enable tracking of the large plastic deformations. The quantitative effects of coupled ductile damage and adaptive remeshing on the sheet metal forming are studied, and qualitative comparison with some available experimental data are given. As illustrated in the presented examples this overall strategy ensures a robust and efficient remeshing scheme for finite element simulation of sheet metal-forming processes.

Keywords: continuum damage mechanics; 3D adaptive remeshing; sheet metal forming

---

1. Introduction

The commercial finite element software has integrated various material models to describe the thermal-visco-plastic behaviors of sheet metal in different forming processes (deep-drawing, hydroforming, incremental forming, blanking). However, when materials are formed by these processes, they experience large plastic deformations leading to the onset of internal or surface micro-defects as voids and micro cracks. When micro-defects initiate and grow inside the plastically deformed metal, the thermo-mechanical fields are deeply modified, leading to significant modifications in the deformation process. On the other hand, the coalescence of micro-voids defects during the deformation can lead to the initiation of macro-cracks or damaged zones, inducing irreversible damage inside the formed part and consequently its loss. Taking into account the damage defect in sheet metal forming necessitates not only the development of a continuum damage theory, but also its coupling with the other mechanical fields. This is useful to avoid damage initiation to obtain a non-damaged work-piece (hot forging, stamping, deep-drawing and hydroforming) and develop the damage initiation and propagation to simulate the machining processes (orthogonal cutting, blanking, guillotining).

During the past decades, constitutive models of ductile damaged materials in the finite deformation range have received considerable attention. An important point in such phenomenological constitutive models is the appropriate choice of the physical nature of mechanical variables realistically describing the damage state of materials. Two main methods exist to predict the macro-defect in sheet metal forming:

(a) The first uncoupled approach aims to calculate the damage (initiation and growth) distribution without taking into account its effect on the other mechanical fields (elastic, thermal, plastic, and hardening). This approach is used to predict zones where local failure has taken place inside the deformed work piece. Generally, this is achieved by post-processing the finite element analysis for a given time step to calculate the damage distribution using the stress and strain fields [1–3].

(b) In the second fully coupled approach, the damage effect is directly introduced into the overall constitutive equations and affects all the mechanical fields. In this case, the damage field assumes that the degradation of structure is due to nucleation and growth of micro defects and their coalescence into macro-cracks. The fully coupled approach has shown their ability to optimize the process plane, not only to avoid the damage occurrence, but also to enhance the damage in order to simulate any metal cutting processes [4–6].

Damage mechanics in metal assumes that the degradation of material due to nucleation and growth of micro defects (voids and cracks), and their coalescence into macro-cracks [1–3]. McClintock [4] firstly develops the relationship between micro defects and ductile failure. After, three main approaches based on micro-defects [5] are extensively used to describe the damage mechanics: fracture mechanics [6], micro-based damage mechanics [7–10], and continuum damage mechanics (CDM). This somewhat fully coupled approach accounts for the direct interactions between the plastic flow, including different kinds of hardening, and the ductile damage initiation and growth. In CDM, the damage is assumed to be one of the internal state variables which relates to material behavior induced by the irreversible deterioration of microstructure. The function of damage variable works with effective stress. Kachanov [11] is the pioneer to characterized ductile damage by a scalar to define the effective stress. Without a clear physical meaning for damage, he introduced a scalar internal variable to model the creep failure of metal under uniaxial loads. A physical significance for the damage variable was given later by Rabotnov [12] who proposed the reduction of the cross sectional area due to micro-cracks as a suitable measure of the state of internal damage.

Lemaitre and Chaboche developed the continuum damage mechanic for ductile damage later [13]. The constitutive equations of damage variables are derived from specific damage potentials by using the effective state variables. These are defined from the classical state variables using one of the three following hypotheses: strain equivalence, stress equivalence or energy equivalence. The coupled constitutive equations of the damaged domain are generally deduced from the same state and dissipation potentials in which the state variable are replaced by the effective state variables [13–16].

In recent years, sets of constitutive equations for elasticity, plasticity and thermo-visco-plasticity coupled with ductile damage are given [15]. The works of Bouchard [17], Brünig [18], and Wang [19] summarized and compared various damage models.

The Johnson–Cook (JC) hardening model is the most attractive among well-known visco-plastic strain flow. This model takes into account both kinematic strengthening and adiabatic heating of the material undergoing strains and can describe the dynamic behavior of materials, which works in different thermal environments. For these advantages, the JC model has been modified in parameters and various forms to fit the different material behaviors. Peirs [20] used the advanced experiment method and finite element simulation tools to verify the material parameters in JC model, especially the strain rate hardening and thermal softening parameters. Through enhancing the thermal softening effects, the simulation results using corrected parameters agreed with the experiments and some strain localization phenomena happened. Calamaz [21] directly changed the JC model to the form of TANH model. A new term, which is controlled by temperature, was added to simulate the serrated chips formation in orthogonal cutting process. On the other hand, Zerilli and Armstrong proposed dislocation-mechanics-based constitutive relations for different crystalline structures, in which the effects of strain hardening, strain rate hardening, and thermal softening based on the thermal activation analysis were incorporated into constitutive relations [22–24]. Holmquist et al. [25] and Hor et al. and [26] propose a comparison of the models to be made independent of the material constants and procedure for which constants can be determined for different constitutive models using the same test data base.

In these models, the damage generates and evolves during tensile, shearing and cutting process had never introduced. This point is not identical to the practical situation. Actually, the stiffness of the material is deteriorating until to losing the abilities of loading which is following with the evolution of damage. Johnson and Cook [27] have given out a threshold, which is a function of stress state (stress triaxiality) for equivalent plastic strain. The damage generates when equivalent plastic strain reaches to this threshold. The limitation of this damage model is that, it can only predict the onset of the damage and the material stiffness will reduce to zero directly without any evolution process. Some phenomena (like strain localization) are hard to obtain and it will also lead the instability into the simulation system. Therefore, it is necessary to integrate a constitutive damage evolution into the JC model. Based on this aspect, a constitutive equation which couple fully the ductile damage into JC isotropic hardening model, is developed. These fully coupled constitutive equations have been implemented into a Dynamic Explicit finite element code (Abaqus/Explicit) using user subroutine. The local integration of the plastic-damage constitutive equations is performed using an asymptotic implicit scheme applied to solve the nonlinear local equations.

Numerical errors are intrinsic in Finite Element Analysis (FEA) of sheet metal forming processes and possess additional difficulties related to the large inelastic deformations with damage imply a severe distortion of the computational domain [28,29]. In fact, the deformed domain undergoes geometrical variation (large displacements and rotations) and are characterized by inhomogeneous spatial distribution of thermo-mechanical fields with evolving localized zones (stress, plastic strain, damage, temperature, etc.). In fact, the time and space discretization of the continuous differential equations governing the physical equilibrium events lead inevitably to numerical errors. In this case, frequent remeshing of the deformed domain during computation is necessary to obtain an accurate solution and complete the computation until the termination of the numerical simulation process. Accordingly, several remeshing have to be performed during the simulation in order to preserve the reliability of the obtained results by minimizing errors generated by either the geometrical transformations or the heterogeneous thermo-mechanical fields.

To decide when the remeshing is required during the analysis, some appropriate criteria are needed and should be automatically executed during the FEA. These are generally based on a priori and a posteriori error estimators. The main goal of any error estimator is the evaluation of the absolute global error as an addition of the estimated local error for each element. For metal forming, the error criteria is classified into three classes:

(1) Geometrical: estimation of the element distortion due to the large transformation of the domain. Distortion criteria are based on the large variation of the geometry of finite elements with respect to their reference state.

(2) Curvature: estimation of the element size needed to avoid inter-penetration between the deformed domain and complex tools. This geometric estimator is based on the curvature of the tools angles inside the contact zones.

(3) Physical: adaptation of the element size to the local or global variation of some physical fields as temperature, displacement gradient, stress, plastic strain etc.

Accordingly, the damage growth induces a decrease in the stress-like variables generated by a decrease in physical properties of the material. When all Gauss points within a given finite element are fully damaged, the corresponding stiffness matrix is zero. Consequently, this element has no more contribution in the global tangent stiffness matrix and should be removed. Accordingly, there is problem for elements lying in boundaries of the deformed domain need special attention when this boundary is concerned with the contact zone between different domains (tools and deformable parts). The best way to treat the fully damaged elements consists in remeshing the domain after dropping the fully damaged elements and smoothing the newly created boundaries of the deformed part.

In this work the damage potential, introduced by Lemaitre [13], is used and coupled into an elasto-visco-plastic material model through defining the effective stress and plastic strain like a Johnson–Cook formulation [27]. 3D adaptive remeshing scheme using linear tetrahedral finite element is developed in order to simulate the large plastic deformations and crack propagation after damage occurring [28,29]. This scheme is established to simulate to predict when and where ductile damage zones may take place inside the deformed part during tensile, compressive, and shearing tests. The localization phenomenon of damage was illustrated clearly. The formation of the cracks and its propagation to the final fracture of the specimen are also illustrated. Four various sheet metal forming processes are proposed to prove that the numerical methodology is an advanced and a reliable tool to simulate various metal forming processes in order to avoid damage in incremental forming, deep drawing, and multi-point drawing or to enhance damage in order to simulate some sheet metal cutting operations.

2. Methods and Constitutive Model

2.1. Visco-Elastoplastic Model Fully Coupled to Isotropic Ductile Damage

This section provides a brief description of the major conception for coupling the ductile damage into the material elasto-visco-plastic behavior. The ductile damage is presented in the framework of irreversible processes with state variables. An isotropic ductile damage variable D $(0 < D < 1)$ is measured in a macroscopic scale way through the surface density of intersection of micro-cracks and micro-cavities at a representative finite elementary volume. In order to perform the effect of this damage variable on the mechanical behavior, the effective state variables are introduced [11–16].

To another consideration, the damage caused by the micro-cracks and micro-cavities has a different evolution processes in tensile and compressive load conditions. In one aspect, the micro-cracks are opened in the tensile state and the module of elasticity reduces gradually until to zero. In another aspect, the micro-cracks are closed in compressive state and the module of elasticity could be able to restore to their initial values before the damage accumulates. This recovery effect of physical properties after closure of micro-cracks is called the quasi-unilateral effect [30–37]. It demands that the definition of effective state variable $(\widetilde{\sigma}, \widetilde{\varepsilon})$ should be in the unilateral condition.

According to the theory of energy equivalence, we define the effective variables, which consider the isotropic ductile damage into state variables [38–42] as follows:

$$\widetilde{\underline{\sigma}} = \frac{\underline{\sigma}}{\sqrt{1-D}}, \ \widetilde{\underline{\varepsilon}}^{\,e} = \underline{\varepsilon}^{e}\sqrt{1-D}, \ \underline{S} = \underline{\sigma} - \sigma_H \underline{1} \text{ and } \sigma_H = \frac{1}{3}\mathrm{tr}\underline{\sigma} \tag{1}$$

where $\underline{\varepsilon}^{e}$ is the small elastic strain tensor representing the elastic flow associated with the Cauchy stress tensor $\underline{\sigma}$, $(\underline{S}, \sigma_H)$ are the deviatoric and hydrostatic Cauchy stresses respectively and D is the ductile damage associated with the potential Y.

The damaged elastoplastic behavior is described in the framework of the thermodynamics of irreversible processes with state variables. The Helmholtz free energy in which elasticity and plasticity are uncoupled gives the law of elasticity coupled with damage. Following the 2nd principle of thermodynamics, non-negativity of the mechanical dissipation, the stress like variables $(\underline{\sigma}, Y)$ are derived from the state potential taken as the classical free energy $\Psi(\underline{\varepsilon}^{e}, D)$ in deformation space [15,16], as follows:

$$\begin{cases} \underline{\sigma} = (1-D)\left(2\mu_e \underline{\underline{I}} + \lambda_e \underline{1} \otimes \underline{1}\right) : \underline{\varepsilon}^e = (1-D)\underline{\underline{\Lambda}} : \underline{\varepsilon}^e \\[2mm] Y = \tfrac{1}{2}\underline{\varepsilon}^e : \underline{\underline{\Lambda}} : \underline{\varepsilon}^e = \dfrac{J_2(\underline{\sigma})^2}{2E(1-D)^2}\left[\tfrac{2}{3}(1+\nu) + 3(1-2\nu)\left(\dfrac{\sigma_H}{J_2(\underline{\sigma})}\right)^2\right] \end{cases} \tag{2}$$

where (λ_e, μ_e) are the classical Lame's constants which are a function of Young modulus E and Poison's coefficient ν.

In order to couple the damage behavior, a single appropriate dissipation potential $F(\underline{\sigma}, Y, D)$ is defined to govern the evolution law for internal variables in the stress space [39–42]:

$$F(\underline{\sigma}, \underline{\varepsilon}^P, Y, D) = f_p + F_Y \leq 0 \quad \begin{cases} f_p(\underline{\sigma}, \underline{\varepsilon}^P, D) = \dfrac{J_2(\underline{\sigma})}{\sqrt{1-D}} - R(\overline{\varepsilon}^P) \\[3mm] F_Y(Y, D) = \dfrac{\gamma}{(1-D)^\beta}\dfrac{1}{(\alpha+1)}\left(\dfrac{Y-Y_0}{\gamma}\right)^{\alpha+1} \end{cases} \tag{3}$$

The parameters $(\alpha, \beta \text{ and } \gamma \text{ and } Y_0)$ are used to control the evolution of damage potential, $R(\overline{\varepsilon}^P)$ is the JC isotropic yield stress in various visco-plastic flow and $J_2(\underline{\sigma}) = \sqrt{\tfrac{3}{2}\underline{S}:\underline{S}}$ is the second invariant of the deviatoric Cauchy stress $\underline{S}$.

For the case of time independent plasticity, the plastic strain rate tensor $\underline{\dot{\varepsilon}}^P$ and the rate damage $\dot{D}$ are obtained from the stationarity conditions as in which the plastic multiplier $\dot{\delta}$ is deduced from the consistency condition [13]:

$$\begin{cases} \underline{\dot{\varepsilon}}^P = \dot{\lambda}\dfrac{\partial f_p}{\partial \underline{\sigma}} = \dfrac{3}{2}\dfrac{\dot{\lambda}}{\sqrt{1-D}}\dfrac{\underline{S}}{J_2(\underline{\sigma})} \\[3mm] \dot{D} = \dot{\lambda}\dfrac{\partial F_Y}{\partial Y} = \dot{\lambda}\left(\dfrac{Y-Y_0}{\gamma}\right)^\alpha \dfrac{1}{(1-D)^\beta} \end{cases} \tag{4}$$

where

$$\dot{\overline{\varepsilon}}^P = \sqrt{\dfrac{2}{3}\underline{\dot{\varepsilon}}^P : \underline{\dot{\varepsilon}}^P} \tag{5}$$

is the effective plastic strain rate.

Failure is assumed to initiate when the damage at a material point reaches the critical damage value D_c. When this happens, the stiffness of the failed element is significantly reduced and consequently incapable of carrying any load. The value of D_c for any material must be acquired through experimental tests. Assuming a fully isotropic material behavior, the plastic multiplier $\dot{\lambda}$ is obtained by the consistency condition $\left(\dot{f}_p = f_p = 0\right)$ associated with the loading–unloading condition [13–15]. It is a strictly positive scalar, which plays the role of Lagrange multiplier for dissipative phenomena:

$$\dot{\lambda} = \dfrac{1}{H_p}\left\langle 3\mu_e\sqrt{1-D}\dfrac{1}{J_2(\underline{\sigma})}\underline{S} : \underline{\dot{\varepsilon}}\right\rangle \tag{6}$$

The non-symmetric fourth order tangent elastoplastic operator $\underline{\underline{L}}_T$ defining the stress rate $\underline{\dot{\sigma}} = \underline{\underline{L}}_T : \underline{\dot{\varepsilon}}$ is defined as [28,29,36,40]:

$$\underline{\underline{L}}_T = 2\mu_e(1-D) - \dfrac{(1-D)}{H_p}\left(3\mu_e\dfrac{\underline{S}}{J_2(\underline{\sigma})}\ddot{A}\,3\mu_e\dfrac{\underline{S}}{J_2(\underline{\sigma})}\right) - \dfrac{1}{H_p(1-D)^{\beta+\frac{1}{2}}}\left(\dfrac{Y-Y_0}{\gamma}\right)^\alpha\left(3\mu_e\dfrac{\underline{S}}{J_2(\underline{\sigma})}\right)\ddot{A}\underline{S} \tag{7}$$

where H_p is the tangent plastic hardening module given by:

$$H_p = 3\mu_e - \dfrac{J_2(\underline{\sigma})}{2(1-D)^{\beta+\frac{3}{2}}}\left(\dfrac{Y-Y_0}{\gamma}\right)^\alpha + \dfrac{1}{\sqrt{1-D}}\dfrac{\delta R}{\delta\overline{\varepsilon}^P} \tag{8}$$

In this paper, the thermal effects are ignored [30–40] and the visco-plastic hardening yield stress is written as:

$$R(\overline{\varepsilon}^P) = \left[A + B(\overline{\varepsilon}^P)^n\right]\left[1 + C\ln\dfrac{\dot{\overline{\varepsilon}}}{\dot{\varepsilon}_0}\right] \tag{9}$$

where $\bar{\varepsilon}^P$ is the equivalent plastic strain, $\dot{\bar{\varepsilon}}$ is the equivalent plastic strain rate. The initial strain rate $\dot{\bar{\varepsilon}}_0$ is determined by experimental conditions. The parameters A, B and n represent the isotropic hardening evolution and C represent the material viscosity. The fully isotropic rate formulation assumes the small strain hypothesis in sheet forming processes where relative slow speeds and inertia effect can be neglected and dynamic phenomena not occur during the process.

For the constitutive equations, this hypothesis is justified by the fact that the applied load increments are still very small during remeshing procedure of sheet metal forming. Accordingly, the total strain rate tensor $\underline{\dot{\varepsilon}}$ is additively partitioned $\underline{\dot{\varepsilon}} = \underline{\dot{\varepsilon}}^e + \underline{\dot{\varepsilon}}^P$ with $\underline{\dot{\varepsilon}}^e$ and $\underline{\dot{\varepsilon}}^P$ are respectively the elastic and plastic strain rate components.

2.2. Local Time Integration of the Constitutive Equations

The fully coupled constitutive equations presented above together with an iterative implicit procedure for the time integration have been implemented using the user's subroutine. By combining the Equations (2) and (4) and saving the damage evolution equation one may obtain the following system of two scalar equations [28,29]:

$$
\begin{cases}
f_1\left(\Delta D_{n+1}, \Delta\lambda_{n+1}\right) = \dfrac{J_2(\underline{\sigma})}{\sqrt{1-D_{n+1}}} - \dfrac{3G\Delta\lambda_{n+1}}{\sqrt{1-D_{n+1}}} - R_{n+1} = 0 & \text{(a)} \\[2ex]
f_2\left(\Delta D_{n+1}, \Delta\lambda_{n+1}\right) = \Delta D^{n+1}_{n+1} - \dfrac{\Delta\lambda_{n+1}}{(1-D_n)^\beta}\left(\dfrac{Y_n-Y_0}{\gamma}\right)^\alpha = 0 & \text{(b)}
\end{cases}
\tag{10}
$$

This simple system is iteratively solved thanks to Newton–Raphson scheme to determine the two unknowns at the time t_{n+1}. The knowledge of $(\Delta\lambda_{n+1}, D_{n+1})$ allows the updating of the hardening and damage variables at the end of the time step. The so-called elastic prediction-return mapping algorithm with an operator splitting methodology is used. For the calculation of stress and plastic strain tensors, accumulated plastic strain and ductile damage, we use the fully implicit Euler method since it contains the property of absolute stability and the possibility of appending further equations to the existing system of nonlinear equations [42,43].

The local numerical integration scheme is known to have important advantages for the constitutive models with a single yielding surface together with a fully implicit global resolution scheme. This approach within the coupled problem, consists in splitting it into two parts:

(1) Damaged elastic prediction, where the problem is assumed to be purely elastic affected by the last damage value.
(2) Damaged-plastic corrector, in which the system of equations includes the damaged elastic relation as well as the damaged-plastic consistency condition. Newton–Raphson iteration algorithm is then used to solve the discretized constitutive equations in the damaged plastic corrector stage around the current values of the state variables (plasticity, hardening, damage) [43].

Two procedures related to the damage coupling have been investigated. The first is the one discussed above and called the strong coupled procedure is implemented into Abaqus subroutines. The second one weak approach solves only the equation $f_1\left(\Delta D_{n+1}, \Delta\lambda_{n+1}\right) = 0$ without damage effect in order to obtain plastic multiplier $\Delta\lambda_{n+1}$. After the convergence the $\Delta\lambda_{n+1}$ is used to calculate the damage increment without any iteration procedure $\Delta D_{n+1} = \dfrac{\Delta\lambda_{n+1}}{(1-D_n)^\beta}\left(\dfrac{Y_n-Y_0}{\gamma}\right)^\alpha$.

If the ductile damage variable reaches its maximum value $D \geq D_{\max}$ at a given integration point, the correspondent elastic modulus is set to zero giving zero stresses and no contribution to the elementary stiffness matrix. This fully damaged integration point is excluded from the integration domain of the element and has no more contribution in the elementary stiffness matrix. However, when a node is found to be connected with fully damaged elements, giving a singular global stiffness matrix, the calculation is terminated. In fact, this situation can be avoided by dropping the corresponding terms from the stiffness matrix. A new mesh is then generated after removing the fully damaged elements.

2.3. Global Resolution Strategy

The principal of virtual power written on the current damaged domain configuration with the volume V and boundary Γ can be written as:

$$\varrho \int_V \ddot{u} \cdot \delta u \, dV = -\int_V \underline{\sigma} : \delta\underline{\dot{\varepsilon}} \, dV + \int_V f \cdot \delta u \, dV + \int_{\Gamma_u} t \cdot \delta u \, d\Gamma + \int_{\Gamma_c} t_c \cdot \delta u_c d\Gamma \tag{11}$$

where $\delta \dot{u}$ (kinematically admissible) and $\delta \ddot{u}$ are the virtual velocity and acceleration fields respectively and $\delta \dot{u}_c$ is the virtual velocity vector of contact nodes.

The deformed domain at each time is supposed to be discretized on isoparametric finite elements (C^0). By using the classical nodal approximation using displacement based Finite Element Analysis FEA, Equation (11) can be easily written, on the overall part, under the following nonlinear algebraic system:

$$I = \sum_e I_e = \left[\sum_e \left([\mathbf{M_e}]\{\ddot{u}_e^N\} + \{F_e^{int}\} - \{F_e^{ext}\} \right) \right] \{\delta u\} = 0 \tag{12}$$

where $[\mathbf{M_e}]$ is the consistent mass matrix and $\{F_e^{ext}\}$ and $\{F_e^{int}\}$ are the vector of external and internal forces defined as:

$$\begin{cases} [\mathbf{M_e}] = \int_{V_e} \varrho [\mathbf{N_N}]^T \cdot \mathbf{N_N} dV_e \\[2mm] \{F_e^{int}\} = \int_{V_e} [\mathbf{B_e^N}]^T : \underline{\sigma} dV_e \\[2mm] \{F_e^{ext}\} = \int_{V_e} [\mathbf{N_N}] \cdot \{\vec{f}\} dV_e + \int_{\Gamma_u} [\mathbf{N_N}] \cdot \{\vec{t}\} d\Gamma_e + \int_{\Gamma_c} [\mathbf{N_N}] \cdot \{\vec{t}_c\} d\Gamma_e \end{cases} \tag{13}$$

$[\mathbf{B_e^N}]$ is the geometric or strain-displacement matrix in the current configuration and $[\mathbf{N_N}]$ is the matrix of the nodal interpolation functions. The index e refers to the eth element.

The system (12) defines a highly nonlinear system expressing the mechanical equilibrium of the work-piece and the tool at each time step. It can be solved either by iterative static implicit methods or by explicit methods [44–51].

(1) The static implicit iterative procedure requires, at each time step, the calculation of the consistent stiffness matrix in order to preserve the quadratic convergence property of the Newton method. When the inertia effect is neglected, the system (Equation (12)) reduces to:

$$\{R_{n+1}\} = \left\{F_e^{int}\right\}_{n+1} - \left\{F_e^{ext}\right\}_{n+1} = \{0\} \tag{14}$$

The nonlinear problem to be solved over the time increment as follows:

$$R_{n+1}^i + \frac{\partial R_{n+1}^i}{\partial U_{n+1}^h}(\Delta U)_{n+1}^{i+1} + \ldots = 0 \tag{15}$$

where the global tangent stiffness matrix at the time t_{n+1} and iteration (i) is defined by:

$$[K]_{n+1}^i = \left[\frac{\partial R_{n+1}^i}{\partial U_{n+1}^i}\right] = \left[K_{\text{Structural}}\left(U_{n+1}^i\right)\right]_{n+1}^i + \left[K_{\text{Contact}}\left(U_{n+1}^i\right)\right]_{n+1}^i + \left[K_{\text{Force}}\left(U_{n+1}^i\right)\right]_{n+1}^i \tag{16}$$

represents the contribution of the elastoplastic behavior to the structural stiffness; the contact and friction stiffness and the external applied body forces.

Note that the tangent matrix are generally non-symmetric and nonlinear because the material Jacobian matrix are themselves non-symmetric. Due to its quadratic rates of asymptotic convergence, this method tends to produce relatively robust and efficient incremental nonlinear finite element schemes. However, the presence of the damage leads to a softening behavior and poses some difficulties for the calculation of the consistent matrix. This, together with the evolving contact conditions, induces some difficulties in the convergence of the iterative procedure. On the other hand, the Newton type implicit iterative resolution strategies are unconditionally stable and allow using large time or loading increments.

(2) The discretized dynamic explicit procedure is formulate as:

$$[\mathbf{M}]_{n+1}\left\{\ddot{U}\right\}_{n+1} + \left\{F_{\mathrm{e}}^{\mathrm{int}}\right\}_{n+1} - \left\{F_{\mathrm{e}}^{\mathrm{ext}}\right\}_{n+1} = \{0\} \Rightarrow [\mathbf{M}]_{n+1}\left\{\ddot{U}\right\}_{n+1} = \{R_{n+1}\} \tag{17}$$

where the degrees of freedom $\{U_{n+1}\}$ are computed as:

$$\begin{cases} \ddot{U}_{n+1} = [\mathbf{M}]_{n+1}^{-1} R_{n+1} \\[2mm] \dot{U}_{n+1} = \dot{U}_n + \frac{\Delta t_n}{2}\left(\ddot{U}_n + \ddot{U}_{n+1}\right) \\[2mm] U_{n+1} = U_n + \Delta t_n \dot{U}_n + \frac{(\Delta t_n)^2}{2}\ddot{U}_n \end{cases} \tag{18}$$

The dynamic explicit procedure avoids the iteration procedure by performing directly a solution of linearized algebraic system. It is extremely robust since there is no iterative procedure in solving the global equilibrium problem and there is no need to construct any consistent tangent matrix. This will reduce greatly the incremental size and generate a large number of increments to calculate the applied loading. The computing cost then will increase sharply for calculating the tangent matrices in each iteration. However, explicit procedure needs to control efficiently and automatically the time step size in order to satisfy the accuracy and stability requirements [43]. The central difference operator is conditionally stable according to the time increment Δt and the stability limit for the operator (with no damping) [47]. Instead, it can be estimated by determining the maximum element dilatational mode of the mesh, and to estimate the time step by:

$$\Delta t \leq \min\left(\frac{\hbar}{C_{\mathrm{d}}}\right), C_{\mathrm{d}} = \sqrt{\frac{E(1-\nu)}{\varrho(1+\nu)(1-2\nu)}} \tag{19}$$

where $\hbar$ is the mesh dependent stability factor and C_{d} is the current dilatational wave speed of the material function of material density ϱ, Young module E and Poisson ratio ν.

The unilateral contact with friction has a capital influence in metal forming processes in general and particularly in cutting operations. In fact, evolving contact with friction takes place between the formed metal sheet and the tools. The most widely used friction models implemented in FE codes are supposed isotropic and time independent (Tresca or Coulomb model) or time dependent (Norton–Hoff model). In this study, we limit ourselves to briefly describe the Coulomb isotropic model available in Abaqus where finite sliding contact with arbitrary rotation of the surfaces of two contacting bodies exist in sheet metal forming [22]. The Coulomb's friction model is defined by:

$$\begin{cases} \tau_{\mathrm{eq}} < \eta P \Rightarrow \dot{u}_t = 0 & \text{Sticking} \\[2mm] \tau_{\mathrm{eq}} = \eta P \Rightarrow \exists \chi \geq 0 / \dot{u}_t = -\chi \tau_{\mathrm{eq}} & \text{Sliding} \end{cases} \tag{20}$$

where η is the temperature dependent friction coefficient, $\dot{u}_t$ is the relative tangential velocity at the contact point lying on the contact boundary Γ_{c}; P is the normal contact pressure and $\tau_{\mathrm{eq}} = \sqrt{\tau_1^2 + \tau_2^2}$ is the equivalent tangential stress in tangential sheet plane. The so-called Signorini unilateral contact conditions where governs the contact between the master surface (representing the tool) and the slave surface (representing the metal sheet deforming plastically):

$$u_n \leq 0, \qquad F_n \leq 0 \qquad \text{and} \qquad u_n \cdot F_n = 0 \tag{21}$$

where $u_n = \vec{u} \cdot \vec{n}_c$ and $F_n = \left(\underline{\sigma} \cdot \vec{n}_c \right) \vec{n}_c$ are the normal components of the displacement and force vectors expressed in a local orthogonal triad and $\vec{n}_c$ is the normal between the bodies at the contact node.

Note that conditions in Equation (21) are similar to the Kuhn–Tucker loading–unloading conditions in classical plasticity. The inequality $u_n \leq 0$ expresses the non-penetration condition, $F_n \leq 0$ expresses the fact that, at each contact point, the normal force is negative in the local triad. Finally $u_n \cdot F_n = 0$ is valid for two cases:

- Case 1: There is contact $u_n \leq 0$ but $F_n < 0$
- Case 2: There no more contact $u_n < 0$ but $F_n = 0$

In the present work, the Dynamic Explicit resolution procedure is used within the general-purpose FE code ABAQUS/Explicit. The fully coupled constitutive equations presented above together with an iterative implicit procedure for the time integration were implemented using the user subroutine VUMAT.

2.4. 3D Adaptive Remeshing Procedure

In sheet forming, the blank shape, the tools geometry and the forming process parameters define the final product shape after metal forming. An incorrect design of the tools and blank shape or an incorrect choice of material and process parameters can yield a product with a deviating shape or with failures. A deviating shape is caused by spring-back after forming and retracting the tools. The most frequent types of failure are wrinkling (high compressive strains) and necking (high tensile strains). During the numerical simulation of sheet metal forming processes, the large plastic deformations imply a severe distortion of the computational mesh of the domain. In this case, frequent remeshing of the deformed domain during computation is necessary to obtain an accurate solution and complete the computation until the termination of the numerical simulation process. In this field, Borouchaki made great contributions in both 2D and 3D numerical simulations [48–52].

The advent of fast computers over the last few years has reduced the solution time once a mesh with an acceptable quality is provided as input. Hence, to obtain a cost and time effective solution to the forming problem is incremental remeshing of the work-piece at each step deformation. 'When to remesh?', and, 'how to remesh?' are the two high level issues that must be considered when automating process simulations. The criteria used to trigger an automatic remesh are collectively called the remeshing criteria. Four sources of errors that influence the remesh criteria are:

(i)   Geometric approximation errors;
(ii) Element distortion errors;
(iii) Mesh discretization errors;
(iv) Mesh rezoning or physical errors.

The impact of the different types of errors encountered based on metrics to measure them will key a remeshing step. The process of remeshing focuses on controlling these errors so that the simulation can continue.

- Until recently, the remeshing process was performed manually and potentially took several days for each remeshing (several are typically needed to model the entire process) of 3D domain. In addition, manual remeshing can potentially smooth the geometry thus preventing boundary defects from being detected, or, introduce constraints that result in false prediction of surface defects from process modeling.

- Hence, a 3D modeling system, that would automatically generate a new mesh on the deformed domain and continue the analysis, can dramatically reduce the overall modeling time and result in this technology being widely used in the design of industrial forming processes.

This section presents 3D adaptive remeshing scheme based on the linear tetrahedral element. The application environment for this scheme was established by python script, which integrates the 3D adaptive mesher, the Abaqus/Explicit solver and the point-to-point field transfer algorithm to new mesh. In order to control the mesh size adaptively and optimize the element quality automatically, both the geometrical and physical error estimates criteria are developed in our scheme [47–49].

We consider computational deformable domain of R^3, each domain Ω being defined from its boundary Γ which is expressed analytically by $G_0(\Gamma)$. We assume that domains of tool Π_k are rigid. Let us denote by $\Omega_{j,k}$ the subset of deformable domains Ω which are in contact with a given rigid domain Π_k. To construct an initial mesh of each deformable domain Ω, at first each boundary of domain Γ is discretized and then the mesh of domain Ω is generated based on this boundary discretization. The discretization of boundary Γ is obtained from its analytical definition $G_0(\Gamma)$. The method proposed in [25] is used to construct the initial "geometric" discretization $T_0(\Gamma)$ of the boundary Γ.

Based on this discretization, an initial tetrahedral coarse mesh $T_0(\Omega)$ of deformable domain Ω is generated. This initial mesh can become invalid after a mechanical computation involving large deformations (zero or a negative Jacobian in one or more elements). The new mesh representation must preserve the original topology of the mesh and must form a "good" geometric approximation to the original mesh with respect the criterion related to the shape of the elements.

In the classical Euclidean space, a popular measure for the shape quality of a mesh element K in three dimensions is:

$$Q(K) = \min_{1 \leq i \leq 4} [Q_i(K)], \quad Q(K) = \xi \frac{V(K)}{[\Sigma^2(e(K))]^{3/2}} \tag{22}$$

where $V(K)$ denotes the volume of element K, i the vertex of K, $e(K)$ the edges of K and ξ is the coefficient such that the quality of a regular element is valued by 1. From this definition, we deduce $0 \leq Q(K) \leq 1$ and that a nicely shaped element has a quality close to 1 while an ill shaped element has a quality close to 0.

The final deformation after the whole simulation is assumed to be obtained iteratively by "small" deformations (which is the case in the framework of an explicit integration scheme to solve the problem). After such a small deformation, rigid domains are moved and deformable domains are slightly distorted (assuming that each mesh element is still valid).

The new geometry $G_j(\Gamma)$ of boundary Γ can be defined in two ways, either by preserving a geometry close to the one before deformation, or by defining a new "smoother" geometry.

(1) In the first case, the new geometry is simply defined by the current discretization $T_{j-1}(\Gamma)$ of the boundary, and the new mesh nodes of $T_j(\Gamma)$ are placed on the elements of this discretization.

(2) In the second case, the new geometry is defined by a smooth curve interpolating the nodes and/or other geometric features of the current boundary discretization $T_{j-1}(\Gamma)$. The new nodes are then placed on this curve. The advantage of this second approach (which seems more complicated) is that the geometry of domain Ω remains smooth during its deformation.

The new geometry $G_j(\Gamma)$ of the domain includes two types of deformations:

(1) Free node deformations: this type concerns the deformations due to mechanical constraints (for instance equilibrium conditions), freely in the space. In this case, the new geometry of the domain after deformation is only defined by the new position of the boundary nodes as well as their connections.

(2) Bounded node deformations: these are the deformations limited by a contact with another domain (deformable or rigid tool domains for example). In this case, domain Ω locally takes the geometric shape of domains in contact Ω_k.

Based on the above classification of deformations, the free and bounded boundary nodes can be identified using the Hausdorff distance (δ). It consists in associating with each surface of part a region centered at this surface and in examining the possible intersection between the regions of the considered domain and those of the other domains. A node of the considered domain is classified as bounded if it belongs to one of the regions associated with the other domains or vice versa.

Formally, the region $R_\delta(e)$ associated with an edge e is defined by:

$$R_\delta(e) = \left\{ X \in R^3, \, d(X,e) \leq \delta \right\} \tag{23}$$

where $d(X,e)$ is the distance from point X to edge e, and δ is the maximum displacement step of domain Ω.

The above node identification allows us to define the new mesh size of boundary nodes.

1. If the node is free, the size is proportional to the curvature radius of the new domain boundary or the new geometry $G_j(\Gamma)$.
2. If the node is bounded, the size is proportional to the curvature radius of the neighboring part of the related domain Ω_k in contact.

The following remeshing scheme is applied to each deformable domain Ω_i after each step increment load j:

(a) Definition of the new geometry $G_j(\Gamma)$ after computation of the field solution S associated to mesh T.
(b) A posteriori geometrical error estimation from S including mesh gradation control to define a new discrete metric map: gap between the new geometry $G_j(\Gamma)$ and the current boundary discretization $T_{k-1}(\Gamma)$. Definition of a geometric size map $h_{g,j}(\Gamma)$ necessary to rediscretize the boundary Γ of the domain Ω.
(c) A posteriori physical error: gap between the physical solution $S_{j-1}(\Omega)$ obtained in Ω and an ideal "smooth" solution considered as the reference solution. Definition of a physical size map $h_{\varphi,j}(\Omega)$ necessary to govern the remeshing of domain Ω.
(d) Calculation of size map $h_j(\Omega_j)$ = minimum ($h_{g,j}(\Gamma)$ and $h_{\varphi,j}(\Omega)$).
(e) Definition of a unique size map with size gradation control parameter (fixed by the users between 1.2 and 2.15) resulting in a modified size map $h_j(\Omega_j)$.
(f) Adaptive rediscretization $T_j(\Gamma)$ of the domain boundary with respect to the size map $h_j(\Omega_j \mid \Gamma)$.
(g) Adaptive remeshing $T_j(\Omega)$ of the domain with respect to the size map $h_j(\Omega)$.
(h) Interpolation of mechanical fields $F_{j-1}(\Omega)$ of the old mesh on the new mesh $T_j(\Omega)$.
(i) Loop if necessary.

The overall adaptive methodology is implemented in the Optiform mesher package (see Figure 1). It includes the remeshing strategy, the interpolation error and the field transfer from the old mesh to the new one. For the simulations of sheet metal forming processes, a special procedure has been developed in order to execute Abaqus software [47] step by step. At each load increment, and after the convergence has been reached, the overall elements are tested in order to detect the fully damaged elements (elements where the damage variable has reach its critical value in all Gauss points). If so, the fully damaged element is removed from the structure and a new adaptive meshing of the part is worked out. The physical fields are interpolated from the old to the new mesh and the next loading step is worked out.

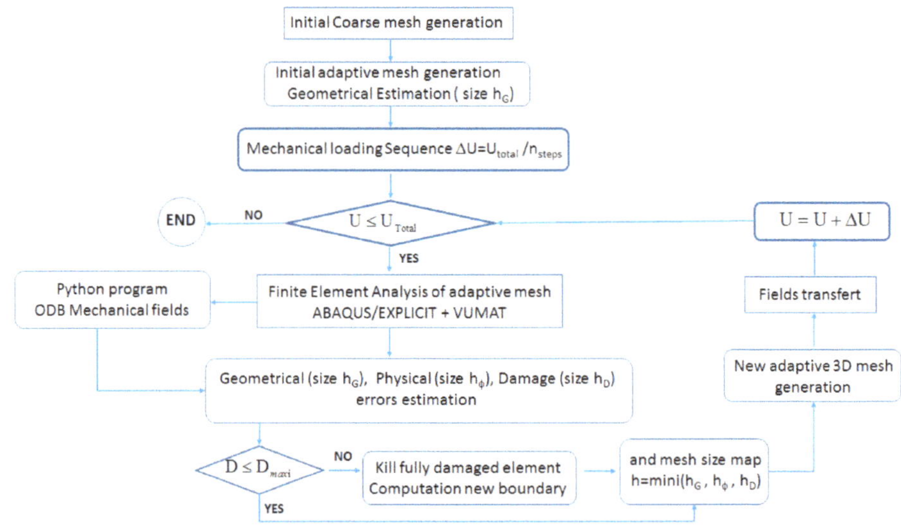

Figure 1. Flowchart of the 3D adaptive numerical methodology.

3. Results and Discussion

This section is dedicated to the validation of the proposed numerical methodology to simulate tensile, shearing, compression, and sheet metal forming processes using Abaqus/Explicit coupled with adaptive remeshing procedure. The characterization of the behavior of a given structure (titanium, copper, steel, and aluminum alloys), needs the knowledge of the material parameters. The difficult and not yet satisfactorily solved problem consists to compute automatically the material parameters under concern, by comparison with the available experimental database.

From a theoretical point of view, this defines a mathematical optimization problem using appropriated inverse analysis termed here as an identification procedure. Two different approaches (deterministic and statistical) exist to relate the problem of parameter identification to a least squares problem. In the deterministic approach, the inverse problem is expressed in a relaxed form and one just trying to minimize a distance between the data from a model and the experimental measurements. In the statistical approach, the inverse problem is seen as the search for the set of parameters which maximizes the probability of carrying out the experimental measurement.

In this study, the inverse constitutive parameter identification using Nelder–Mead Simplex algorithm is used [53,54]. The importance of this identification procedure is proportional to the increasing of the so-called advanced constitutive equations describing many coupled physical phenomena. The identification of the isotropic, isothermal elastoplastic constitutive equations accounting for isotropic hardening and ductile damage is based on the following steps [22,55]:

- The used database contains only uniaxial tensile tests conducted on a given specimen until the final fracture;
- The plastic parameters (A, B, C, n, $\dot{\bar{\varepsilon}}_0 = 1$) are determined on the hardening stage when damage effect is very small and can be neglected. The strain velocity is supposed fixed;
- The damage parameters ($Y_0, \alpha, \beta, \gamma, D_c$) are determined using the softening stage of the stress-strain curve;

The inverse identification procedure is performed by integrating the above constitutive equations on a single material point submitted to the tensile loading path using Matlab Software (Nelder–Mead Simplex algorithm) and the user's subroutine. Using material parameters determined above, the real specimen in tensile, shear or compression test is simulated by FEA and the global force-displacement curve is compared to the experimental one. If needed, the material parameters are adjusted and new FEA simulations are performed until the experimental and numerical force-displacement curves compares well.

Some obvious phenomena, like strain localization and damage evolutions, were presented in order to test the capability of the proposed fully coupled model and adaptive remeshing scheme to simulate the sheet metal forming process like blanking, multi-point drawing, single incremental forming and deep-drawing. All the numerical simulation are performed on the Dell Precision T7600 Workstation, $2\times$ Intel Xeon E5-2670 2.6 GHz 4 CPU Cores Processors; 128 GB Memory, Ubuntu Linux 64Mbit.

3.1. Uniaxial Tensile Test of XES Steel Sheet

The proposed constitutive equations are used to predict the stress-strain curve of XES steel which is used in the tensile experiment [20]. The fully coupled damage constitutive equations are implemented to predict the maximum stress $\bar{\sigma}_{max} = 383$ MPa, the plastic strain at damage initiation $\bar{\varepsilon}_c^p = 0.22$ and the plastic strain to fracture $\bar{\varepsilon}_{max}^p = 0.26$. The damage evolves from the damage initiation to the fracture $D_{max} = 0.99$ and the material stiffness degrades from maximum tensile force to zero. The best parameters found to fit the experiment stress-strain curve are shown in Table 1. The validation focused on the tensile specimen with the dimension of 8.9 mm $\times$ 3.2 mm $\times$ 0.5 mm subject to displacement load of $V = 5$ mm/s (see Figure 2).

The tensile tests are simulated firstly without remeshing for both coupled and uncoupled damage with plasticity cases. The 3D hexahedral finite elements with reduced integration (C3D8R) are used and a fine mesh (size is $h_{min} = 0.01$ mm in the region of plastic strain localization) is applied. Secondly, the remeshing procedure is applied, with the parameters given in Table 2, to simulate the localization of the equivalent plastic strain and the ductile damage. The iso-values of the damage in the specimen with/without remeshing procedure using tetrahedral finite elements are shown in Figure 3:

(1) In the coupled model without remeshing procedure (Figure 3b), the damage localization appears in the middle at a displacement of $U = 0.5$ mm. Then, two shear bands are formed quickly at a displacement of $U = 1.24$ mm. The damage variable reaches to the maximum value $D_{max} = 0.99$ in the cracked zones at a displacement of $U = 1.5$ mm.

(2) In the uncoupled model (Figure 3a), no damage and shear bands localization exist.

(3) In the coupled model with remeshing procedure the damage focused in the center of the specimen in Figure 3c and the shear band extended from center to the two sides along the direction of a 45°. Then, the crack generates in the center at a displacement of $U = 1.0$ mm and propagates along the shear band when the tensile displacement increased from $U = 1.586$ mm to $U = 1.587$ mm (see Figure 4).

The predicted force-displacement curves are compared with the experiment result in Figure 5. From these figures, one can observe the predicted results fit well the experiment values in the plastic stage and the effect of the damage-induced softening are clearly in the simulation using fully coupled damage constitutive equations.

The mesh size sensibility was also studied by comparing the tensile force-displacement curves for three minimum values $h_{min} = 0.5$, 0.1 and 0.05 mm as shown in Figure 6. It is clear that the damage evolution is sensitive to the element size, and the damage variable accumulates more quickly in the smaller mesh size. Table 3 presents the time performance (CPU time, number of elements) of numerical simulation with different mesh size $h_{min} = 0.05$, 0.1 and 0.5 mm). It is possible to confirm remeshing with small mesh (156,847 elements) advantage even for higher adaptive refinement compared with the coarse mesh (35,128 elements). A significant reduction in the overall error and CPU time (170%) in tensile stresses force is observed for fine meshes. These results prove that the proposed numerical methodology using elastoplastic fully coupled ductile damage model and adaptive remeshing procedure is reliable to predict the material behavior.

Table 1. Material parameters of the used XES steel [55,56].

E (GPa)	ν	A (MPa)	B (MPa)	n	C	Y_0 (MPa)	α	β	γ (MPa)	$\dot{\bar{\varepsilon}}_0$
210	0.29	150	448	0.406	0.025	0	2	1	0.37	1

Table 2. Adaptive remeshing parameters for the tensile test of the XES steel.

$h_{\min}$ (mm)	$h_{\max}$ (mm)	Physical Adaptive	Critical Value D_c	$D_{\max}$
0.05	2.0	Equivalent plastic strain	0.48	0.99

Table 3. Remeshing time performance of tensile test.

	Element Number	CPU	Global Error Estimation
Coupled model with remeshing: $h_{\min} = 0.05$	156,847	1 h 13 min	2.1%
Coupled model with remeshing: $h_{\min} = 0.1$	88,168	55 min	8.5%
Coupled model with remeshing: $h_{\min} = 0.5$	35,128	41 min	17%
Coupled model without remeshing	29,889	22 min	-
Uncoupled model without remeshing	15,289	15 min	-

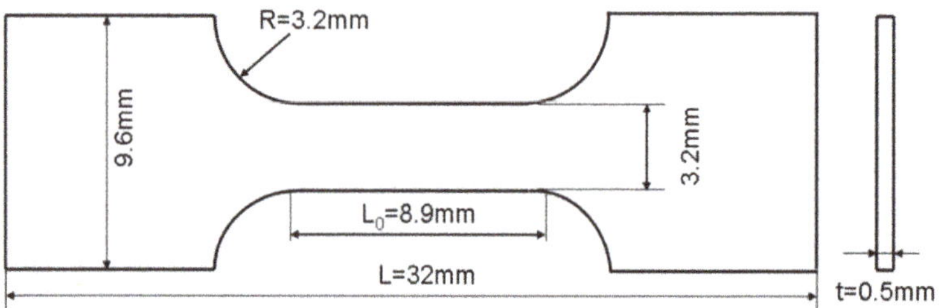

Figure 2. Geometry and dimensions for tensile specimen.

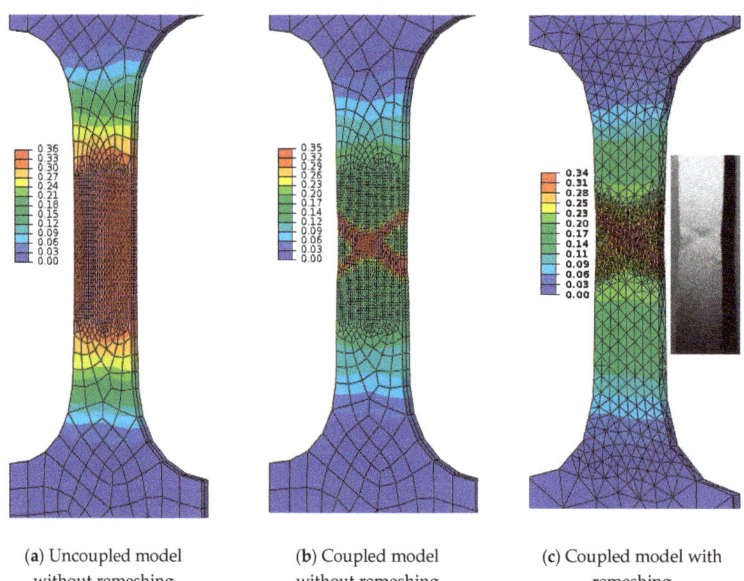

(**a**) Uncoupled model without remeshing

(**b**) Coupled model without remeshing

(**c**) Coupled model with remeshing

Figure 3. Ductile damage localization in various cases.

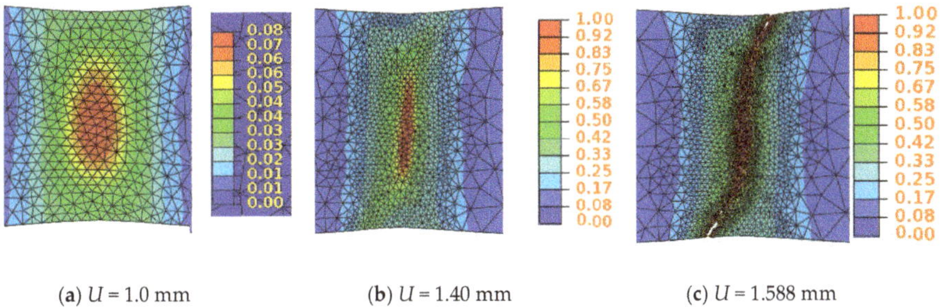

(**a**) $U = 1.0$ mm   (**b**) $U = 1.40$ mm   (**c**) $U = 1.588$ mm

Figure 4. Micro-crack initiation, growth and propagation in shear band localization [57].

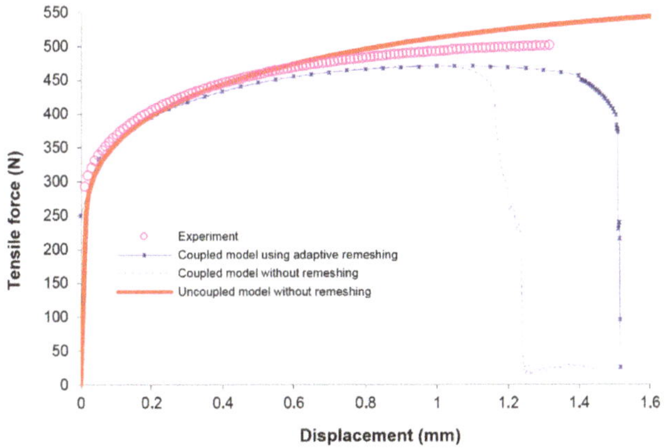

Figure 5. Tensile force versus displacement of the XES Steel with and without remeshing.

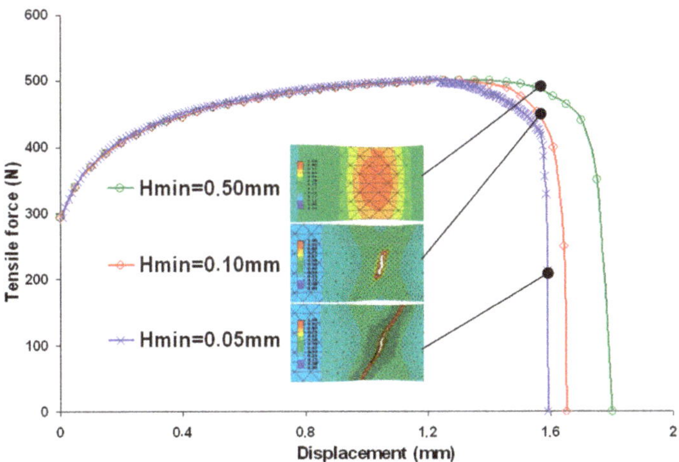

Figure 6. Mesh size effects on the response of force versus displacement.

3.2. Pure Shear Test of Titanium Alloy Sheet

In the above tensile simulation, a short shear band was already observed after damage localization and propagation. With respect to the tensile test, there is no sectional reduction in the shear sample and pure shear stress conditions are imposed. The application to shearing test is compared with the Peirs' work in literature [20]. The shear specimen of Peirs was recently studied and compared with other shear geometries proposed by Abedini [58] in which its capabilities for constitutive and fracture characterization of sheet metals was discussed. The low stress triaxiality in shear reduces the damage-accumulating rate, which can lead to a large strain than that in tensile test. The used shearing specimen dimension is shown in Figure 7.

The specific center shape of the specimen is adapted in order to concentrate shear stress in this zone. The titanium alloy was used, but the plastic hardening and damage parameters are estimated from the experimental shear curve (see Table 4). With these parameters, the maximum equivalent stress is about $\bar{\sigma}_{max} = 1340$ MPa, and a ductility of $\bar{\varepsilon}^P_{max} = 0.297$ [55,56]. As shown in Figure 8, the coupled material model had fit the experiment value well when equivalent plastic strain less than $\bar{\varepsilon}^P = 0.21$ and the material damage accumulate rapidly to $D_{max} = 0.99$ when equivalent plastic strain $\bar{\varepsilon}^P = 0.297$. The parameters of remeshing procedure are given in Table 5 and the initial coarse mesh for adaptive remeshing scheme with 3000 linear tetrahedral finite elements is illustrated in Figure 7.

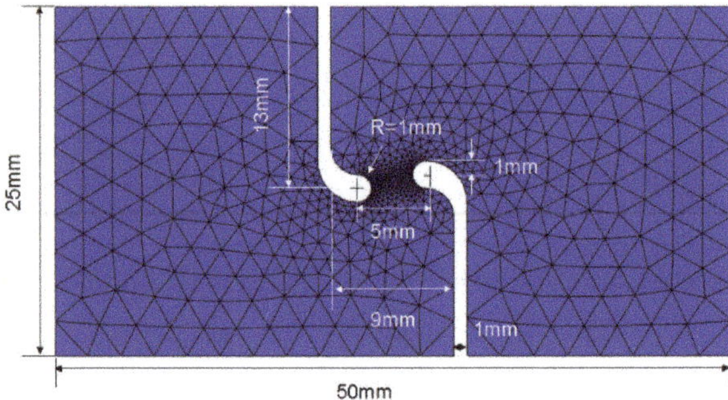

Figure 7. The dimension of shearing specimen and the initial mesh for adaptive remeshing scheme.

The simulation of shear test is also under a load of constant velocity ($V = 10$ mm/s). Firstly, the predicted ductile damage generated not in the center of shearing specimen but on both sides where near to the shear band for $U = 0.40$ mm (Figure 9a). Afterwards, the damage is initiated in the center of the specimen for $U = 0.60$ mm (Figure 8b) and propagated fast until $U = 0.72$ mm (Figure 9c). Then, the micro-crack is localized along the shear band when the specimen is completely damaged for $U = 0.74$ mm. As seen in Figure 9d, the size of finite mesh was refined adaptively to the minimum value in the region where micro-crack propagated. Finally, the specimen fractured when the tensile displacement was between $U = 0.74$ mm and $U = 0.75$ mm. The damage variable and the 3D damage section were illustrated in Figure 9e.

This simulation process described a very clear phenomenon of the shear band formation and the crack propagation. The shear band and the crack propagation direction in pure shearing stress condition are in a straight line, which is obviously different comparing with the tensile test. Figure 10 shows the comparison of the predicted force versus displacement curves for this fully coupled model obtained with adaptive remeshing under a controlled displacement with the constant velocity. According to this figure, can note that the strong effect of the softening induced by the damage occurrence giving a final fracture of the specimen around $U = 0.75$ mm. The maximum force is $F_{max} = 1504$ N reached for $U = 0.68$ mm.

Table 4. Material parameters used for shear test of titanium alloy [55,56].

Plastic Hardening Parameters				Ductile Damage Parameters			
A (MPa)	B (MPa)	C	n	Y_0 (MPa)	α	γ (MPa)	β
951	892	0.027	0.37	0.15	2	9.3	1

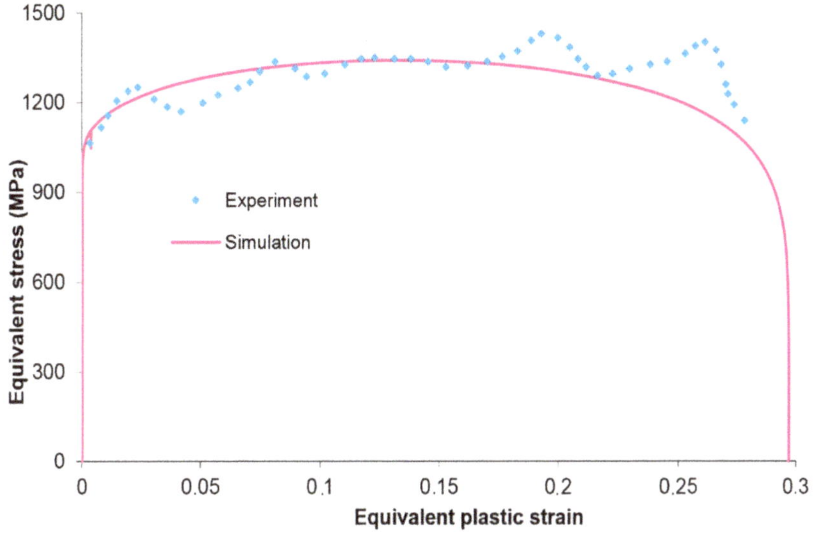

Figure 8. Experimental and predicted stress–strain curve of the TC4 titanium alloy.

Table 5. Adaptive remeshing parameters for the pure shearing test.

$h_{\min}$ (mm)	$h_{\max}$ (mm)	Physical Adaptive	Critical Value D_c	$D_{\max}$
0.05	2.5	Equivalent plastic strain	0.3	0.999

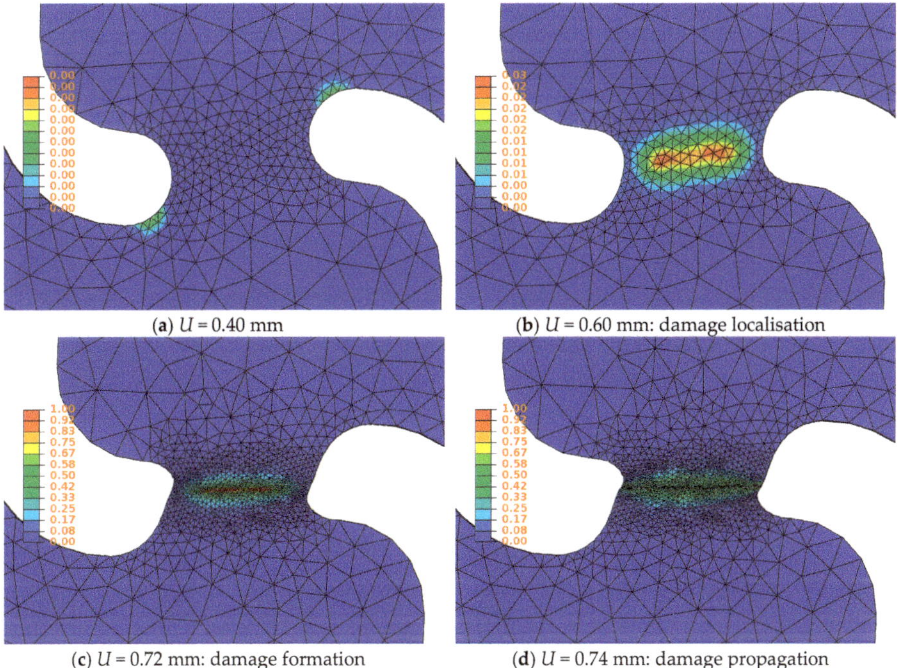

(**a**) U = 0.40 mm

(**b**) U = 0.60 mm: damage localisation

(**c**) U = 0.72 mm: damage formation

(**d**) U = 0.74 mm: damage propagation

Figure 9. *Cont.*

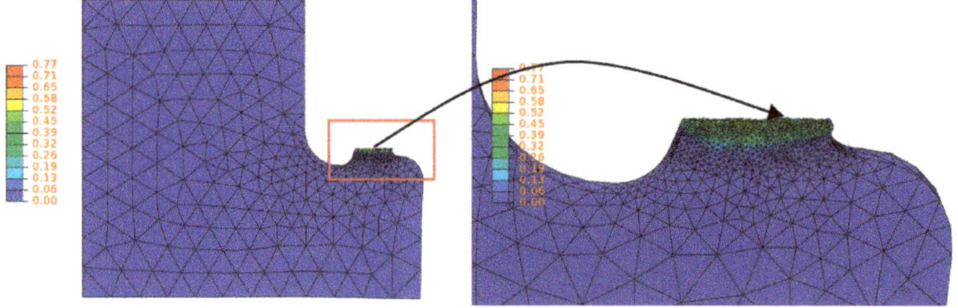

(**e**) U = 0.75 mm: Fully damaged specimen

Figure 9. The localization of damage, the formation and propagation of crack in the coupled model with remeshing.

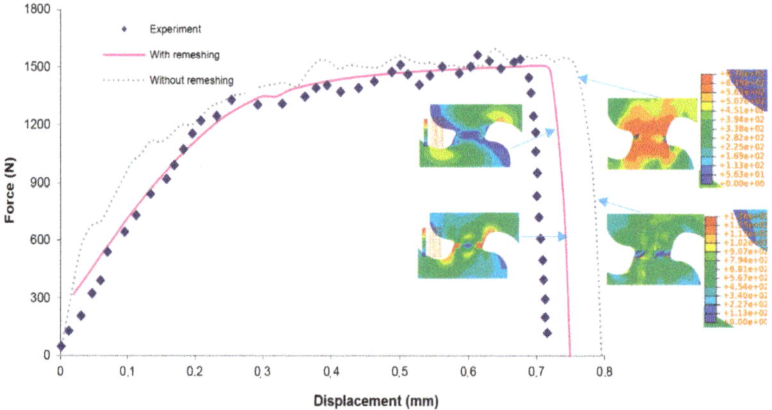

Figure 10. Comparison of predicted force versus displacement with experimental results.

3.3. Compression of Thin Truncated Aluminum Sheet Cylinder

The third test concerns the study of truncated aluminum alloy 5086 sheet cylinder compression. The thick truncated cylinder sheet of thickness t = 5 mm and radius R = 22.5 mm is compressed between rigid flat tools that tend to flatten it a displacement of U = 10 mm (see Figure 11). Following with the move of the top tool, the localization phenomenon of damage, the formation of a crack and its propagation to the final fracture of the truncated cylinder was clearly simulated. The aluminum alloy 5086 is primarily alloyed with magnesium and is a high strength structural alloy. It is not strengthened by heat treatment, instead becoming stronger due to strain hardening, or cold mechanical working of the material. The chemical composition of the used alloy is: Al 95.4%, Mg 4%, Cr 0.15%, and Mn 0.4%. The elastic-plastic and damage parameters of the aluminum alloy (5086) are given in Table 6. The plastic parameters of the material are obtained, from the experimental curves, in which the maximum stress is about $\overline{\sigma}_{max}$ = 366 MPa for equivalent plastic strain $\overline{\varepsilon}^P$ = 0.17 and then the material stiffness deteriorates to zero when equivalent plastic strain equal to $\overline{\varepsilon}^p_{max}$ = 0.27 (Figure 12) [55,56]. The numerical meshing parameters of the compression test are presented in Table 7. The proposed fully coupled methodology will be applied to improve compression test by avoiding the ductile damage occurrence in order to obtain a final part free from any defect.

Firstly, the damage generated at the four corners of the specimen, which contact with the compressive tools. After the accumulation of the plastic deformation, the damage in the center of the specimen was appearing as in Figure 13a when the top tool moved to U = 6.0 mm. Initially the damage is initiated at the four corners of the cylinder that come into contact with the top and bottom flat tools. After an accumulation of the plastic deformation is located in the center of the cylinder and the beginning of damage appeared for a displacement of the upper tool of $U = 6.0$ mm as shown in Figure 13a. The damage accumulated is propagated more rapidly from the center to the four corners of the cylinder at each compression stage. When the top tool move to $U = 8.1$ mm, the fully damaged element was generated and deleted firstly in the center as shown in Figure 13b. Two shear bands along the direction of 45° and 135° generated quickly after the center elements lost their stiffness. The crack propagated from center to the four corners along these two shear bands, and one main crack elongated faster than another as shown in Figure 13c. Finally, the specimen fractured when the top tool moved from $U = 8.7$ mm and $U = 8.8$ mm. The damage variable in the cross section of damaged specimen illustrated in Figure 13d. The final experiment fracture is presented in Figure 13f and two obviously cracks elongated from the center to the four corners. It is clear that the macro-crack initiation and propagation after damage localization were well simulated in the model using adaptive remeshing procedure for compressive test.

The predicted force-displacement curves using the fully coupled model and adaptive remeshing procedure under a controlled displacement was compared to the experiment values in Figure 14. From this figure, the effect of the softening induced by the damage occurrence giving a final fracture of the specimen around $U = 8.7$ mm which is a little earlier than that in experiment of $U = 9.7$mm. The predicted maximum force for coupled model is $F_{max} = 105.7$ kN which is also a slightly smaller than that of experiment value $F_{max} = 106$ kN. It is remarkable to see that the maximum force and the displacement corresponding to the final specimen cutting are predicted with a good precision (0.8% error) but over estimates the decrease of the compression-force in the softening stage.

However, for uncoupled approach, the total force increased monotonously and had non-degradation of the structure and the specimen is not completely damaged (see Figure 13e). Under the help of the adaptive remeshing procedure, only the coupled constitutive equation can predict the damage behavior clearly for compressive test.

Table 6. Material parameters of aluminum alloy 5086 [55,56].

Elastic Parameters	E (GPa)	v		
	85	0.38		
Plastic Parameters	A (MPa)	B (MPa)	C	n
	150	460	0.025	0.41
Damage Parameters	Y_0 (MPa)	α	γ (MPa)	β
	5	2	0.8	0

Table 7. Adaptive remeshing parameters for the compressive test of aluminum alloy 5086.

h_{min} (mm)	h_{max} (mm)	Kill Element	Physical Adaptive	Critical Value
0.5	1.0	$D_{max} = 0.99$	$\bar{\varepsilon}^p$	0.27

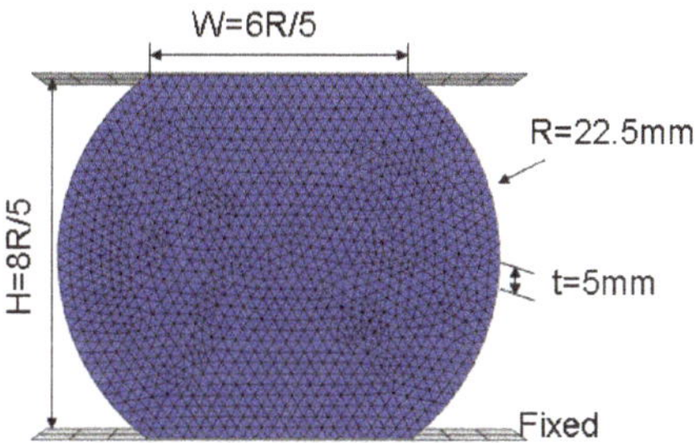

Figure 11. Dimension of truncated cylinder thick sheet and initial mesh.

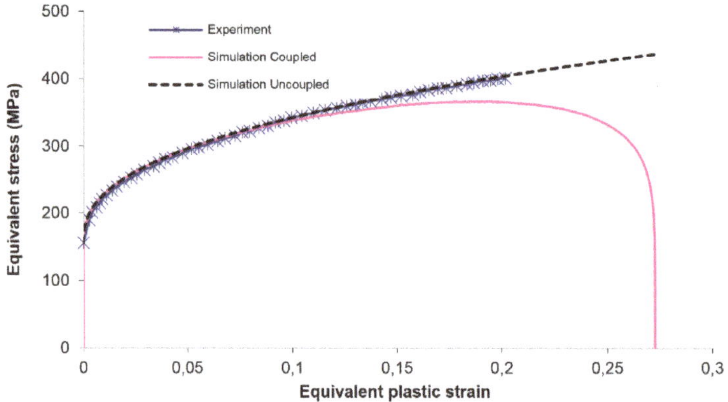

Figure 12. The material stress–strain curve obtained by experiment and simulation models.

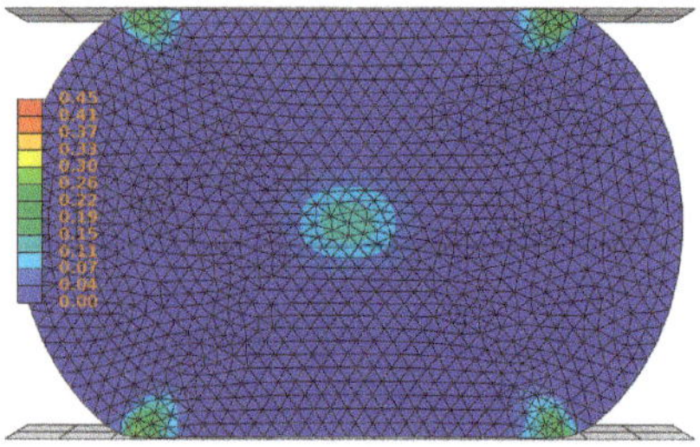

(**a**) U = 6.0 mm damage localisation

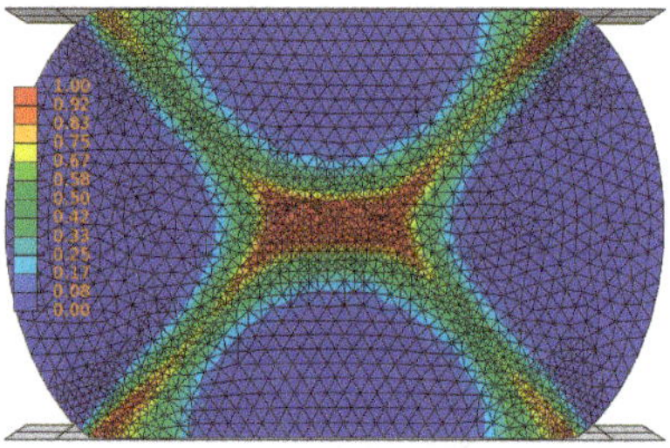

(**b**) U = 8.1 mm damage formation

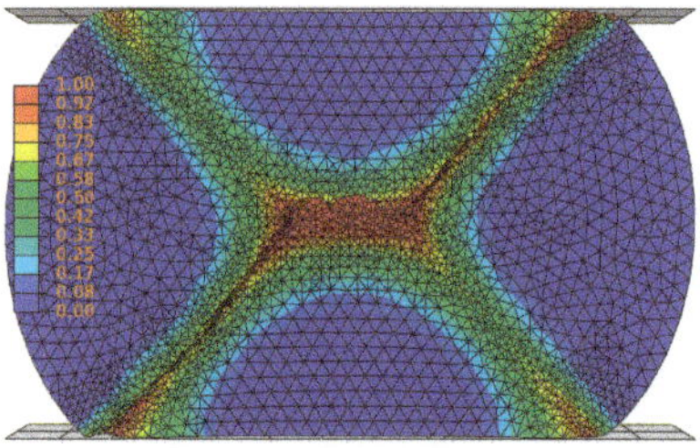

(**c**) U = 8.7 mm damage propagation

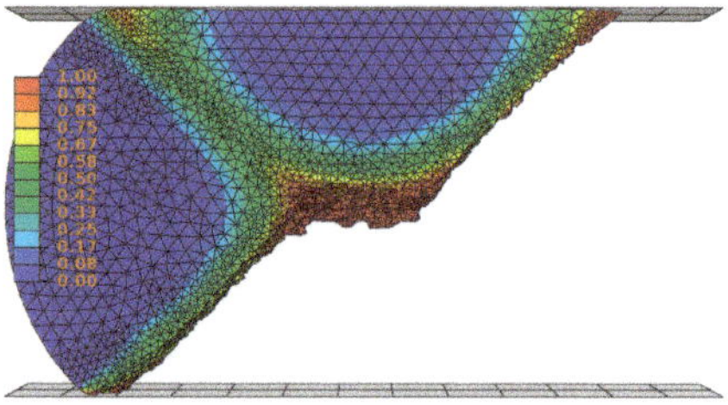

(**d**) U = 8.8 mm Final fracture (coupled with remeshing)

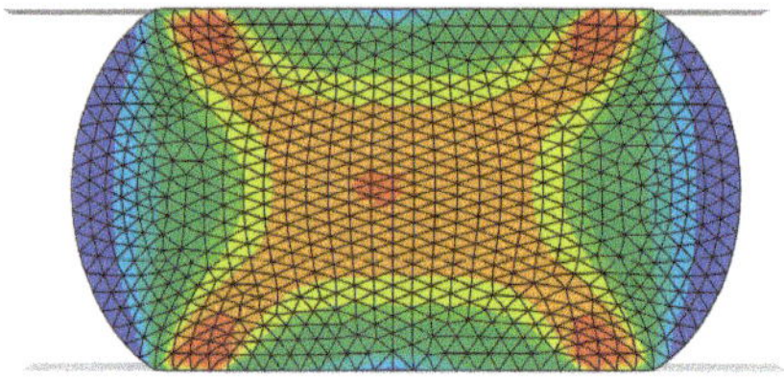

Coupled model

Figure 13. *Cont.*

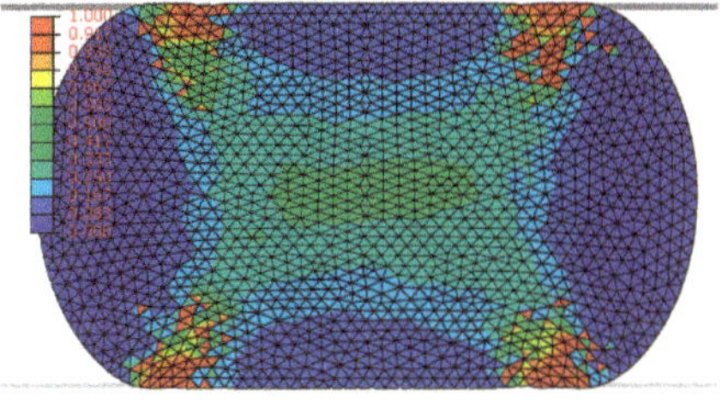

Uncoupled model

(**e**) U = 10 mm uncoupled and coupled models without remeshing

(**f**) Experimental compression before and after compression for U = 9.68 mm

Figure 13. The localization of damage, the formation and propagation of crack in compressive test.

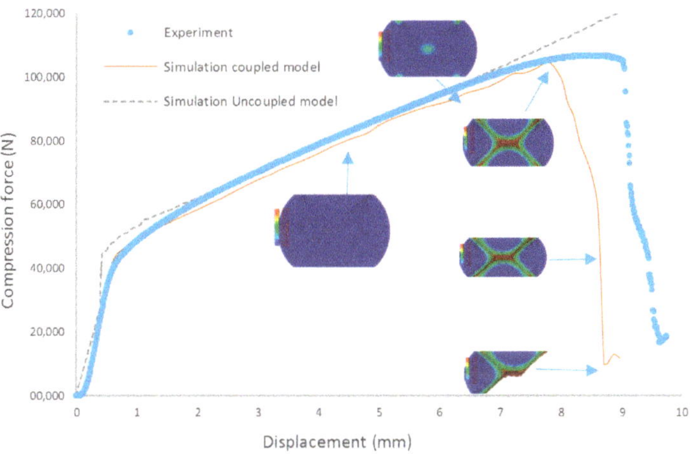

Figure 14. Comparison of predicted compression force response and experimental values.

3.4. Blanking of Oxygen-Free-High-Conductivity (OFHC) Copper Sheet Metal

The main purpose of the numerical simulation is to "virtually" predict a final sheared edge free from any defect (ratio of the burnished edge, burr height), to estimate the residual stress profile through the thickness and the maximum blanking force versus displacement. The punch force and penetration at fracture are important for tool and machine dimensioning. The shape of the cut edge and the burr height are crucial for the final product quality. The prediction of these parameters by numerical adaptive simulation may be very helpful in blanked or sheared part design.

This section focuses on simulating the sheet metal blanking process using the fully coupled model using the numerical methodology. The sheet metal is oxygen-free-high-conductivity (OFHC) copper.

The numerical model will be validated through comparing the blanking force and the profiles of sheared edges with the experimental values [59]. The geometrical model and the boundary conditions of cylindrical sheet metal blanking are described in Figure 15. For the friction of contact, the surface to surface contacts is defined between tools and piece; the self-contact is defined in the piece when some fully damage elements are removed. The friction coefficient between the blank and tools is taken as representing 'dry' friction without lubrication is used in both surface to surface contact (punch/die and work-piece) and self-contact (fractured surface).

The blanking parameters used to blank the sheet of OFHC copper sheet are presented in Table 8. The material parameters used in this model are listed in Table 9 and it accurately predict the material behavior with a maximum stress of $\bar{\sigma}_{max} = 350$ MPa, $\bar{\varepsilon}^P = 0.77$ and $\bar{\varepsilon}^P_{max} = 0.9$. In order to save the computing cost, quarter of the round sheet is considered. The round sheet is initially discretized by a very coarse mesh (453 linear tetrahedral elements as shown in Figure 16a). Geometrical error estimation and the contact region with tools (punch and die) with respect the minimum size h_{min} then control the adaptive mesh (Table 10). In the following steps, the mesh regenerates adaptively according to the plastic strain and the damage variable. When the damage accumulates $D_{max} > 0.99$, the element will be removed and a new boundary will be generated. The macro-crack initiation and propagation during the sheet blanking were analyzed at each punch penetration in Figure 16b for $U = 0.01$ mm and in Figure 16c for $U = 0.2$ mm. The copper sheet was fully cut for $U = 0.33$ mm.

The predicted force versus displacement with/without adaptive remeshing is compared with the experiment results in Figure 17. Note that the predicted result without adaptive remeshing deviates to the experiment values more and more severely with the increasing of the punch displacement. These differences are due essentially to the qualities of finite elements, despite the used large number of elements (230,150) and to the convergence of the computational procedure. Also, note that the overall error is very large (19.0%) in the case without remeshing with significant time calculations (3 h 48 min). In the case of adaptive remeshing procedure, the predicted curve is in agreement with the experiment values with a maximum force of $F_{max} = 3235$ N (experimental value $F_{max} = 3353$ N) and displacement at fracture $U_f = 0.34$ mm (experimental value $U_f = 0.36$ mm). Also note that in this case using a number of elements not high (54,519), the overall error is small (1.23%) with relatively correct time calculation (2 h 17 min) (see Table 11).

The depth of Rollover (L_R), Burnish (L_B), Fracture (L_F) and Burr (L_{Burr}) define the morphology of the sheared edge of sheet. Compared to the experimental results given by Husson [60], the predicted sheared edges with $L_R = 0.5$ mm, $L_B = 40$ μm, $L_F = 268$ μm and $L_{Burr} = 260$ μm using both the models with and without adaptive remeshing scheme are shown in Figure 18. It is clear that the morphology of sheared edge predicted by adaptive remeshing scheme (middle) is similar to experiment one (right). The rollover depth $L_r = 55$ mm and burnish depths $L_B = 271$ mm are a little bigger than the experimental ones ($L_r = 40$ mm and $L_B = 268$ mm) and there is no burr generate in this blanking condition. The fracture angle α and rollover radius R are almost the same. Against in the case of a model without remeshing procedure the sheared edge profile obtained remains unrepresentative of the experimental result and is not in agreement with experimental profile. For these results, we can nothing that only the adaptive remeshing scheme is able to predict the sheared edge profile of sheet cutting (see Table 12).

Figure 19 illustrates the damage evolution at the sheared edge during blanking process. We observe that the damage starts from two sides which contact with the tool tips ($U = 0.10$ mm) and extends to the center of sheared edge ($U = 0.2$ mm). When the punch achieves $U = 0.31$ mm, the elements fully damage at the sheared region are removed and a macro-crack is formed.

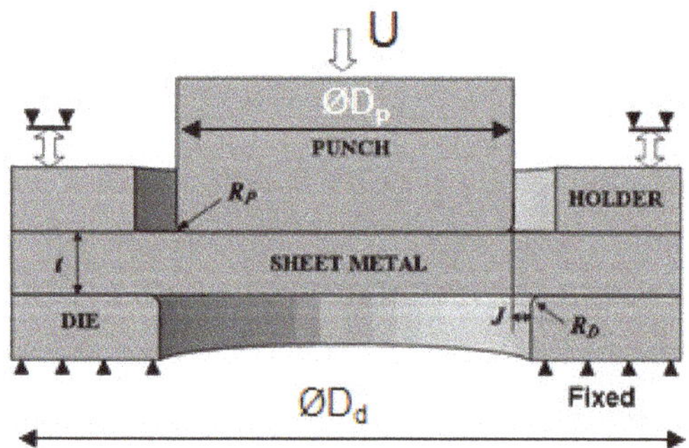

Figure 15. Geometry parameters and boundary of blanking process.

Table 8. Geometrical parameters of sheet blanking.

t (mm)	$R_P = R_D$ (mm)	J (mm)	D_P (mm)	D_D (mm)	Blanking Frequency	Friction Coefficient Tools/Sheet μ
0.58	0.01	0.0145	3.5	3.53	700 (stroke/min)	0.02

Table 9. Elasto-plastic and damage parameters of the oxygen-free-high-conductivity copper [55,56].

E (GPa)	ν	A (MPa)	B (MPa)	n	C	Y_0 (MPa)	α	β	γ (MPa)	$\dot{\bar{\varepsilon}}_0\,(\text{s}^{-1})$
124	0.34	90	292	0.4	0.025	0.38	0.2	0.3	0.37	1

Table 10. Adaptive remeshing parameters for the blanking of the OFHC sheet.

$h_{\min}$ (mm)	$h_{\max}$ (mm)	D_c	$D_{\max}$	δ (mm)
0.015	5	0.5	0.99	0.026

Table 11. Adaptive remeshing time performance of blanking process simulation.

	Element Number	CPU (hours)	Global Error Estimation
Without remeshing	230,150	3 h 48 min	19%
With remeshing	54,519	2 h 17 min	1%

Table 12. Predicted blanking performance and sheet characterize with and without remeshing.

	$F_{\max}$ (N)	U_f (mm)	α	R (mm)	L_R (µm)	L_B (µm)	L_F (µm)	L_{Burr} (µm)
Without remeshing	6640	0.53	-	0.65	56.5	270	248	55
With remeshing	3235	0.35	7°	0.53	55	271	242	0
Experiment	3353	0.36	5°	0.50	40	268	260	0

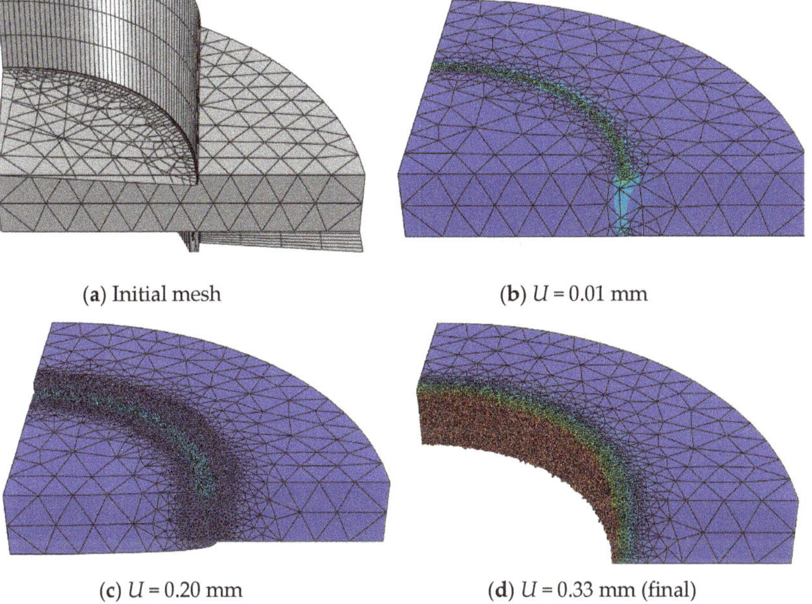

(**a**) Initial mesh (**b**) U = 0.01 mm

(**c**) U = 0.20 mm (**d**) U = 0.33 mm (final)

Figure 16. Iso-values of damage variable at different blanking step.

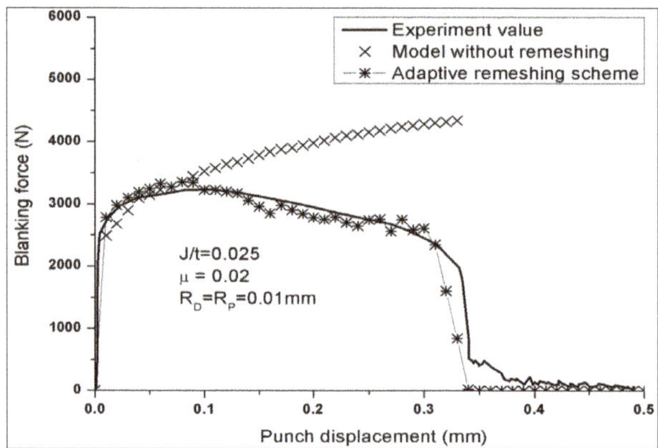

Figure 17. Predicted blanking force versus punch displacement with and without remeshing.

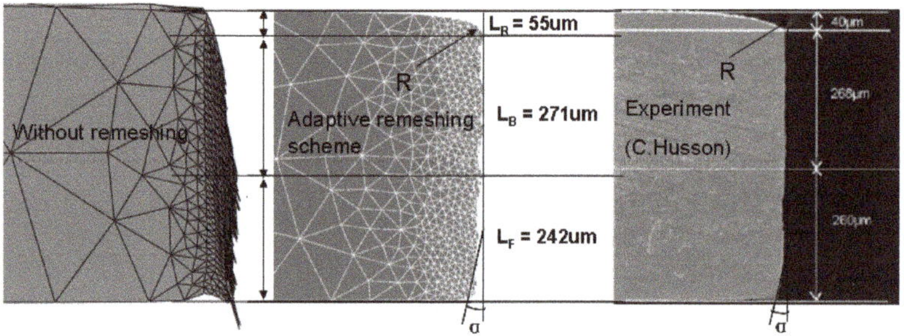

Figure 18. Sheared edges morphologies of sheet: Experimental–simulation with and without remeshing.

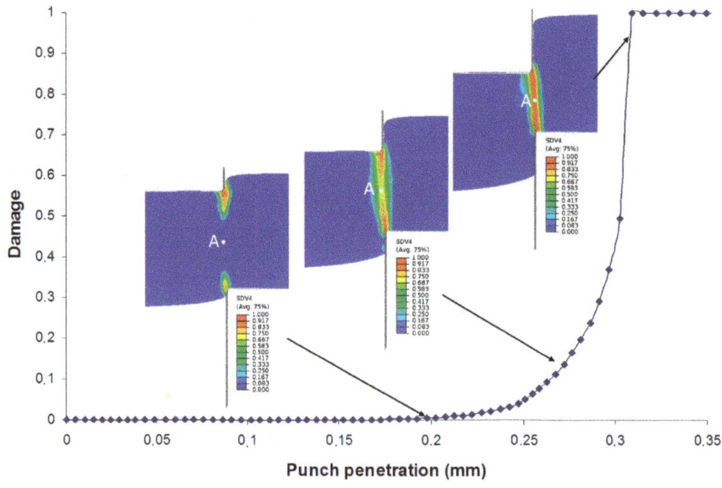

Figure 19. Damage initiation and growth in the center of OFHC copper sheet (Point A) during the blanking.

3.5. Multi-Point Forming of Thick Steel Sheet

Curved sheet metal parts are widely used in automotive and aerospace industries. Traditional deep-drawing or hydroforming process is difficult to deform plastically thick sheet. The fundamental component in the Multi-Point Forming (MPF) press is a pair of matrices of punches, which approximates to a continuous forming surface of dies [61]. MPF is a flexible forming technology, which is suitable to produce customized products. The factors influencing the shape accuracy of the final product are analyzed in order to minimize the spring-back.

The objective of the Multi-Point forming of thick steel sheet simulation is to highlight the abilities and capacities of numerical procedure to simulate complex shapes and its relative robustness of the quality of the final product obtained by discrete of the multipoint tool contact. According to the fact that the MPF operation is axisymmetric, only half of the billet section is modelled. Figure 20a shows the upper and lower matrices of square punches with curved ends in the MPF press, 100 punches for an effective forming surface of 1700 mm × 1700 mm. The initial thick sheet of steel material is 1500×1500 mm^2 and 40 mm of thickness is discretized with linear tetrahedral finite element. Tools (punch and die) are supposed as rigid surfaces. The output of the simulation is given in terms of the sheet profile, the strain history as well as the punch forming forces. Comparison between the simulation results and the experimental data is made.

The materials parameters of the used steel should be determined using the experimental results obtained from classical tensile tests conducted until the final fracture of the specimens. The damage parameters have been estimated in order to have a maximum stress $\overline{\sigma}_{max} = 1200$ MPa, for $\overline{\varepsilon}^p = 12\%$ and a rupture of $\overline{\varepsilon}^p_{max} = 35\%$. A master–slave contact approach is used in the analysis where the tools are considered as the master surfaces, and the top and the bottom surfaces of the blank (surface facing the punch and the die) constitutes the slave surfaces. Friction (with coefficient $\eta = 0.12$) is introduced using Coulomb model and the interaction between the sheet and the billets is formulated using finite sliding approach, which allows for the possibility of separation between the two surfaces during sliding.

The adaptive remeshing simulation is performed under a load of constant velocity ($V = 2$ mm/s) and the small punch displacement $U = 10$ mm of the sheet were loaded by numerous small time steps. The simulation consists, in the present case, of 230 computational steps such an adaptive remeshing process ensures a maximum error in the whole computational domain to be limited by the threshold excepted in the zone where the element size $h_{min} = 0.1$ mm and $h_{max} = 5$ mm is achieved (see Table 13).

Figure 20b–g illustrates the predicted deformed sheet ate each step with respect to the punch displacement. Noting that the mesh is only adapted when the punches are close to the sheet for $R_\delta(e) \leq 1$ mm. It can also be noted that the plastic strain and damage ($D_{max} < 0.12$) remains very low. Quantitative comparison with experimental result provides the performance of the proposed approach (Figure 20h).

The curves of forming force versus punch are presented in Figure 21 with and without remeshing. It can be seen clearly that the forming force depends on the contact state. During the initial forming step the punch force is small estimated at $F_{punch} < 500$ kN, and it increases very slowly since only a few punches contact with sheet ($F_{punch} < 750$ kN). After the displacement of punch reaches a value about the 3/4 of the total tool move more and more punches get into contact with sheet and the forming force increases sharply. At the end of MPF process the forming force reaches its maximum since all punches contact with sheet ($F_{punch} = 3500$ kN).

The formed work-piece profiles, before and after the spring-back, are shown in Figure 22. From these results, it is concluded that for thick sheet the spring-back during the MPF is very small. Table 14 presents the time performance (CPU time, number of elements) of numerical simulation performed on Dell Precision T7600 Workstation, 2× Intel Xeon E5-2670 2.6 GHz 4 CPU Cores Processors; 128 GB Memory with and without remeshing. It is possible to confirm remeshing advantage even for higher adaptive refinement compared with the very fine mesh used here as reference (700,143 tetrahedral elements). A reduction of CPU time about approximately 75% has been obtained with the remeshing procedure (4 h 17 min) compared to the simulation without remeshing obtained with very fine mesh (700,143 elements).

Table 13. Adaptive remeshing parameters for Multi-Point Forming (MPF) process.

h_{min} (mm)	h_{max} (mm)	$\overline{\varepsilon}_C^P$	$\overline{\varepsilon}_{max}^P$	δ (mm)
0.1	5	0.1	0.35	2.5

Table 14. Adaptive remeshing time performance of MPF process simulation.

	CPU Time	Number of Steps/Increments	Initial n° of Elements	Final n° of Elements
Without remeshing	29 h 48 min	100 increments	700,143	700,143
With remeshing	4 h 17 min	230 remeshing steps	15,000	284,700

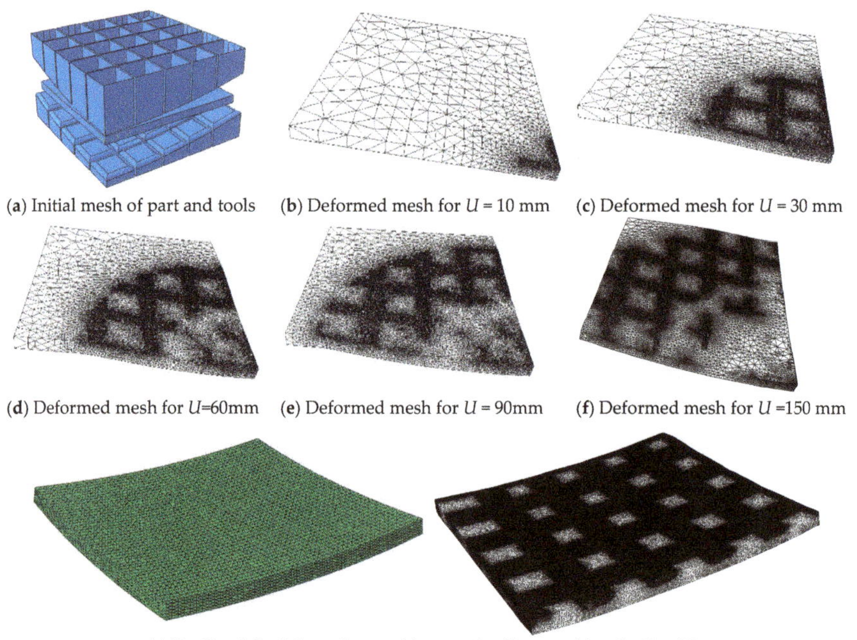

(**a**) Initial mesh of part and tools (**b**) Deformed mesh for U = 10 mm (**c**) Deformed mesh for U = 30 mm

(**d**) Deformed mesh for U=60mm (**e**) Deformed mesh for U = 90mm (**f**) Deformed mesh for U =150 mm

(**g**) Predicted final sheet shape without and with remeshing for U = 180 mm

Figure 20. Predicted sheet profile at each punch displacement [61].

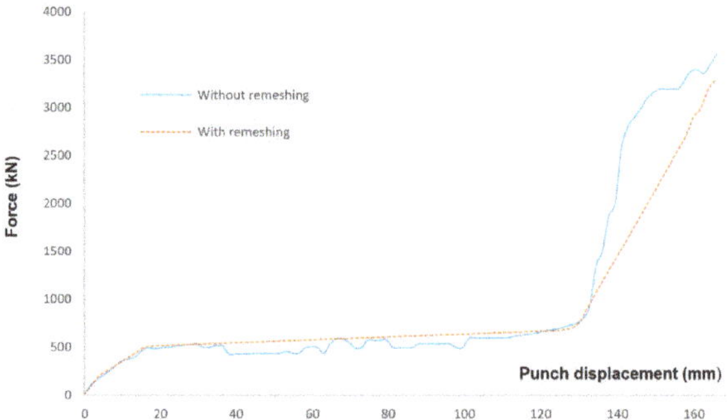

Figure 21. MPF punch force versus punch displacement after spring-back.

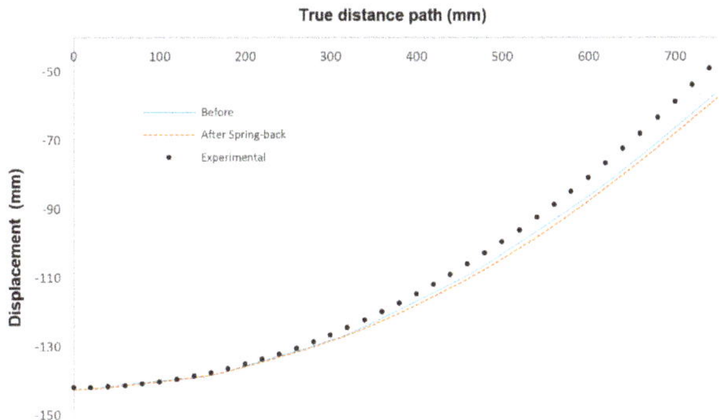

Figure 22. Sheet profile before and after spring-back.

3.6. Incremental Sheet Forming of Titanium Conical Shape

The study will be focused on Numisheet'2014 benchmark for the manufacturing of a truncated cone shape made by single point incremental forming process (SPIF). The incremental step-down size (Δz) is the amount of material deformed for each revolution of the forming tool (similar to depth of cut in machining) which does not require dies [62]. The advantages of this process are the flexibility of forming process for the manufacturing of complex shapes. Several applications of this process exist in the medical field for the manufacturing of personalized titanium prosthesis (cranial plate, knee prosthesis, etc.) due to the need of product customization to each patient. The CAD package CATIA is used to create solid models of parts and the tool path is generated according to the obtained profile. A 3-axis CNC machine is generally used to control the movement of the spherical tool. Tool trajectory is created and connected using a step or a spiral transition method [63–68].

The internal roughness of SPIF shaped component is influenced primarily by the tool (dp) and step (ΔZ) sizes, as other factors such as, sheet thickness, material properties, forming angle, feed rate and etc., may affect both the internal and external roughness [54–58]. The influence of forming speed, both rotational spindle speed N and feed rate V_f had an important effect on the heat generation due to the friction with the sheet. Too high friction generates an increase of temperature and sheet damage [47,48]. In this study, the forming conditions (see Table 15) are adopted for the SPIF with rotational spindle speed $N = 40$ revolutions per minute (RPM) and feed rate $V_f = 6000$ mm/mm. Truncated cone that was formed using continuous spiral tool paths (Figure 22) also in order to achieve smaller surface roughness and to avoid tool entry and exit marks.

This section presents simulation results of the SPIF to produce a truncated conical height $h = 40$ mm titanium Ti-6Al-4V sheet. The effective working area of the initial circular specimen was $\varnothing D = 130$ mm and thickness $t_i = 0.5$ mm. The forming was made with semi hemispherical punch X160CrMoV12 steel in order to obtain prototypes according to the desired profiles. The plastic parameters of Ti-6Al-4V sheet presented in Table 16 was obtained following uniaxial tensile tests [59].

After incremental forming, visual control is performed first with goal to detect potential defects and estimate surface finish. On the both plates, the only problem identified was a slight increase in roughness of internal surface in the zone of the maximum deformation. The rest of internal surface was smooth and without tool marks. Dimensional control of the obtained parts was performed by coupling with the original one. For that purpose, both replicas were scanned applying the same measuring procedure. The obtained clouds of points are then transformed to CAD model and compared with the desired CAD profile.

Table 15. Single point incremental forming process (SPIF) process parameters.

t_i	rc	h	d_p	ΔZ	ØD	α	N	V_f
0.5 mm	5 mm	40 mm	5 mm	0.5 mm	130 mm	50°	40 RPM	600 mm/mm

Table 16. Elasto-plastic parameters of Ti-6Al-4V sheet [60,61].

A (MPa)	B (MPa)	C	n	$\dot{\bar{\varepsilon}}_0(\text{s}^{-1})$
968	380	0.0197	0.421	1

The FEA simulation of conical forming design of single point incremental sheet forming SPIF presented in Figure 23 is performed on Dell Precision T7600 Workstation, $2\times$ Intel Xeon E5-2670 2.6 GHz 4 CPU Cores Processors; 128 GB Memory, with same experimental working conditions. The friction of contact with coefficient $\eta = 0.14$ and the surface-to-surface contact formulation are defined to characterize the interaction between the punch and the sheet.

The main outputs results are the final shape of the sheet and the evolution of the thickness in a cross-section along the symmetric axis. The adaptive remeshing simulation is performed with fully coupled model under a load of constant velocity and the small punch displacement loaded by numerous incremental path (Table 17). During the SPIF simulation many elements are refined and coarsened, so as a result 1500 computational steps such an adaptive remeshing process ensures a maximum error in the whole computational domain to be limited by the threshold excepted in the zone where the element size $h_{min} = 0.5$ mm and $h_{max} = 10$ mm is achieved.

Table 17. Adaptive remeshing parameters for SPIF.

h_{min} (mm)	h_{max} (mm)	$\bar{\varepsilon}_c^p$	$\bar{\varepsilon}_{max}^p$	δ (mm)
0.15	10	0.012	0.30	2.5

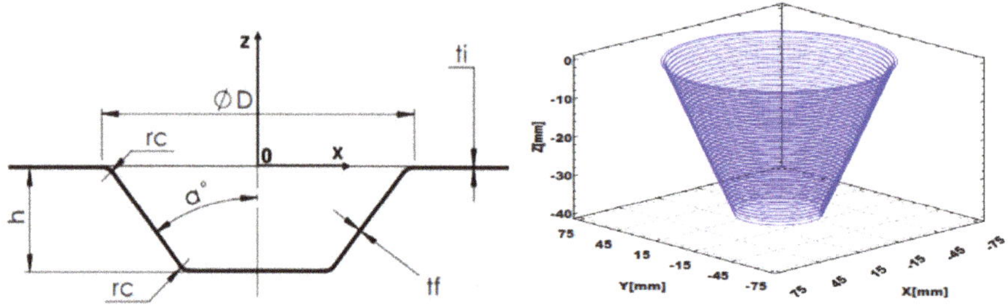

Figure 23. Tool path trajectory of incremental process.

Table 18 presents the adaptive remeshing time performance for different level of refinement (CPU time, final sheet shape profile, convergence and strain thickness distribution). It is possible to confirm remeshing advantage even for higher adaptive refinement compared with the very fine mesh used here as reference (300,477 tetrahedral elements). The results are quite similar between the reference case and the remeshing process using high number nodes per edge.

They are higher with remeshing because the mesh used is coarser than the reference one. It is notable that the results (CPU time, convergence and sheet strain) are sensitive to the variation of the number of elements. A reduction of CPU time about approximately 85% has been obtained and at the same time good simulation results are achieved and compared to reference results. Figure 24b–j illustrates the deformed sheet at each step with respect to the spherical punch travel and the number of element used at the step forming. Noting that the mesh is only adapted when the punches are close to the sheet for $R_\delta(e) \leq 0.12\,\text{mm}$. It can also be noted that, the plastic strain ($\bar{\varepsilon}^p_{max} < 0.3$) remains low and no damage occurs during forming process. Quantitative comparison with experimental result provides the performance of the proposed approach. Figures 25 and 26 indicate a good prediction of the conical shape and the cross section thickness along central X-axis for compared with the experimental and the desired profiles. A lower thickness reduction is observed on the bottom surface of deformed sheet.

Table 18. Adaptive remeshing time performance of SPIF.

	CPU Time	N° Steps/Increments	Initial n° of Elements	Final n° of Elements
Without remeshing (coarse mesh)	06 days and 19 h	Stopped, convergence problem	100,000	100,000
Without remeshing (fine reference mesh)	09 days and 15 h	2460 increments	300,477	300,477
With remeshing	18 h and 20 min	1500 remeshing steps	2,853	128,558

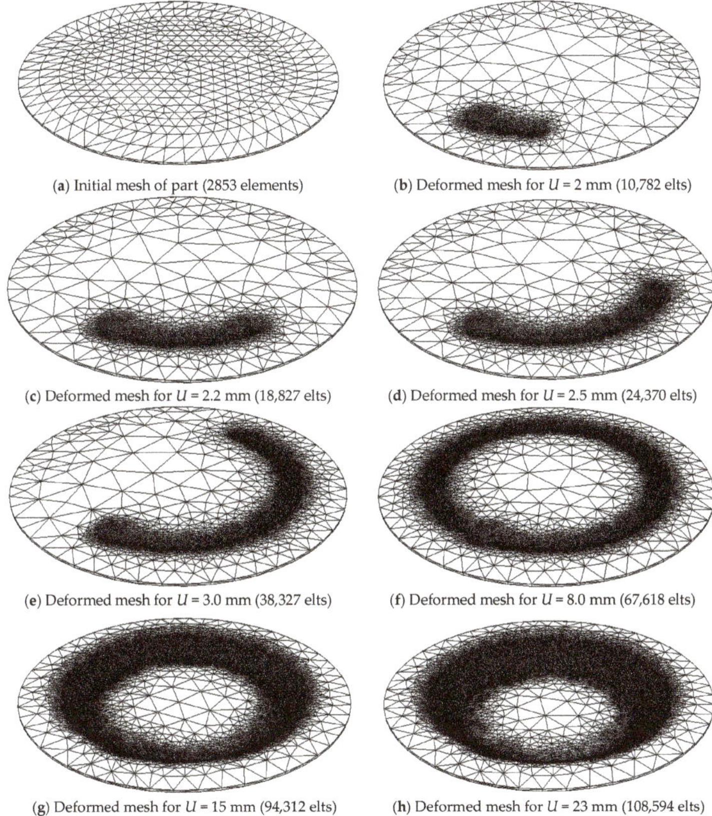

(a) Initial mesh of part (2853 elements)

(b) Deformed mesh for U = 2 mm (10,782 elts)

(c) Deformed mesh for U = 2.2 mm (18,827 elts)

(d) Deformed mesh for U = 2.5 mm (24,370 elts)

(e) Deformed mesh for U = 3.0 mm (38,327 elts)

(f) Deformed mesh for U = 8.0 mm (67,618 elts)

(g) Deformed mesh for U = 15 mm (94,312 elts)

(h) Deformed mesh for U = 23 mm (108,594 elts)

Figure 24. *Cont.*

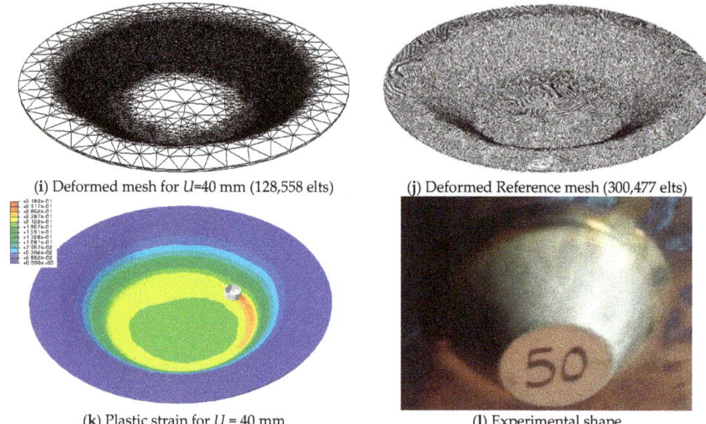

(**i**) Deformed mesh for U=40 mm (128,558 elts)

(**j**) Deformed Reference mesh (300,477 elts)

(**k**) Plastic strain for U = 40 mm

(**l**) Experimental shape

Figure 24. Predicted conical profile at each punch displacement and experimental deformed sheet.

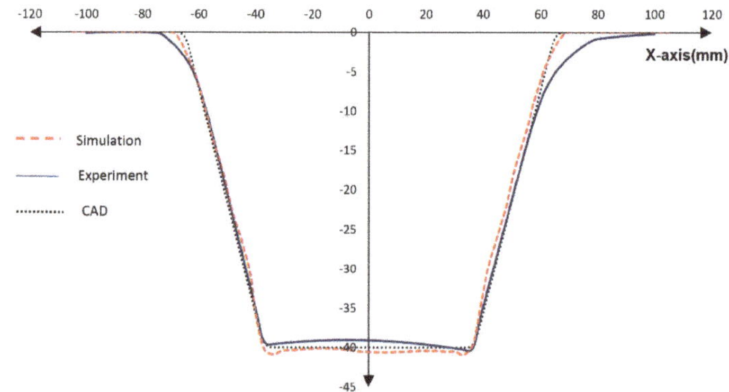

Figure 25. Comparison predicted and experimental profiles along the central X-axis.

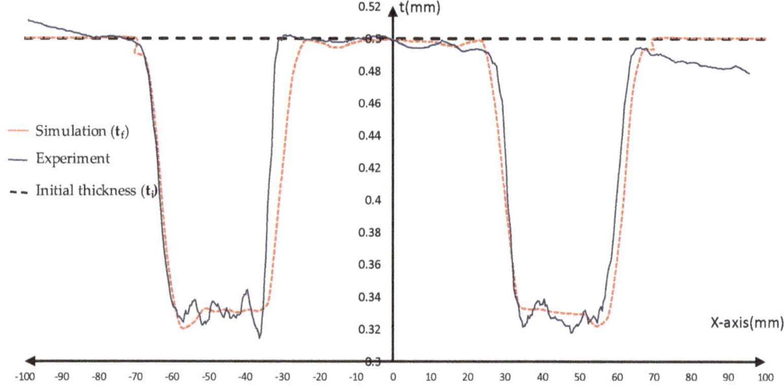

Figure 26. Comparison predicted and experimental final thickness **t$_f$** distribution along X-axis.

3.7. Deep Drawing Sheet Forming of a Front Door Panel

The last example concerns the numerical simulation of a front door panel proposed at Numisheet'2002 (Figure 27). The main goal is to illustrate the capability of the proposed fully coupled model and the numerical methodology to simulate complex shape part. The material properties used in the simulation curved mild steel with the initial thickness of $t_i = 0.78$ mm are shown in Table 19.

Solid linear tetrahedral finite elements (2628 elements) are used to discretize initially the curved blank (Figure 28a). The tools (lower punch, upper die, and blank-holder) whose geometry is presented on Figure 27 are discretized by rigid surfaces. A master–slave contact approach is used in the analysis where the tools are considered as the master surfaces, and the top and the bottom surfaces of the blank (surface facing the punch and the die) constitutes the slave surfaces. Friction ($\eta = 0.15$) is introduced using Coulomb model and the interaction between the blank and the tools is formulated using finite sliding approach, which allows for the possibility of separation between the two surfaces during sliding. The punch moves with the total displacement $U = 130$ mm, die and blank-holder tools are supposed fixed.

Meshes adapted to the lower punch curvature surface corresponding to different punch displacement ($U = 30, 40, 50, 60, 80, 100, 120,$ and 130 mm) are given in Figure 28. The simulation is done using an initial coarse mesh (2628 elements), the deformed mesh is again refined uniformly at each punch displacement and the refinement is activated in regions of large curvature of tools using the remeshing parameters (Table 20). The final front door panel sheet is plastically deformed with small plastic strain $\bar{\varepsilon}_{max}^{p} < 15\%$ (see Figure 28j) and no damage and wrinkling are occurred.

Figure 27. Geometry of front door panel tools.

Table 19. Elasto-plastic parameters of mild steel [60,61].

A (MPa)	B (MPa)	C	n	$\dot{\varepsilon}_0\,(\mathrm{s}^{-1})$
203.39	645.24	0.010	0.25177	1

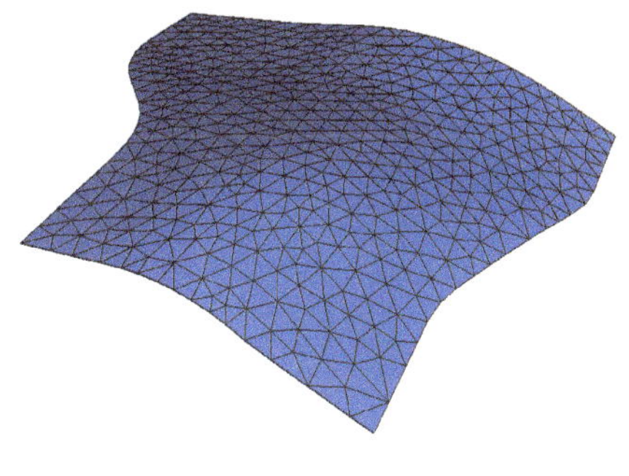

(**a**) Initial sheet

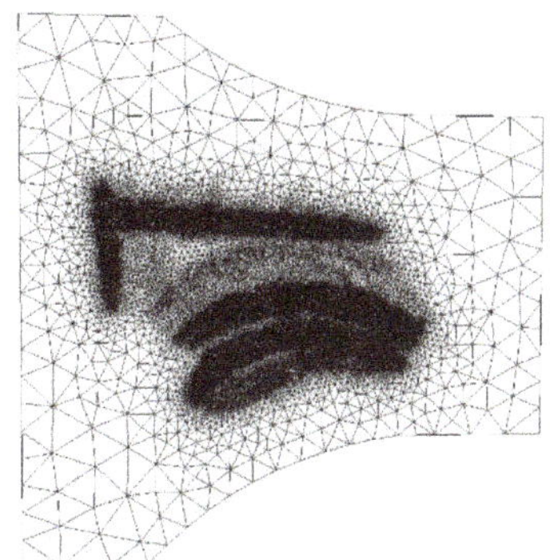

(**b**) Deformed shape for $U = 30$ mm

Figure 28. *Cont.*

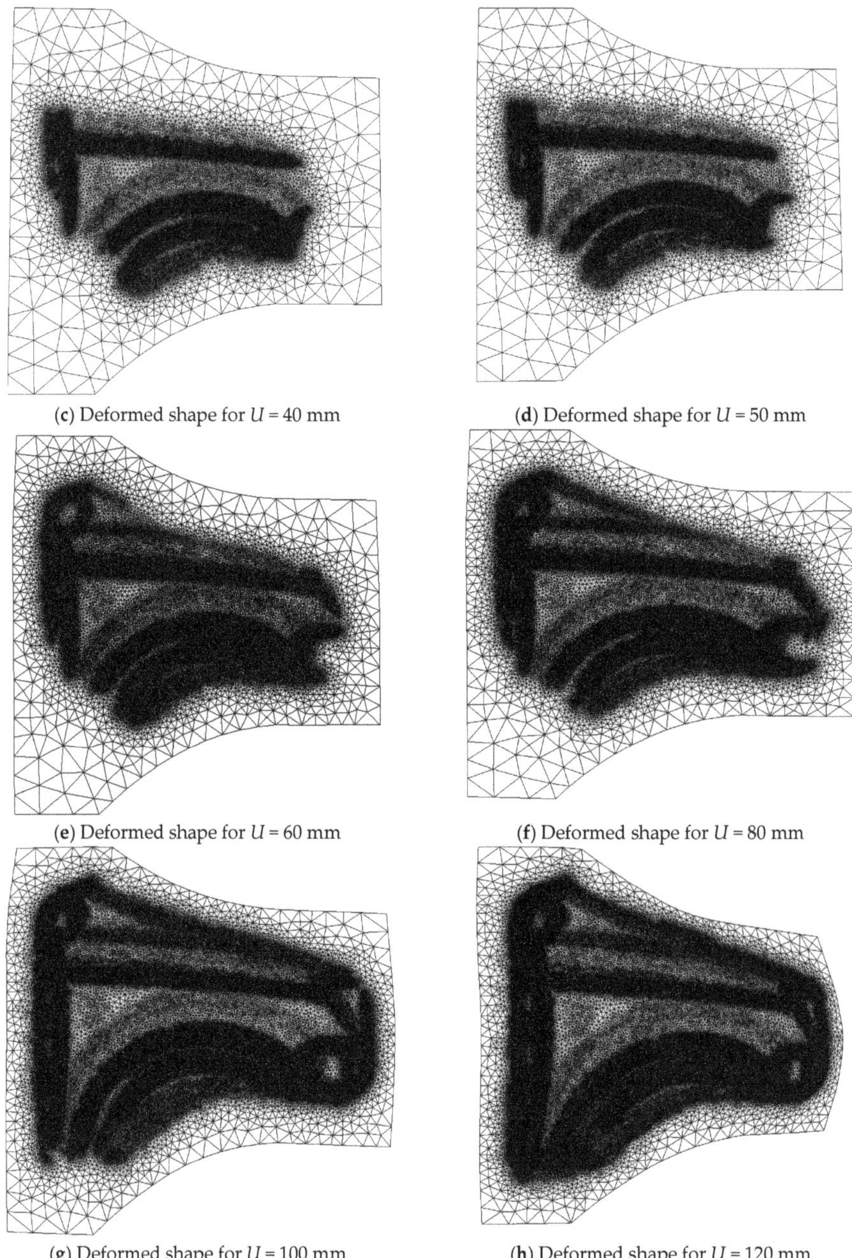

(**c**) Deformed shape for $U = 40$ mm

(**d**) Deformed shape for $U = 50$ mm

(**e**) Deformed shape for $U = 60$ mm

(**f**) Deformed shape for $U = 80$ mm

(**g**) Deformed shape for $U = 100$ mm

(**h**) Deformed shape for $U = 120$ mm

Figure 28. *Cont.*

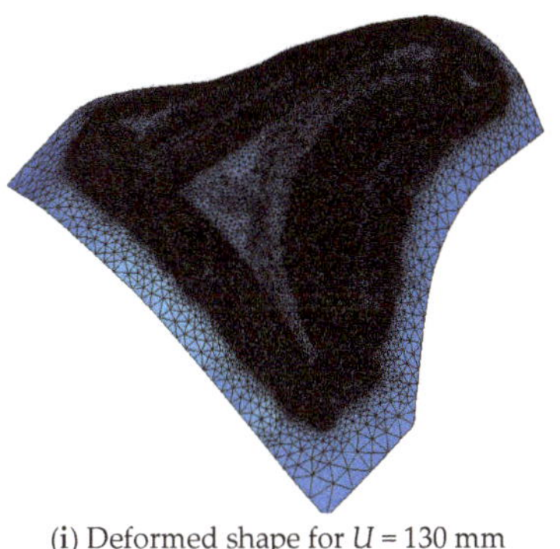

(i) Deformed shape for $U = 130$ mm

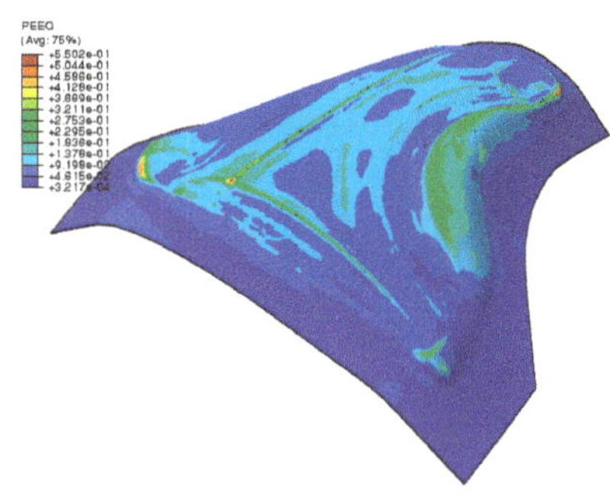

(j) Plastic strain plotted on the final shape of front door panel for $U = 130$ mm

Figure 28. Deformed front door panel at different punch displacement.

Table 20. Adaptive remeshing parameters for deep-drawing.

h_{min} (mm)	h_{max} (mm)	$\bar{\varepsilon}_c^p$	$\bar{\varepsilon}_{max}^p$	δ (mm)
1.0	15.0	10%	42%	2.5

4. Conclusions

The paper is dedicated to the study of the multi-physical coupling in sheet metal forming. The standard framework of the thermodynamics of irreversible processes with state variables is used to derive a fully coupled elasto-visco-plastic equations accounting for nonlinear hardening and ductile damage. First, the proposed model had been validated by comparing predicted force-displacement curves in tensile, compressive and shearing tests. During the simulation of tensile, shearing and compression tests, the coupled damage constitutive equations were validated and the damage behavior in these various stress conditions were well simulated under the help of adaptive remeshing scheme. In practice, it is not easy to get the convergence of the problem without remeshing, and the elements that distort will lose their accuracies of reflecting the actual situation. In contrary, the adaptive remeshing scheme constantly optimized the element quality and refined the mesh size in the whole model. The finest finite elements (size and quality) were obtained in the front of crack, and the crack can propagate easily and accurately. The numerical analysis based on geometrical and physical criteria is integrated in a computational environment using the ABAQUS solver and OPTIFORM mesh. Applications are made to complex sheet metal forming.

It has been demonstrated that only the strong coupling between elasticity, plasticity and ductile damage effect at large plastic strain with contact/friction is able to predict the mechanical field in some dynamic metal forming or machining processes. In addition, the 3D solid finite element adaptive remeshing procedure of deformable sheet is capable to optimize the element quality and refined the mesh size in the whole model or in the crack zone.

Author Contributions: Conceptualization, A.C. and Z.J.; Methodology, A.C. and H.B.; Software, A.C., H.B. and Z.J.; Validation, A.C., H.B. and Z.J.; Formal Analysis, A.C.; Resources, A.C. and Z.J.; Writing-Original Draft Preparation, A.C.; Writing-Review & Editing, A.C.; Visualization, A.C.; Supervision, A.C.; Project Administration, A.C.

Funding: This research received no external funding.

Conflicts of Interest: The authors declare no conflict of interest.

References

1. Lin, J.; Liu, Y.; Dean, T.A. A Review on Damage Mechanisms, Models and Calibration Methods under Various Deformation Conditions. *Int. J. Damage Mech.* **2005**, *14*, 299–319. [CrossRef]
2. Krajcinovic, D. Damage Mechanics: Accomplishments, Trends and Needs. *Int. J. Damage Mech.* **2000**, *37*, 267–277. [CrossRef]
3. Ibijola, E.A. On Some Fundamental Concepts of Continuum Damage Mechanics. *Comput. Meth. Appl. Mech. Eng.* **2002**, *191*, 1505–1520. [CrossRef]
4. McClintock, F.A. A criterion for ductile fracture by the growth of holes. *J. Appl. Mech.* **1968**, *35*, 363–371. [CrossRef]
5. Bonora, N.; Gentile, D.; Pirondi, A.; Newaz, G. Ductile damage evolution under triaxial state of stress: theory and experiment. *Int. J. Plast.* **2005**, *21*, 981–1007. [CrossRef]
6. Racy, J.R.; Tracy, D.M. On ductile enlargement of voids in triaxial stress fields. *J. Mech. Phys. Solids* **1969**, *17*, 201–217.
7. Gurson, A.L. Continuum theory of ductile rupture by void nucleation and growth and flow rules for porous ductile media. *J. Eng. Mater. Technol.* **1997**, *99*, 2–15. [CrossRef]
8. Needleman, A.; Tvergaard, V. An analysis of ductile rupture in notched bars. *J. Mech. Phys. Solids* **1984**, *32*, 461–490. [CrossRef]
9. Tvergaard, V.; Niordoson, C. Non local plasticity effects on interaction of different size voids. *Int. J. Plast.* **2004**, *20*, 107–120. [CrossRef]
10. Bonfoh, N.; Lipinski, P.; Carmasol, A.; Tiem, S. Micromechanical modeling of ductile damage of polycrystalline material with heterogeneous particles. *Int. J. Plast.* **2004**, *20*, 85–106. [CrossRef]
11. Kachanov, L.M. *Introduction to Continuum Damage Mechanics*; Martinus Nijhoff Publisher: Dordrecht, The Netherlands, 1986.
12. Rabotnov, Y. Creep rupture. In Proceedings of the XII International Congress of Applied Mechanics, Stanford, CA, USA, 26–31 August 1968; pp. 342–349.
13. Lemaitre, J. A continuous damage mechanics model for ductile fracture. *J. Eng. Mater. Technol.* **1985**, *107*, 83–89. [CrossRef]
14. Chaboche, J.L. On some modifications of kinematic hardening to improve the description of ratcheting effects. *Int. J. Plast.* **1991**, *7*, 661–678. [CrossRef]
15. Chaboche, J.L. Anisotropic creep damage in the framework of the continuum damage mechanics. *Nucl. Eng. Des.* **1984**, *79*, 309–319. [CrossRef]
16. Soyarslan, C.; Tekkaya, A.E. Finite deformation plasticity coupled with isotropic damage: Formulation in principal axes and applications. *Finite Elem. Anal. Des.* **2010**, *46*, 668–683. [CrossRef]
17. Bouchard, P.O.; Bourgeon, L.; Fayolle, S.; Mocellin, K. An enhanced Lemaitre model formulation for materials processing damage computation. *Int. J. Mater. Form.* **2011**, *4*, 299–315. [CrossRef]
18. Brünig, M.; Chyra, O.; Albrecht, D.; Driemeier, L.; Alves, M. A ductile damage criterion at various stress triaxialities. *Int. J. Plast.* **2008**, *24*, 1731–1755. [CrossRef]
19. Wang, T.J. A General Ductile Damage Model for Engineering Materials. In *Fracture of Engineering Materials and Structures*; Teoh, S.H., Lee, K.H., Eds.; Springer: Dordrecht, The Netherlands, 1991; pp. 798–803.
20. Peirs, J.; Verleysen, P.; Van Paepegem, W.; Degrieck, J. Determining the stress-strain behaviour at large strains from high strain rate tensile and shear experiments. *Int. J. Impact Eng.* **2011**, *38*, 406–415. [CrossRef]
21. Calamaz, M.; Coupard, D.; Girot, F. A new material model for 2D numerical simulation of serrated chip formation when machining titanium alloy Ti-6Al-4V. *Int. J. Mach. Tools Manuf.* **2008**, *48*, 275–288. [CrossRef]
22. Sun, Z.C.; Yang, H.; Han, G.J.; Fan, X.G. A numerical model based on internal-state-variable method for the microstructure evolution during hot-working process of TA15 titanium alloy. *Mater. Sci. Eng. A* **2010**, *527*, 3464–3471. [CrossRef]
23. Zerilli, F.J. Dislocation mechanics–based constitutive equations. *Metall. Mater. Trans. A* **2004**, *35*, 2547–2555. [CrossRef]
24. Zhang, H.; Wen, W.; Cui, H.; Xu, Y. A modified Zerilli–Armstrong model for alloy IC10 over a wide range of temperatures and strain rates. *Mater. Sci. Eng. A* **2009**, *527*, 328–333. [CrossRef]
25. Holmquist, T.; Johnson, G. Determination of constants and comparison of results for various constitutive models. *J. Phys. IV* **1991**, *1*, 853–860.

26. Hor, A.; Morel, F.; Lebrun, J.L.; Germain, G. Modelling, identification and application of phenomenological constitutive laws over a large strain rate and temperature range. *Mech. Mater.* **2013**, *64*, 91–110. [CrossRef]
27. Johnson, G.R.; Cook, W.H. Fracture characteristics of three metals subjected to various strains, strain rates, temperatures and pressure. *Eng. Fract. Mech.* **1985**, *21*, 31–48. [CrossRef]
28. Cherouat, A.; Moreau, L.; Borouchaki, H. Advanced numerical simulation of metal forming processes using adaptive remeshing procedure. *Mater. Sci. Forum* **2009**, *614*, 27–33. [CrossRef]
29. Cherouat, A.; Borouchaki, H.; Saanouni, K.; Laug, P. Numerical methodology for metal forming processes using elastoplastic model with damage occurrence. *J. Mater. Sci. Technol.* **2006**, *22*, 279–283.
30. Steinberg, D.J.; Cochran, S.G.; Guinan, M.W. A constitutive model for metals applicable at high-strain rate. *J. Appl. Phys.* **1980**, *51*, 1498–1504. [CrossRef]
31. Hoge, K.G.; Mukherjee, A.K. The temperature and strain rate dependence of the flow stress of tantalum. *J. Mater. Sci.* **1977**, *12*, 1666–1672. [CrossRef]
32. Zerilli, F.J.; Armstrong, R.W. Dislocation-mechanics-based constitutive relations for material dynamics calculations. *J. Appl. Phys.* **1987**, *61*, 1816–1825. [CrossRef]
33. Follansbee, P.S.; Kocks, U.F. A constitutive description of the deformation of copper based on the use of the mechanical threshold. *Acta Metall.* **1988**, *36*, 81–93. [CrossRef]
34. Preston, D.L.; Tonks, D.L. Model of plastic deformation for extreme loading conditions. *J. Appl. Phys.* **2002**, *93*, 211–220. [CrossRef]
35. Zhu, Y.Y.; Cescotto, S.; Habraken, A.M. A fully coupled elastoplastic damage modeling and fracture criteria in metal forming processes. *J. Mater. Process. Technol.* **1992**, *32*, 197–204. [CrossRef]
36. Badreddine, H.; Labergère, C.; Saanouni, K. Ductile damage prediction in sheet and bulk metal forming. *C.R. Mec.* **2016**, *344*, 296–318. [CrossRef]
37. Chaboche, J.L.; Cailletaud, G. Integration methods for complex plastic constitutive equations. *Comput. Meth. Appl. Mech. Eng.* **1996**, *133*, 125–155. [CrossRef]
38. Kim, J.; Gao, X.; Srivatsan, T.S. Modeling of void growth in ductile solids: effects of stress triaxiality and initial porosity. *Eng. Fract. Mech.* **2004**, *71*, 379–400. [CrossRef]
39. Malcher, L.; Andrade Pires, F.M.; César de Sá, J.M.A. An assessment of isotropic constitutive models for ductile fracture under high and low stress triaxiality. *Int. J. Plast.* **2012**, *30–31*, 81–115. [CrossRef]
40. Xue, L.; Wierzbicki, T. Ductile fracture initiation and propagation modeling using damage plasticity theory. *Eng. Fract. Mech.* **2008**, *75*, 3276–3293. [CrossRef]
41. Lemaitre, J. Coupled elasto-plasticity and damage constitutive equations. *Comput. Meth. Appl. Mech. Eng.* **1985**, *51*, 31–49. [CrossRef]
42. Boudifa, M.; Saanouni, K.; Chaboche, J.L. A micromechanical model for inelastic ductile damage prediction in polycrystalline metals for metal forming. *Int. J. Mech. Sci.* **2009**, *51*, 453–464. [CrossRef]
43. Simo, J.C.; Taylor, R. Consistent tangent operators for rate independent elastoplasticity. *Comput. Meth. Appl. Mech. Eng.* **1985**, *48*, 101–118. [CrossRef]
44. Nagtegaal, J.C.; Taylor, L.M. Comparison of implicit and explicit finite element methods for analysis of sheet forming processes. In *FE-Simulation of 3D Sheet Metal Forming Processes in Automotive Industry*; VDI Verlag: Dusseldorf, Germany, 1991.
45. Yang, D.Y.; Jung, D.W.; Song, I.S.; Yoo, D.J.; Lee, J.H. Comparative investigation into implicit, explicit, and iterative implicit/explicit schemes for the simulation of sheet-metal forming processes. *J. Mater. Process. Technol.* **1995**, *50*, 39–53. [CrossRef]
46. Chen, L. Comparisons of explicit and implicit finite element methods for sheet metal forming. *Adv. Mater. Res.* **2014**, *936*, 1836–1839. [CrossRef]
47. *ABAQUS*, version 6.14; Abaqus Inc.: Palo Alto, CA, USA, 2015.
48. Borouchaki, H.; Cherouat, A.; Laug, P.; Saanouni, K. Adaptive remeshing for ductile fracture prediction in metal forming. *C.R. Mec.* **2002**, *330*, 709–716. [CrossRef]
49. Borouchaki, H.; Grosges, T.; Barchiesi, D. Enhancement of the accuracy of numerical field computation using an adaptive three-dimensional remeshing scheme. *C.R. Mec.* **2010**, *338*, 127–131. [CrossRef]
50. Borouchaki, H.; Grosges, T.; Barchiesi, D. Improved 3d adaptive remeshing scheme applied in high electromagnetic field gradient computation. *Finite Elem. Anal. Des.* **2010**, *46*, 84–95. [CrossRef]
51. Borouchaki, H.; George, P.L.; Hecht, F.; Laug, P.; Saltel, E. Delaunay mesh generation governed by metric specifications. Part I. Algorithms. *Finite Elem. Anal. Des.* **1997**, *25*, 61–83. [CrossRef]

52. Borouchaki, H.; George, P.L.; Mohammadi, B. Delaunay mesh generation governed by metric specifications Part II. Applications. *Finite Elem. Anal. Des.* **1997**, *25*, 85–109. [CrossRef]

53. Broggiato, G.B.; Campana, F.; Cortese, L. Identification of Material Damage Model Parameters: an Inverse Approach Using Digital Image Processing. *Meccanica* **2007**, *42*, 9–17. [CrossRef]

54. Springmann, M.; Kuna, M. Identification of material parameters of the Rousselier model by non-linear optimization. *Comput. Mater. Sci.* **2003**, *26*, 202–209. [CrossRef]

55. Yue, Z.M.; Soyarslan, C.; Badreddine, H.; Saanouni, K.; Tekkaya, A.E. Identification of fully coupled anisotropic plasticity and damage constitutive equations using a hybrid experimental–numerical methodology with various triaxialities. *Int. J. Damage Mech.* **2015**, *24*, 683–710. [CrossRef]

56. Feng, F.; Li, J.; Yuan, P.; Zhang, Q.; Huang, P.; Su, H.; Chen, R. Application of a GTN damage model predicting the fracture of 5052-O aluminum alloy high-speed electromagnetic impaction. *Metals* **2018**, *8*, 761. [CrossRef]

57. Zhang, J.; Cherouat, A.; Borouchaki, H. 3D Thermo-Mechanical Simulation Coupled with Adaptive Remeshing for Metal Milling. *Adv. Mater. Res.* **2013**, *698*, 11–20. [CrossRef]

58. Abedini, A.; Butcher, C.; Worswick, M.J. Fracture characterization of rolled sheet alloys in shear loading: studies of specimen geometry, anisotropy, and rate sensitivity. *Exp. Mech.* **2017**, 57–75. [CrossRef]

59. Zhang, J. Multi-axial Damage Model for Numerical Simulation of Metal Forming Processes Using 3D Adaptive Remeshing Procedure. Ph.D. Thesis, University of Technology of Troyes, Troyes, France, April 2005.

60. Husson, C.; Correia, J.P.M.; Daridon, L.; Ahzi, S. Finite elements simulation of thin copper sheets blanking: Study of blanking parameters on sheared edge quality. *J. Mater. Process. Technol.* **2008**, *199*, 74–83. [CrossRef]

61. Cherouat, A.; Ma, X.; Borouchaki, H.; Zhang, Q. Numerical study of Multi-Point Forming of thick sheet using remeshing procedure. In Proceedings of the ESAFORM 2018, Palermo, Italy, 23–25 April 2018.

62. Saidi, B.; Giraud-Moreau, L.; Cherouat, A. Optimization of the Single Point Incremental Forming Process for titanium sheets by using response surface. *MATEC Web Conf.* **2016**, *80*, 10011. [CrossRef]

63. Sena, J.I.V.; Alves de Sousa, R.J.; Valente, R.A.F. Single point incremental forming simulation with an enhanced assumed strain solid-shell finite element formulation. *Int. J. Mater. Form.* **2010**, *3*, 963–966. [CrossRef]

64. Liu, Z.; Daniel, W.J.T.; Li, Y. Multi-pass deformation design for incremental sheet forming: Analytical modeling, finite element analysis and experimental validation. *J. Mater. Process. Technol.* **2014**, *214*, 620–634. [CrossRef]

65. Honarpisheh, M.; Jobedar, M.; Alinaghian, I. Multi-response optimization on single-point incremental forming of hyperbolic shape Al-1050/Cu bimetal using response surface methodology. *Int. J. Adv. Manuf. Technol.* **2018**, *96*, 3069–3080. [CrossRef]

66. Duflou, J.; Tunçkol, Y.; Szekeres, A.; Vanherck, P. Experimental study on force measurements for single point incremental forming. *J. Mater. Process. Technol.* **2007**, *189*, 65–72. [CrossRef]

67. Li, Y.; Chen, X.; Liu, Z. A review on the recent development of incremental sheet-forming process. *Int. J. Adv. Manuf. Technol.* **2017**, *92*, 2439–2462. [CrossRef]

68. Ambrogio, G.; De Napoli, L.; Filice, L. Application of Incremental Forming process for high-customized medical product manufacturing. *J. Mater. Process. Technol.* **2005**, *162*, 156–162. [CrossRef]

© 2018 by the authors. Licensee MDPI, Basel, Switzerland. This article is an open access article distributed under the terms and conditions of the Creative Commons Attribution (CC BY) license (http://creativecommons.org/licenses/by/4.0/).

Article

Springback Prediction of Aluminum Alloy Sheet under Changing Loading Paths with Consideration of the Influence of Kinematic Hardening and Ductile Damage [†]

Zhenming Yue [1], Jiashuo Qi [1], Xiaodi Zhao [1,2], Houssem Badreddine [3], Jun Gao [1] and Xingrong Chu [1,*]

[1] School of Mechanical, Electrical and Information Engineering, Shandong University at Weihai, Weihai 264209, China; yuezhenming@sdu.edu.cn (Z.Y.); qjs@mail.sdu.edu.cn (J.Q.); zxd.xiaodi@foxmail.com (X.Z.); shdgj@sdu.edu.cn (J.G.)

[2] Pinggao Group Weihai High Voltage Apparatus CO., LTD., Weihai 264209, China

[3] ICD/LASMIS-CNRS-FRE-2848-University of Technology of Troyes, 12, rue Marie Curie, CS 42060, 10004 Troyes CEDEX, France; houssem.badreddine@utt.fr

* Correspondence: xrchu@sdu.edu.cn; Tel.: +86-133-8630-3657

† This paper is an extend version of our paper published in 17th International Conference on Metal Forming, Metal Forming 2018, Toyohashi, Japan, 16–19 September 2018.

Received: 20 October 2018; Accepted: 13 November 2018; Published: 14 November 2018

Abstract: Springback prediction of sheet metal forming is always an important issue in the industry, because it greatly affects the final shape of the product. The accuracy of simulation prediction depends on not only the forming condition but also the chosen material model, which determines the stress and strain redistributions in the formed parts. In this paper, a newly proposed elastoplastic constitutive model is used, in which the initial and induced anisotropies, combined nonlinear isotropic and kinematic hardenings, as well as isotropic ductile damage, are taken into account. The aluminum alloy sheet metal AA7055 was chosen as the studied material. In order to investigate springback under non-proportional strain paths, three-point bending tests were conducted with pre-strained specimens, and five different pre-straining states were considered. The comparisons between numerical and experimental results highlighted the hard effect of both kinematic hardening and ductile damage on the springback prediction, especially for a changed loading path case.

Keywords: springback; non-proportional loading paths; mixed hardening; ductile damage; plastic anisotropy

---

1. Introduction

In recent years, with the rapid development of the need for lightweight materials, more high-strength aluminum alloy sheets have been widely used in the products of automotive, aerospace, and medical health fields. However, some undesirable problems during the forming of high strength sheet metal have been observed. Springback is one of these important issues, and it will affect the final appearance and accuracy of the workpiece. Sheet metal forming usually induces important and complicated plastic strain state, particularly under complex loading paths and large strain. Understanding these behaviors of sheet metal forming becomes more important, which can give great help during the tool adjustments in the metal forming process, especially for the new lightweight and high-strength materials.

The springback can be predicted using many approaches: using Finite Element Analysis (FEA) [1,2] and some analytical methods [3,4]. Large-scale numerical simulations with FEA are more widely used to predict and consequently avoid the effect of springback, especially for the geometrical complex parts. It can give fast and accurate results. The accuracy of the simulation depends on not only the forming conditions [5] but also on the material models [6,7], which are used to describe the forming behavior of the material under different forming conditions.

Li et al. have investigated the influence of element type and size and other physical effects (e.g., friction) [8] and showed the importance of accounting Bauschinger effect on the springback prediction. Oliveira et al. studied the effects of different hardening laws on the final springback prediction [9]. Armstrong and Frederick have ever proposed nonlinear kinematic hardening to improve the accuracy of springback simulation. The modified Chaboche type model with the combined isotropic-kinematic hardenings as well as the non-quadratic anisotropic yield potential Yld2000-2d has been used by Chung et al. to improve the prediction capability [10]. Chun et al. changed the isotropic part of the Chaboche model to do the simulation by explicit/implicit integral algorithm [11]. Besides the Yld2000-2d yield function, the influence of Barlat 89's yield function on springback of sheet metal forming has also been conducted and investigated [12]. Zang et al. show the hard influence of plastic anisotropy through comparison between experiments and simulations [13].

Nowadays, more works have noticed the hard impact of Young's modulus evolution during the unloading phase. Stoughton et al. have pointed out the difficulties and challenges of constitutive modelling of metal forming with considering the modulus variation [14]. Lee et al., based on classical Dafilias/Popov and Krieg concepts, developed a two-surface plasticity model combined with Bauschinger effect, transient behavior, and permanent softening according to the Chaboche model [15]. Recently, Govik et al. have investigated the unloading behavior of dual phase steel with micromechanical FE model exhibited non-linear strain recovery due to local plasticity caused by interaction between two phases [16]. Gau and Kinzel [17] investigated the difference of the springback angle with isotropic and kinematic hardenings. The importance of kinematic hardening on springback prediction has also been proved by the work of Wang et al. [18]. Meanwhile considering the large strain during forming, the influence of ductile damage on springback has also attracted a lot of attention [19]. So, through literature studies, it can be found that the constitutive model plays a critical role in the simulation process, including the equations of the yield surface description, hardening flow, anisotropies and texture evolution, even the ductile damage, etc. There are a lot of studies which have concluded that new and more suitable models need to be proposed in order to improve the numerical prediction accuracy [20,21].

In this paper, a newly proposed model considering initial anisotropy and subsequent yield surface distortion, non-linear combined isotropic and kinematic hardenings, and fully coupling with ductile damage [22] is used to predict the springback of aluminum alloy sheet AA7055. The nonlinear elastic unloading-reloading behavior will not be accounted for, so elastic linear unloading and reloading behavior will be considered in this model. Also, this can help better investigate the influence of kinematic hardening and ductile damage on springback. The model has been implemented in ABAQUS Standard and Explicit through the user subroutines VUMAT/UMAT. The springback phase of AA7055 sheets is investigated through three-point bending with different levels of pre-straining states. Through the comparison between numerical and experimental results, the influence of kinematic hardening and ductile damage on final springback results is discussed.

2. Constitutive Equations

The newly proposed elastoplastic constitutive equations coupled with the isotropic ductile damage, and accounting for non-linear combined isotropic and kinematic hardenings, are used to predict the springback [22]. The initial anisotropy and hardening induced subsequent yield surface distortion are also considered in the model. The following couples of state variables: $(\underline{\varepsilon}^e, \underline{\sigma})$, $(\underline{\alpha}, \underline{X})$, (r, R), (d, Y) are included in the model, and they represent respectively the elastoplastic flow, kinematic hardening, isotropic hardening, and isotropic ductile damage, respectively. Detailed description of the fully coupled relationship can be found in the published research works [22,23].

$$\underline{\sigma} = 2\mu_e\left[(1-d)\langle\underline{e}^e\rangle_+ + (1-hd)\langle\underline{e}^e\rangle_-\right] + k_e[(1-d)\langle tr(\underline{\varepsilon}^e)\rangle - (1-hd)\langle -tr(\underline{\varepsilon}^e)\rangle]\underline{1} \qquad (1)$$

$$\underline{X} = (1-d)\frac{2}{3}C\underline{\alpha} \qquad (2)$$

$$R = (1-d^Y)Qr \qquad (3)$$

$$Y = Y^e + Y^\alpha + Y^r \qquad (4)$$

where $\underline{e}^e$ denotes the deviatoric part of the elastic strain, $\langle\cdot\rangle_+$ and $\langle\cdot\rangle_-$ denote positive and negative parts. μ_e and K_e are the classical shear and bulk elastic moduli. h is the microcracks closure parameter $0 \leq h \leq 1$, C and Q are the hardening moduli for kinematic and isotropic hardening, respectively, γ is the parameter of damage effect on isotropic hardening. Y^e, Y^α, and Y^r are the density energy release rates corresponding respectively to elastoplastic flow, kinematic hardening, and isotropic hardening. Details about the state variables and evolution equations are given in Appendix A.

The given model will be implemented in both ABAQUS/Standard and ABAQUS/Explicit finite element codes through the user subroutines UMAT and VUMAT. The implementation of the developed model is based on purely local implicit integration scheme used with the elastic predictor-plastic corrector approach.

3. Experimental Investigations

The AA7055 aluminum alloy sheet with 1.60 mm thickness was used for the experimental study. In order to investigate the influence of kinematic hardening and ductile damage on springback under the loading path changing, a series of tests were conducted. Firstly, the simple uniaxial tensile test, cyclic shear test, and cyclic loading-unloading tensile tests were conducted separately to determine the material parameters. In order to investigate the springback of AA7055 alloy sheet the tensile specimens were subjected to five different pre-strain states, and then bent with 25 mm depth. With the neutral layer as the boundary, the inner sheet material experiences the cyclic tension-compression loading process during tensile-bending tests. During the bending tests, the internal damage appearing in the sheet increases with plastic strains and causes the decrease of the rigidity of the material. The increase of micro-cracks and the micro-voids in the sheet reduces the stress levels in the sheet, which will affect the final formed shape.

3.1. Uniaxial Tensile Tests

The geometry of the uniaxial tensile test specimen is given in Figure 1. The specimens were cut along the orientations ($0°$, $45°$, $90°$) according to the rolling direction. The tensile velocity was fixed to 2.0 mm/min. To ensure the repeatability of the results, each test was conducted at least three times. The uniaxial tensile test is used to identify the material properties, including the yield function, anisotropy, and hardening laws. The obtained plastic flow curves $0°$, $45°$, and $90°$ according to the rolling direction are given in Figure 2.

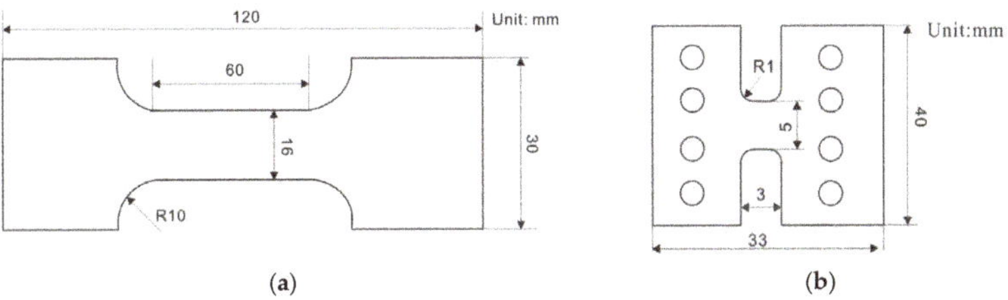

(a)   (b)

Figure 1. Specimen geometry for the tensile test (**a**) (1.6 mm thickness) and cyclic shear tests (**b**).

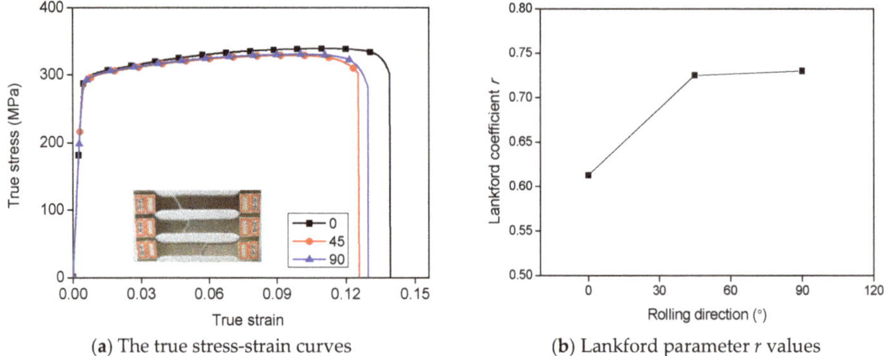

(**a**) The true stress-strain curves   (**b**) Lankford parameter *r* values

Figure 2. Tensile tests of AA7055 aluminum alloy sheet.

3.2. Cyclic Shear Loading Tests

The geometry of the cyclic shear loading test specimen is given in Figure 1. In the proposed model, the kinematic hardening parameters should be determined through the cyclic loading tests. Through the literature study, the cyclic shear test can be one valid method to investigate the Bauschinger phenomenon of sheet metal [24]. In this study, the cyclic shear loading test was conducted with a butterfly clamp on the universal tensile machine. The Digital image correlation (DIC) system was used to capture the vertical displacement between point A and point B shown in Figure 3. The engineering strain and stress can be calculated by:

$$\gamma_e = \frac{\Delta h}{h_0}, \quad \tau_e = \frac{F}{t_0 h_0} \tag{5}$$

where Δh denotes the vertical displacement; h_0 denotes the initial height of the connect zone; t_0 denotes the initial thickness of the sheet; F denotes the loading force during the tests.

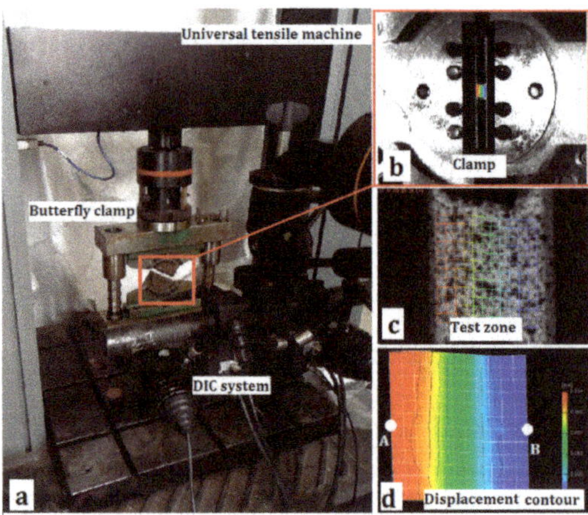

Figure 3. Experimental setup for the cyclic shear test.

With the obtained stress-strain curves in uniaxial and cyclic shear tests, the material parameters suited to aluminum alloy sheet AA7055 can be determined through the use of an appropriate inverse hybrid numerical-experimental methodology [25]. The relevant numerical and experimental result are the displacement-load responses. The inverse method here is to search the minimum error value between simulation and experiment responses. The Trust Region reflective method is used, which suits itself to the nonlinear least squares optimization problem. The optimization process involves the approximate solution of a large linear system based on the method of preconditioned conjugate gradients. The optimization algorithm is coded within MATLAB and Python script in conjunction with the FE commercial software ABAQUS.

Note that the use of only these two tests cannot allow the accurate determination of all model parameters such as the determination of micro-cracks closure parameter h and distortional parameters (X_{11}^c, X_{12}^c, X_{11}^p). In this study, the micro-cracks closure parameter is assumed to be $h = 0.2$. The distortional effect in this study will be ignored by considering $X_{11}^c = X_{12}^c = X_{11}^p = \infty$. The set of material parameters for the AA7055 is summarized and given in Table 1.

Table 1. The material parameters of the AA7055 aluminum alloy.

Material	t (mm)	ρ (g/cm^3)	E (MPa)	ν	σ_y (MPa)	σ_s (MPa)	$F = F^*$	$G = G^*$	$H = H^*$	$L = L^*$	$M = M^*$
AA7055	1.60	2.86	68439	0.3	290.98	369.19	0.52	0.62	0.38	1.5	1.5
$N = N^*$	Q (MPa)	b	C (MPa)	a	S	s	β	Y_0	γ	h	d_c
1.4	720	14.5	720	14.5	0.75	1.0	4	1	3	0.2	0.99

In this study, the decomposition of kinematic hardening and isotropic hardening was determined through cyclic shear tests (Figure 4). But they also can be determined with the help of the springback angle after the bending test with different pre-stretch strain [13]. The initial yield criteria (Hill 48) in the yield function f and potential equation F are assigned the same values ($F = F^*$, $G = G^*$, $H = H^*$, $L = L^*$, $M = M^*$). In the coming studies, this decomposition ratio will be changed in order to study the influence of kinematic hardening on the prediction of the final springback values.

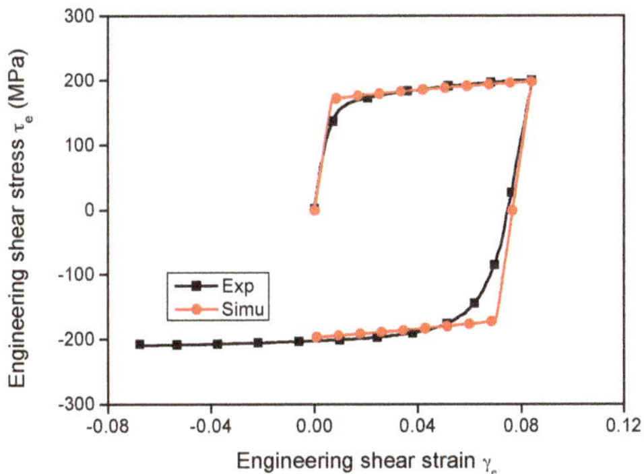

Figure 4. Experimental and numerical cyclic loading stress-strain curve.

3.3. Cyclic Loading-Unloading Uniaxial Tension Tests

In order to better investigate the damage initiation and growth during the plastic flow, uni-axial tensile test with loading-unloading cycles was conducted, and the evolution of elastic modulus was recorded during the unloading phases (Figure 5). The specimen is prepared following the rolling direction. The strain levels of unloading are chosen (1.6%, 2.7%, 3.7%, 4.7%, 5.7%, 6.7%). The average value of elastic moduli during the unloading process and the decrease ratios (E_i/E_0) are given in Figure 6. In this figure, we can observe clearly the decreases of elastic modulus according to the tensile strain which can be related to the damage evolution during the tensile loading. The decrease ratio of E can be regarded as the damage value in a way. During the simulation on the prediction of the springback phenomenon, the E decrease can be taken into account through user subroutine, and this result can compared with ones obtained with ductile damage model (marked as Simu-E in Figure 13).

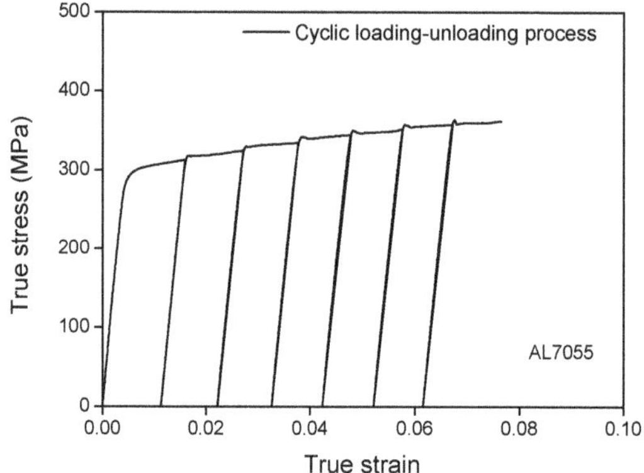

Figure 5. Stress-strain curve during the cyclic loading-unloading process.

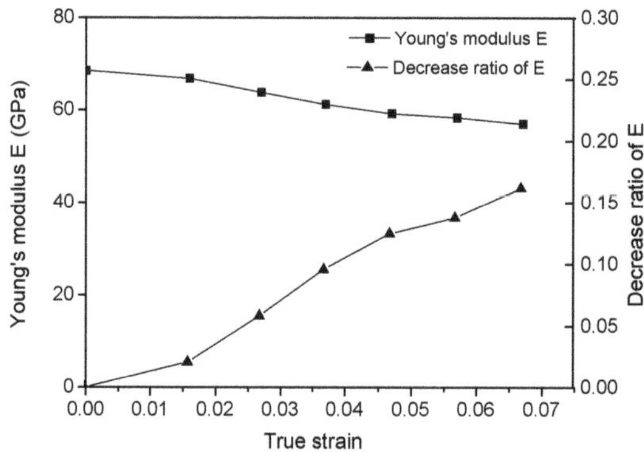

Figure 6. Variations of elastic moduli during the cyclic loading-unloading process.

3.4. Bending Test with Pre-Stretch Strain

Three-point bending tests were conducted to investigate the springback phase of AA7055 specimens. The test design and device are given in Figure 7, and also the specimen is prepared following the rolling direction. Considering the maximum strain of 0.12 (where the local necking happens after this point) under uniaxial tensile loading path, five pre-straining states are used of 4.6% (2.4 mm), 6.4% (3.3 mm), 8.0% (4.2 mm), 9.7% (5.1 mm), and 11.3% (6.0 mm) were selected to assign on the bending specimens. The bending depth is about 25 mm, and the fixed velocity of the upper die is about 5.0 mm/min. The geometry sizes of the bending dies are shown in Figure 7. The springback angle is measured for the different specimens including the pre-straining states.

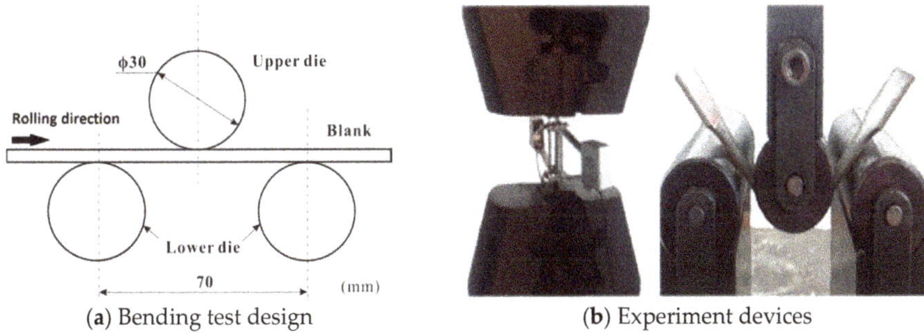

(**a**) Bending test design (**b**) Experiment devices

Figure 7. Three-point bending test device.

The test design and part of final obtained samples are showed in Figure 8. In Figure 9 the obtained average springback angles after bending tools unloading are shown. From the comparison of the results it can be seen that plastic pre-straining affects greatly the resulting springback angles. We observe that the increase of tensile pre-strain induces an increase of springback angle. On the other hand, when the plastic pre-strain increases, the increases of hardening stress and plastic strain will be obtained on one side accompanying the increase of ductile damage. This damage evolution allows a significant reduction of rigidity of the material so that when the bending tools are unloaded the critical parts are submitted to the high level of stress with the low rigidity which will contribute to maximizing the springback phase.

Test design	Pre-tensile plastic strain	Stroke(mm)
1	0.0%	25
2	1.6%	25
3	6.4%	25
4	8.0%	25
5	9.7%	25
6	11.3%	25

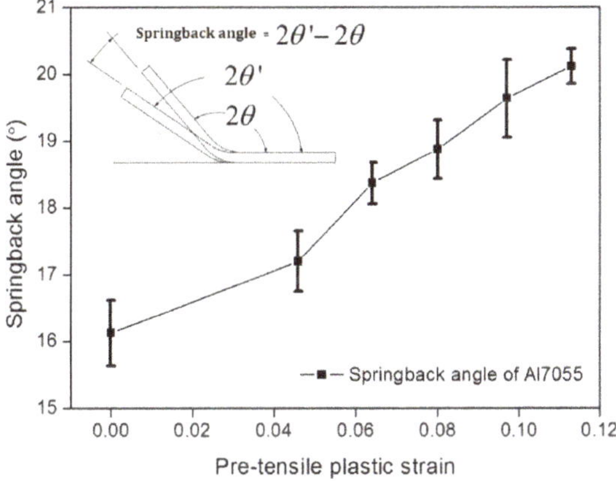

(**a**) Test design (**b**) Test samples

Figure 8. Comparison of three-point bending specimens with different pre-strains.

Figure 9. Springback angles after bending tools unloading versus pre-straining states.

Note that when the bending depth is the same, under different pre-strain levels, the stress and strain distributions across the critical deformed zone situated under the punch radius shall be quite different. Without considering pre-strains, the normally neutral layer is situated in the central side of the sheet thickness. Tensile pre-strains shall be added to the bending strain state and causes the sliding of the neutral layer in the direction of the punch. The movement of the neutral layer will reduce the volume ratio of the compression stress state in the inner thickness section. This fact will contribute also to spring back increase.

4. Simulation Results

In the simulation process, the influences of the kinematic hardening and the ductile damage on the springback were studied. The loading processes include pre-straining and bending phases that were simulated using ABAQUS/Explicit® with user subroutine (VUMAT), which takes into account the dynamic process and avoid the non-convergence problem. The unloading process was simulated using ABAQUS/Standard® with user subroutine (UMAT) in order to reflect the final static shape after unloading. The decrease of Young's modulus during the deformation process can be defined using two approaches: the direct assignment according to the decrease ratios of E given in Figure 6. The second one is obtained automatically through the full coupling with ductile damage. In the coming sections, the comparison of these two approaches is analyzed. In order to better investigate the effect of the stress gradient along thickness, the 2D plane strain geometrical model was chosen in the simulation process. Constant mesh size 0.1 mm is used in the critical deformation zone, as shown in Figure 10.

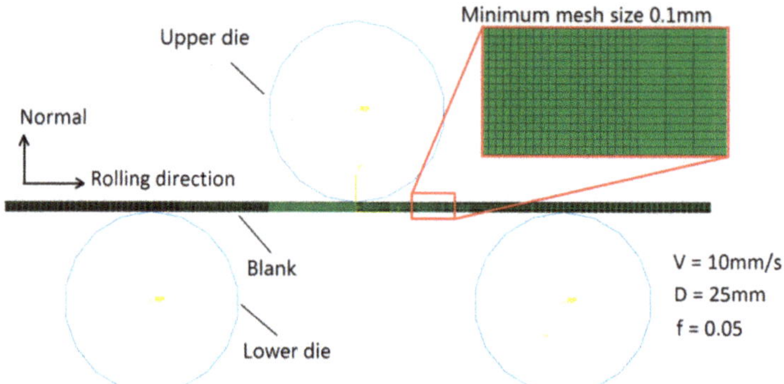

Figure 10. Finite element model for bending tests.

4.1. Effect of Kinematic Hardening Fraction on Springback Prediction

The hardening stresses obtained through the uniaxial tensile test combines isotropic and kinematic hardenings. Without the cyclic loading tests, it is difficult to separate these two hardenings and determine the exact fraction of every part. In this study, the kinematic hardening fraction is considered through a parametric study. The fraction ratio of kinematic hardening with respect to total hardening is defined by:

$$t = X_{\text{sat}}/(X_{\text{sat}} + R_{\text{sat}}) = C/(C + Q \cdot a/b) \tag{6}$$

where X_{sat} is the kinematic hardening saturation value and R_{sat} is the isotropic hardening saturation value. In the parametric study, the kinematic hardening fraction values chosen are $t = (0.0, 0.1, 0.2, 0.3, 0.4, 0.5, 0.6, 0.7, 0.8, 0.9, 1.0)$. For all these kinematic hardening fractions, the values of hardening modulus Q and C are re-identified in order to obtain the same fit with experimental tensile test curves. In Table 2 are given the obtained sets of hardening parameters corresponding to each considered kinematic hardening fraction.

Table 2. Hardening parameters corresponding to the different kinematic hardening fraction.

Kinematic Hardening Fraction	Isotropic Hardening		Kinematic Hardening	
	Q (MPa)	b	C (MPa)	a
0.0	1441		0	
0.1	1297		144.1	
0.2	1153		288.3	
0.3	1009		432.4	
0.4	864.9		576.6	
0.5	720.8	14.53	720.8	14.53
0.6	576.6		864.9	
0.7	432.4		1009	
0.8	288.3		1153	
0.9	144.1		1297	
1.0	0		1441	

In order to enhance the stress gradient description through the thickness, element size of 0.1 mm was considered along with the thickness direction. The simulation was consistent with the experimental setup, and five different tensile pre-strains 4.6%, 6.4%, 8.0%, 9.7%, and 11.3% were applied during a step before the bending process.

The friction coefficient between the specimen and the dies was fixed to 0.05. The bending depth of the upper die was about 25.0 mm. The pre-tension and three-point bending processes were simulated through the dynamic explicit methodology, while the unloading process after bending was conducted by the static analysis methodology. The state variables after the bending analysis were imported as the initial state of the unloading process. After the unloading process, the resulting values obtained of springback angles are given in Figure 11 with respect to the pre-strain states.

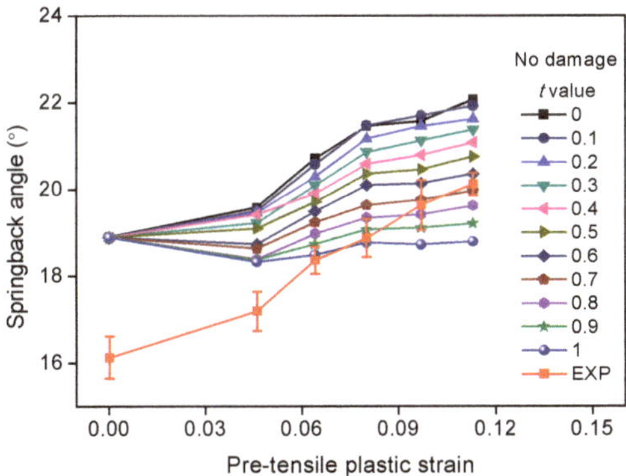

Figure 11. Springback angle obtained after three-point bending test with different considered kinematic hardening fractions and tensile pre-strains (no damage).

In Figure 11 are given the springback angles obtained after the bending tests applied for different considered pre-strained specimens without considering ductile damage. When pre-tensile plastic strain is zero, there is still a gap of 3°, which can be caused by the friction and the thickness variation of the sheet, and also the anisotropy following the thickness direction is ignored during the simulation. In this figure, the results of the simulations using the various kinematic hardening fraction cases ($t = 0.0\sim1.0$) are compared with the experimental results. The case of $t = 0.0$ value fraction corresponds to a purely isotropic hardening model; however, the case of $t = 1.0$ value fraction corresponds to purely kinematic hardening model. As mentioned before, the springback angle increases when the pre-strain increases. In this figure, we observe an important effect of kinematic hardening fraction on the springback predicted angle. It needs to be highlighted that for high kinematic hardening fraction ($t > 0.8$) the increase of pre-strain state induces a decrease of springback angle. For these cases, the difference among numerical results is important. The stress contours before springback are given in Figure 12, which clearly present the big effect of kinematic hardening fraction on the final obtained stress contour on the thickness section during the bending process.

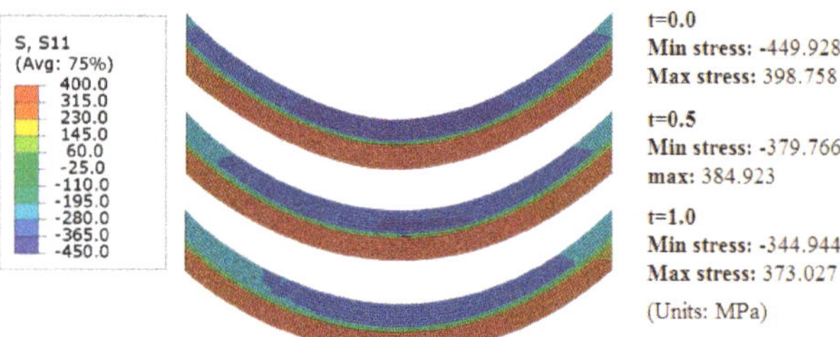

Figure 12. Numerical obtained stress contour with different kinematic hardening fraction.

Through comparing the experimental and numerical obtained springback angles, it is recommended that the kinematic hardening fraction shall be defined between $t = 0.1$ and $t = 0.5$ for Al7050. In the coming section about the ductile damage, the ratio of kinematic hardening in total hardening was assigned to be 0.5.

4.2. Effect of Damage Coupling on Springback Prediction

The Young's modulus plays an important role in the springback prediction. In this section, the decrease of Young's modulus during the deformation process is considered through the two approaches of including Young's moduli variation according to loading strain as discusses in the last section. The direct evolution of Young's moduli with respect to accumulating plastic strain is defined with tabular form according to Figure 6. Linear interpolation is considered between given points. For the ductile damage model, the material parameters were defined according to Table 2. The corresponding obtained numerically predicted springback angles are shown in Figure 13. These two methods can both well predict the tendency of the influence of pre-stretch on the final springback angles. Together with the increase of the pre-stretch displacement, the springback angles almost increase linearly with the same slope. Also, it can be found that the obtained springback angle with the ductile damage model can be closer to the experimental results.

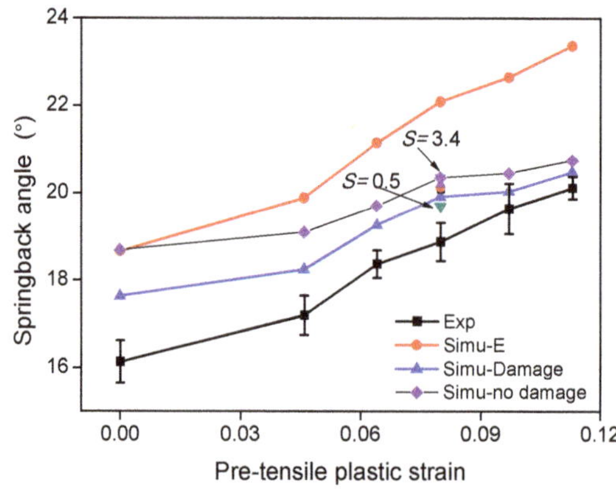

Figure 13. Comparison between the numerical and experimental obtained springback.

The damage parameters S will be assigned different values of (0.5, 0.75, 1.0, 1.7, 3.4). In Figure 14, the effect of the damage parameters S on the strain-stress curves is compared. It is found that with the decrease of the damage parameters S, the ductility of the material highly increased. In Figure 15, the ductile damage contour across the thickness zone after the bending process with pre-strains of 4 mm was plotted and compared. When S is smaller, more ductile damage appears on the section, and the springback angle is smaller due to the smaller stress caused by damage. The comparisons with experimental and other numerical obtained results are also given in Figure 15.

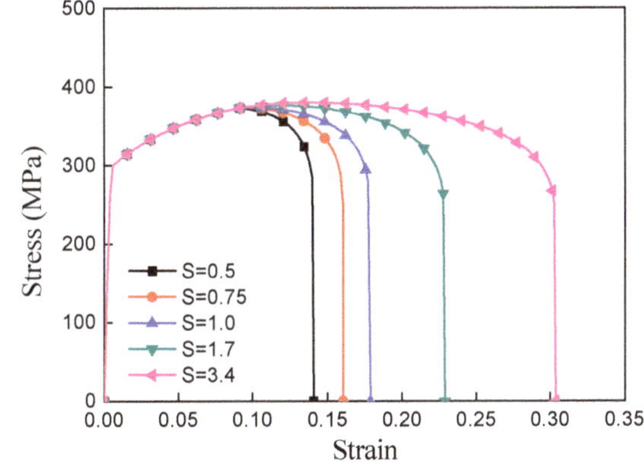

Figure 14. The effect of damage parameter S on the stress-strain curve.

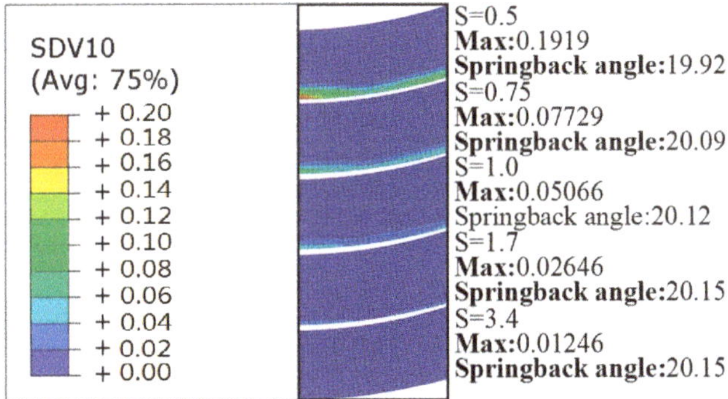

Figure 15. Ductile damage contour across the thickness section.

5. Conclusions

In this study, the springback phenomena of AA7055 alloy sheet metal under three points bending test after pre-stretch were investigated. Under the same bending depth, five different pre-stretch displacements were conducted on the samples, and the obtained springback angles were discussed by comparing the experimental and numerical obtained results. Meanwhile, in the numerical aspect, the influence of the kinematic hardening and ductile damage on the springback was compared and discussed. The conclusions in this study are given as follows:

- Through the experimental observation, it was found that under the same bending depth the springback angle increases with the increase of the pre-stretch displacement. The pre-deformation plays an important role in the final sample shapes.
- The influence of the kinematic hardening on the springback prediction of sheet metal is relatively large. The numerically obtained springback angle decreases with the increase of the ratio of

kinematic hardening in total hardening. With the increase of pre-stretch displacement, the influence of kinematic hardening will be enlarged. Meanwhile, though this methodology, the ratio of kinematic hardening in total hardening can be coarsely determined.

- The ductile damage has a great influence on the springback prediction of sheet metal. It not only affects the Young's modulus but the mechanical state of the formed samples. The numerical simulation with consideration of the ductile damage gives more accurately predicted springback angles.

Author Contributions: Z.Y. and X.Z. conceived and prepared the manuscript; J.Q. performed the experiments; H.B. focused on the constitutive model building; J.G. and X.C. are responsible for simulation and analysis.

Funding: This research was funded by National Natural Science Foundation of China (NO.51605257), the European Funds (FEDER) and the Alsace-Champagne-Ardenne-Loraine region (ESSAIMAGE D201600778).

Conflicts of Interest: The authors declare no conflict of interest.

Appendix A

In the framework of non-associative plasticity, the yield criterion f and plastic potential F are separately defined in Equation (A5). The evolution relations of plastic strain, isotropic and kinematic hardenings, and ductile damage are given by Equations (A1) to (A4), respectively, by the normality rule with respect to plastic potential F.

$$\underline{D}^p = \dot{\lambda}\frac{\partial F}{\partial \underline{\sigma}} = \dot{\lambda}\frac{\underline{\underline{H}}^p : \left(\underline{S}^p_d - \underline{X}\right)}{\sqrt{1-d}\|\underline{S}^p_d - \underline{X}\|_H} : \left[\underline{\underline{I}}^D + \frac{\underline{X} \otimes \underline{S}_0}{X^p_{11}(1-d)(R/\sqrt{1-d^\gamma} + \sigma_y)}\right] = \dot{\lambda}\underline{n}^p \tag{A1}$$

$$\dot{\underline{\alpha}} = -\dot{\lambda}\frac{\partial F}{\partial \underline{X}} = \frac{\dot{\lambda}}{\sqrt{1-d}}(\underline{n}^x - a\underline{\alpha}) \tag{A2}$$

$$\dot{r} = -\dot{\lambda}\frac{\partial F}{\partial R} = \dot{\lambda}(\frac{n^i}{\sqrt{1-d^\gamma}} - br) \tag{A3}$$

$$\dot{d} = \dot{\lambda}\frac{\partial F}{\partial Y} = \frac{\dot{\lambda}}{(1-d)^\beta}\left(\frac{\langle Y - Y_0\rangle}{S(\theta)}\right)^s \tag{A4}$$

$$\begin{cases} f = \frac{\|\underline{S}^c_d - \underline{X}\|_H}{\sqrt{1-d}} - \frac{R}{\sqrt{1-d^\gamma}} - \sigma_y = 0 \\ F = \frac{\|\underline{S}^p_d - \underline{X}\|_H}{\sqrt{1-d}} - \frac{R}{\sqrt{1-d^\gamma}} - \sigma_y + \frac{3a}{4C}\frac{\underline{X}:\underline{X}}{1-d} + \frac{b}{2Q}\frac{R^2}{1-d} + \frac{S}{s+1}\left\langle\frac{Y-Y_0}{S'(\theta)}\right\rangle^{s+1}\frac{1}{(1-d)^\beta} \end{cases} \tag{A5}$$

where σ_y is the initial yield stress, and the parameters a and b represent the non-linearity of the kinematic and isotropic hardening respectively. $\underline{D}^p$ denotes the plastic strain tensor. $S'(\theta)$ is function of lode angle θ [26]. S, s, β, and Y_0 are the parameters governing the ductile damage evolution. The plastic multiplier $\dot{\lambda}$ is defined by condition of consistency with respect to the yield surface $f = \dot{f} = 0$. The equivalent stresses $\|\underline{S}^i_d - \underline{X}\|_H = \sqrt{(\underline{S}^i_d - \underline{X}) : \underline{\underline{H}}^i : (\underline{S}^i_d - \underline{X})}, i = \{c, p\}$ given in yield criterion ($i = c$) and plastic potential ($i = p$) of Hill 48 type. The deviatoric distortional stress tensor $\underline{S}^i_d(i = \{c, p\})$ of François type [27]:

$$\underline{S}^p_d = \underline{S} + \frac{\underline{S}_0 : \underline{S}_0}{2(1-d)X^p_{11}(R/\sqrt{1-d^\gamma} + \sigma_y)}\underline{X} \tag{A6}$$

$$\underline{S}^c_d = \underline{S} + \frac{\underline{S}_0 : \underline{S}_0}{2(1-d)X^c_{11}(R/\sqrt{1-d^\gamma} + \sigma_y)}\underline{X} - \frac{\underline{X}:\underline{X}}{2(1-d)X^c_{12}(R/\sqrt{1-d^\gamma} + \sigma_y)}\underline{S}_0 \tag{A7}$$

with $\underline{S}_0 = \underline{S} - \frac{\underline{S}:\underline{X}}{\underline{X}:\underline{X}} \cdot \underline{X}$. These distortion stresses reduce to classical deviatoric stress $\underline{S}$ for the case of parameters $X_{11}^c = X_{12}^c = X_{11}^p = \infty$ which corresponding to the annealing distortion effect giving the classical Hill 48 yield surface. The outward normal tensors to plastic potential are defined as follows:

$$\underline{n}^p = \frac{1}{\sqrt{1-d}\|\underline{S}_d^p - \underline{X}\|_H}\left[\underline{\underline{I}}^D + \frac{\underline{X} \otimes \underline{S}_0}{X_{11}^p(1-d)(R/\sqrt{1-d^\gamma}+\sigma_y)}\right] : \underline{\underline{H}}^p : \left(\underline{S}_d^p - \underline{X}\right) \tag{A8}$$

$$\underline{n}^x = \frac{1}{\sqrt{1-d}}\left[\begin{array}{c} \frac{\underline{\underline{H}}^p:(\underline{S}_d^p-\underline{X})}{\|\underline{S}_d^p-\underline{X}\|_H} - \frac{\underline{S}_0:\underline{S}_0}{2(1-d)X_{11}^p(R/\sqrt{1-d^\gamma}+\sigma_y)}\frac{\underline{\underline{H}}^p:(\underline{S}_d^p-\underline{X})}{\|\underline{S}_d^p-\underline{X}\|_H} \\ + \frac{(\underline{S}:\underline{X})(\underline{X}:\frac{\underline{\underline{H}}^p:(\underline{S}_d^p-\underline{X})}{\|\underline{S}_d^p-\underline{X}\|_H})}{(1-d)X_1^p(\underline{X}:\underline{X})(R/\sqrt{1-d^\gamma}+\sigma_y)}\underline{S}_0 \end{array}\right] \tag{A9}$$

$$n^r = \frac{1}{\sqrt{1-d^\gamma}}\left[\frac{(\underline{S}_0:\underline{S}_0)(\underline{X}:\frac{\underline{\underline{H}}^p:(\underline{S}_d^p-\underline{X})}{\|\underline{S}_d^p-\underline{X}\|_H})}{2(1-d)^{3/2}X_{11}^p(R/\sqrt{1-d^\gamma}+\sigma_y)^2} + 1\right] \tag{A10}$$

The consistent tangent operator $\underline{\underline{K}}$ is given by:

$$\underline{\dot{\sigma}} = \underline{\underline{K}} : \underline{D} = \left\{\frac{\partial \underline{\sigma}}{\partial \underline{\varepsilon}^e} - \frac{1}{H^P}\left[(\frac{\partial \underline{\sigma}}{\partial \underline{\varepsilon}^e} : \underline{n}^p) \otimes (\frac{\partial \underline{\sigma}}{\partial \underline{\varepsilon}^e} : \underline{n}^c) - \overline{Y}\frac{\partial \underline{\sigma}}{\partial d} \otimes (\frac{\partial \underline{\sigma}}{\partial \underline{\varepsilon}^e} : \underline{n}^c)\right]\right\} : \underline{D} \tag{A11}$$

$$\underline{\underline{K}} = \frac{\partial \underline{\sigma}}{\partial \underline{\varepsilon}^e} - \frac{1}{H^P}\left[(\frac{\partial \underline{\sigma}}{\partial \underline{\varepsilon}^e} : \underline{n}^p) \otimes (\frac{\partial \underline{\sigma}}{\partial \underline{\varepsilon}^e} : \underline{n}^c) - \overline{Y}\frac{\partial \underline{\sigma}}{\partial d} \otimes (\frac{\partial \underline{\sigma}}{\partial \underline{\varepsilon}^e} : \underline{n}^c)\right] \tag{A12}$$

$$H^P = (1-d)\underline{n}^p : \underline{\underline{A}} : \underline{n}^c + \underline{n}^p : (\underline{\underline{A}} : \underline{\varepsilon}^e)\overline{Y} + \frac{n^x}{\sqrt{1-d}} : (\overline{Y}(-\tfrac{2}{3}C\alpha) - \tfrac{2}{3}C(1-d)(a\alpha - \frac{n^x}{\sqrt{1-d}})) +$$
$$\frac{n^i}{\sqrt{1-d^\gamma}}Q(\overline{Y}(-r\gamma d^{\gamma-1}) - (1-d)(br - \frac{n^i}{\sqrt{1-d^\gamma}})) - \tfrac{1}{2}\left[(1-d)^{-\frac{3}{2}}\|\underline{S}_d^c - \underline{X}\| + \gamma d^{\gamma-1}(1-d)^{-\frac{3}{2}}R\right]\overline{Y} \tag{A13}$$

$$\overline{Y} = \frac{1}{(1-d)^{\beta+\frac{1}{2}}}\left\langle\frac{Y-Y_0}{S(\theta)}\right\rangle^s \tag{A14}$$

where $\underline{\underline{A}}$ is the elastic operator defined as $A_{ijkl} = 2\mu_e(\delta_{ik}\delta_{jl} - \tfrac{1}{3}\delta_{ij}\delta_{kl}) + K_e\delta_{ij}\delta_{kl}$.

References

1. Sumikawa, S.; Ishiwatari, A.; Hiramoto, J. Improvement of springback prediction accuracy by considering nonlinear elastoplastic behavior after stress reversal. *J. Mater. Process. Technol.* **2017**, *241*, 46–53. [CrossRef]
2. Hassan, H.U.; Traphöner, H.; Güner, A.; Tekkaya, A.E. Accurate springback prediction in deep drawing using pre-strain based multiple cyclic stress–strain curves in finite element simulation. *Int. J. Mech. Sci.* **2016**, *110*, 229–241. [CrossRef]
3. Chan, K.C.; Wang, S.H. Theoretical analysis of springback in bending of integrated circuit leadframes. *J. Mater. Process. Technol.* **1999**, *91*, 111–115. [CrossRef]
4. Yuen, W.Y.D. A generalised solution for the prediction of springback in laminated strip. *J. Mater. Process. Technol.* **1996**, *61*, 254–264. [CrossRef]
5. Neto, D.M.; Oliveira, M.C.; Santos, A.D.; Alves, J.L.; Menezes, L.F. Influence of boundary conditions on the prediction of springback and wrinkling in sheet metal forming. *Int. J. Mech. Sci.* **2017**, *122*, 244–254. [CrossRef]
6. Liao, J.; Xue, X.; Lee, M.G.; Barlat, F.; Vincze, G.; Pereira, A.B. Constitutive modeling for path-dependent behavior and its influence on twist springback. *Int. J. Plast.* **2017**, *93*, 64–88. [CrossRef]
7. Prates, P.A.; Pereira, A.F.G.; Sakharova, N.A.; Oliveira, M.C.; Fernandes, J.V. Inverse strategies for identifying the parameters of constitutive laws of metal sheets. *Adv. Mater. Sci. Eng.* **2016**, *2016*, 1–18. [CrossRef]
8. Li, K.; Carden, W.; Wagoner, R. Simulation of springback. *Int. J. Mech. Sci.* **2002**, *44*, 103–122. [CrossRef]
9. Oliveira, M.C.; Alves, J.L.; Chaparro, B.M.; Menezes, L.F. Study on the influence of work-hardening modeling in springback prediction. *Int. J. Plast.* **2007**, *23*, 516–543. [CrossRef]

10. Chung, K.; Lee, M.G.; Kim, D.; Kim, C.; Wenner, M.L.; Barlat, F. Spring-back evaluation of automotive sheets based on isotropic-kinematic hardening laws and non-quadratic anisotropic yield functions: Part I: Theory and formulation. *Int. J. Plast.* **2005**, *21*, 861–882. [CrossRef]

11. Chun, B.; Jinn, J.; Lee, J. Modeling the Bauschinger effect for sheet metals, part I: Theory. *Int. J. Plast.* **2002**, *18*, 571–595. [CrossRef]

12. Dongjuan, Z.; Zhenshan, C.; Xueyu, R.; Yuqiang, L. Sheet springback prediction based on non-linear combined hardening rule and Barlat89's yielding function. *Comput. Mater. Sci.* **2006**, *38*, 256–262. [CrossRef]

13. Zang, S.; Lee, M.G.; Sun, L.; Kim, J.H. Measurement of the Bauschinger behavior of sheet metals by three-point bending springback test with pre-strained strips. *Int. J. Plast.* **2014**, *59*, 84–107. [CrossRef]

14. Stoughton, T.; Xia, C.; Du, C.; Shi, M. Challenges for Constitutive Models for Forming of Advanced Steels. In Proceedings of the National Science Foundation Workshop, Arlington, VA, USA, 29–30 March 2006.

15. Lee, M.G.; Kim, D.; Kim, C.; Wenner, M.L.; Wagoner, R.H.; Chung, K. A practical two-surface plasticity model and its application to spring-back prediction. *Int. J. Plast.* **2007**, *23*, 1189–1212. [CrossRef]

16. Govik, A.; Rentmeester, R.; Nilsson, L. A study of the unloading behaviour of dual phase steel. *Mater. Sci. Eng. A* **2014**, *602*, 119–126. [CrossRef]

17. Gau, J.-T.; Kinzel, G.L. A new model for springback prediction in which the Bauschinger effect is considered. *Int. J. Mech. Sci.* **2001**, *43*, 1813–1832. [CrossRef]

18. Wang, Z.; Hu, Q.; Yan, J.; Chen, J. Springback prediction and compensation for the third generation of UHSS stamping based on a new kinematic hardening model and inertia relief approach. *Int. J. Adv. Manuf. Technol.* **2017**, *90*, 875–885. [CrossRef]

19. Nayebi, A.; Shahabi, M. Effect of continuum damage mechanics on springback prediction in metal forming processes. *J. Mech. Sci. Technol.* **2017**, *31*, 2229–2234. [CrossRef]

20. Zajkani, A.; Hajbarati, H. An analytical modeling for springback prediction during U-bending process of advanced high-strength steels based on anisotropic nonlinear kinematic hardening model. *Int. J. Adv. Manuf. Technol.* **2017**, *90*, 349–359. [CrossRef]

21. Jamli, M. Finite element analysis of springback process in sheet metal forming. *Int. J. Adv. Manuf. Technol.* **2017**, *11*, 75–84.

22. Badreddine, H.; Yue, Z.; Saanouni, K. Modeling of the induced plastic anisotropy fully coupled with ductile damage under finite strains. *Int. J. Solids Struct.* **2017**, *108*, 49–62. [CrossRef]

23. Saanouni, K. *Damage Mechanics in Metal Forming: Advanced Modeling and Numerical Simulation*; John Wiley & Sons: Hoboken, NJ, USA, 2012.

24. Souto, N.; Andrade-Campos, A.; Thuillier, S. Material parameter identification within an integrated methodology considering anisotropy, hardening and rupture. *J. Mater. Process. Technol.* **2015**, *220*, 157–172. [CrossRef]

25. Yue, Z.; Soyarslan, C.; Badreddine, H.; Saanouni, K.; Tekkaya, A.E. Identification of fully coupled anisotropic plasticity and damage constitutive equations using a hybrid experimental–numerical methodology with various triaxialities. *Int. J. Damage Mech.* **2015**, *24*, 683–710. [CrossRef]

26. Yue, Z. Ductile Damage Prediction in Sheet Metal Forming Processes. Ph.D. Thesis, University of Technology of Troyes, Troyes, France, 8 September 2014.

27. François, M. A plasticity model with yield surface distortion for non proportional loading. *Int. J. Plast.* **2001**, *17*, 703–717. [CrossRef]

© 2018 by the authors. Licensee MDPI, Basel, Switzerland. This article is an open access article distributed under the terms and conditions of the Creative Commons Attribution (CC BY) license (http://creativecommons.org/licenses/by/4.0/).

Article

Modeling Bake Hardening Effects in Steel Sheets—Application to Dent Resistance

Sandrine Thuillier [1,*], Shun-Lai Zang [2], Julien Troufflard [1], Pierre-Yves Manach [1] and Anthony Jegat [1]

[1] Univ. Bretagne Sud, UMR CNRS 6027, IRDL, F-56100 Lorient, France; julien.troufflard@univ-ubs.fr (J.T.); pierre-yves.manach@univ-ubs.fr (P.-Y.M.); anthony.jegat@univ-ubs.fr (A.J.)

[2] Key Laboratory of Education Ministry for Modern Design and Rotor-Bearing System, Xi'an Jiaotong University, No. 28 Xianning Road, Xi'an 710049, China; shawn@mail.xjtu.edu.cn

* Correspondence: sandrine.thuillier@univ-ubs.fr; Tel.: +33-297-874-537

Received: 4 July 2018; Accepted: 24 July 2018; Published: 30 July 2018

Abstract: This study is dedicated to the experimental characterisation and phenomenological modeling of the bake hardening effect of a thin steel sheet, to predict the static dent resistance and perform an experimental validation on a bulged part. In a first step, rectangular samples are submitted to a thermo-mechanical loading to characterise the bake hardening magnitude in tension. A three-step procedure is considered, involving first a pre-strain in tension up to several values followed by unloading. Secondly, a heat treatment during a fixed time and a given temperature is performed, and finally, a reloading in tension in the same direction as the pre-strain is applied. Then, a specific device is developed to perform dent tests on a bulged specimen, to evaluate the influence of bake hardening on the dent resistance. A three-step procedure is also considered, with a pre-strain applied with a hydraulic bulge test followed by a heat treatment and then static dent test at the maximum dome height. An original phenomenological model is proposed to represent the yield stress increase after the heat treatment and the second reloading. Material parameters are identified from the tensile tests and are input data to a finite element model. The numerical prediction of the load evolution during the dent test is then compared with experimental data and shows an overall good correlation.

Keywords: steel sheet; bake hardening; mechanical modeling; dent resistance

1. Introduction

Outer automotive panels may be subjected during their service life to the high or low velocity impact of a projectile, the size of which is small compared to the panel dimensions. The resistance of the part to such an impact, or dent resistance, is a major industrial concern. It has led to several academic studies, using specific devices to reproduce at the laboratory scale the impact of the projectile either under static or dynamic conditions. In most of these studies, thin sheet structures are subjected locally to the action of an indenter under an applied load of a few hundreds newtons. For example, large and doubly curved drawn parts representative of a roof panel are submitted in their middle to the action of a flat headed indenter [1]; or a trimmed sample and a hemispherical indenter are used in [2]. The dent resistance is defined as the load corresponding to a permanent dent depth of 0.1 mm. In the area in contact with the indenter, the sheet is subjected to stretching and highly localised plastic strain, whereas the remaining zones are mostly drawn and deformed elastically. The characteristics load-deflection curve of dent tests is highly non linear and exhibits a significant hysteresis, highlighting the non-reversibility of the test [2,3]. Such feature is rather difficult to predict numerically within a classical elasto-visco-plastic scheme especially when considering several points [4]. Introducing mixed hardening in the material model provides a better description of the dent resistance of steel panels [5]. However, the highly non-linear evolution during unloading is difficult to capture, though the description can be improved by considering the chord modulus degradation with plastic strain in balanced biaxial tension [6]. Moreover, the dent resistance is affected by the bake hardening effect, though little evidence or discussion can be found in the literature related to this dependency.

Bake hardening (BH) is a phenomenon occurring in steel materials corresponding to an increase of the flow stress after a pre-strain followed by heat treatment (or annealing) within a specific temperature range. Classical BH magnitudes are from 30 up to 60 MPa [7] after a few percent of pre-strain and heating at around 170 °C (443 K) during 20 min. The magnitude of this phenomenon depends on the annealing temperature and time [8]. Specific series of materials are concerned by bake hardening, like series with low carbon contents as E180BH, dual phase and TRIP steels [9–11] and also Mg alloys [12]. Carbon atoms are in solid solution in the iron matrix. Their capacity to diffuse, even at room temperature, especially from close-packed areas to areas less close-packed, is at the origin of the bake hardening. Indeed, after the pre-strain, carbon atoms, forming Cottrell atmospheres, move close to the core of the dislocations generated during this step, due to a decrease of the elastic energy of the crystallographic network, leading to a pinning of the dislocations. Secondly, C-rich clusters lead to the precipitation of coherent precipitates, e.g., [13]. Therefore, upon re-straining after the heat-treatment, a stress increase is necessary in order to free the dislocations, leading to a Lüders-like phenomenon of catastrophic slip in macroscopic bands of localised strain. And eventually to overcome the newly-formed precipitates. After a load stagnation corresponding to the band propagation [10], hardening is resumed at a higher level than before thermal treatment. The bake hardening magnitude results both from strain hardening after the pre-strain and thermal ageing. BH magnitude is defined as the difference between the elastic stress upon reloading, after heat treatment, and the stress level reached at the end of the pre-strain.

The bake hardening effect has practical applications in the automotive industry [14,15]. Indeed, the flow stress increase during paint curing, which is a necessary step in outer automotive panel production, leads to an increase of the dent resistance of the panels without altering the forming properties. Indeed, the deep drawing corresponds to the pre-strain; the equivalent plastic strain resulting from this step can differ significantly from one area to the other, e.g., bent areas near the door handle or windows or the trimming line exhibit highly localised strain whereas the center part is less deformed. Therefore, after paint curing, usually at a temperature close to 170 °C (443 K), the mechanical properties of the material increase [7,10,16], leading to an improvement of the static dent resistance of the panel. Within a complete virtual forming of metallic sheet parts, it seems relevant to add bake hardening and dent resistance to the list of predicted properties, to improve the design of new materials able to avoid non-aesthetic effects such as the one illustrated in Figure 1.

Berbenni et al. [13] propose a micro-mechanical model of the bake hardening effect that relates the flow stress increase to the dislocation density, content of solute atoms, volume fraction of C-rich coherent clusters and time. It depends on three material parameters, which are identified for E180BH steel using only the homogeneous strain range, after removing the transient part of the stress-strain curve corresponding to Lüders phenomenon. A BH magnitude of around 80 MPa is measured, after the two pre-strain values of 0.02 and 0.05 and 20 min annealing at 170 °C (443 K). Ballarin et al. [17] develop also a micro-mechanical model based on the dislocation density and considers also the Piobert-Lüders phenomenon. They obtained a very close description of the bake hardening effect in tension for several annealing times and baking temperatures. This model is further extended to strain path changes between the pre-strain and the subsequent strain path [18]. A neural network approach is proposed in [19–21] and information such as the baking temperature and the carbon contents are input data.

The aim of this study is to characterise and predict the magnitude of the bake hardening effect on the dent resistance via a phenomenological approach. In this study, a low carbon steel referred to as E220BH is considered. Firstly, the magnitude of the bake hardening effect is investigated in tension, for fixed time and temperature conditions and several pre-strain values ranging from 0.02 to 0.08. Then, hydraulic bulging of E220BH circular blank is performed and the dent resistance is investigated experimentally, for two different pre-strain amounts and after a heat treatment. Finally, an original phenomenological model is presented, which represents an increase of the stress level after the thermal treatment. As a first step, the model depends only on the pre-strain magnitude. The numerical simulation of these tests is then performed and compared with experimental data. The comparison on the load level and the strain field gives a thorough validation of the numerical model.

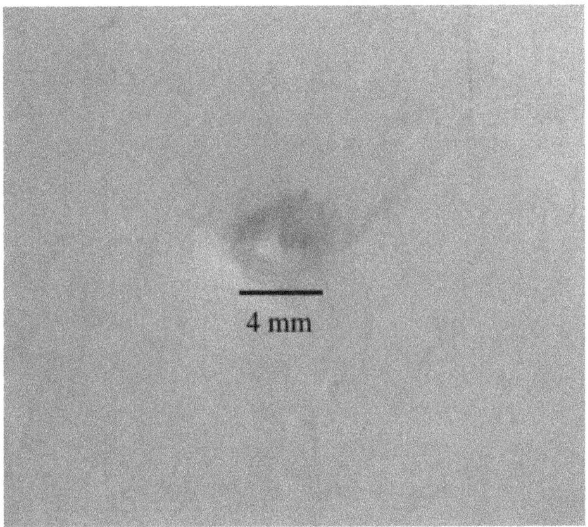

Figure 1. Impact of a hard projectile on a metallic panel, illustrating the importance of an improved dent resistance, in terms of the aesthetics of the product. The impact if highly localised compared to the size of the outer panel.

2. Materials and Experiments

A low carbon steel referred to as E220BH, that exhibits bake hardening effect, is used in this study. The chemical composition, as given by the supplier, is of 0.06% (in weight) of C and 0.7% of Mn. It is provided in sheets with a thickness of 0.74 mm. The mechanical properties were characterised in tension [22] and are simply recalled in Table 1. Though the influence of anisotropy is out of the scope of the present study, normal and in-plane anisotropy coefficients (respectively $\bar{r}$ and Δr) are also given as general characteristics of the material.

Table 1. Measured mechanical properties of E220BH steel. $R_{p0.2\%}$ is the conventional elastic limit and R_m is the tensile strength.

$R_{p0.2\%}$ (MPa)	R_m (MPa)	$\bar{r}$	Δr
220	350	1.75	0.66

2.1. Tensile Tests and Thermal Treatment

The three-steps necessary to investigate the bake hardening effect are first a tensile pre-strain followed by a thermal treatment and finally a second tension in the same direction as the previous one; these steps are detailed below. Tensile tests are performed on samples with a rectangular shape, of width 20 mm and gauge length 150 mm, in the rolling direction. The local longitudinal strain is measured with an extensometer (2630 series, INSTRON, Norwood, MA, USA) having a gauge length of 10 mm. Three pre-strain values, corresponding to a total logarithmic strain of 0.02, 0.04 and 0.08, are considered. Tensile tests up to rupture are also performed, in order to characterise the material in its original state.

Then, the samples are heated at an average value of 165 °C during 15 min in an oven. Three thermocouples are used to record the temperature evolution at different locations in the oven, one in the middle of the samples and two others on each side. Three different batches are heat-treated, each of them made of 3 to 5 tensile samples. It can be seen in Figure 2 that the gradient in the oven is lower than 5 °C, whatever the batch of samples. In a first step, the oven is heated up to 170 °C during a time long enough (above 100 min) to reach a stable temperature. Then, samples are settled inside the oven; the temperature drops down to 80 °C and, as soon as the door is closed, goes rapidly up to 160 °C. The samples are then kept during 15 min at a temperature that evolves from 160 °C to a maximum of 170 °C, with an average value of 165 °C (Figure 2). The sample then slowly cools down to room temperature.

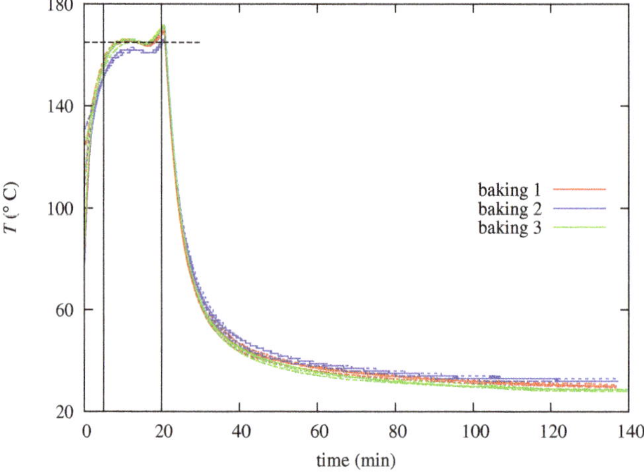

Figure 2. Temperature evolution during heating (15 min at an average temperature of 165 °C) and then slow cooling down to room temperature. Each colour corresponds to a batch of samples and three curves of the same colour represent the signals of the three thermocouples.

The samples are then subjected to a second tensile test in the same direction as the first tension, up to rupture. The reproducibility is investigated over 2 or 3 samples tested in the same conditions. As the global mechanical behavior, before and after the heat treatment, as well as the evolution of BH magnitude with the pre-strain are close, only one test is presented.

2.2. Dent Resistance

The validation test is related to static dent resistance. Indeed, opening parts of cars, like doors and hoods, are first formed and then painted on the outside surface and submitted to baking. Materials exhibiting bake hardening effect are then interesting to use, for the static dent resistance should increase after baking. To reproduce such a multi-step process at an academic laboratory scale, circular blanks are pre-strained by bulging up to several strain levels or pole heights, then submitted to the same thermal treatment as for tension. Finally, a dent test is specifically designed in order to characterise the dent resistance of the bulged blanks after baking. The same procedure is also carried out without the thermal treatment, to investigate the influence of the bake hardening effect on dent resistance. The experiments consist of a three-step process: pre-strain in hydraulic bulging, thermal treatment representative of paint baking and finally static dent test.

2.2.1. Pre-Strain

A hydraulic bulge test, developed at IRDL and already detailed in [23], is used to apply a pre-strain. Circular blanks of diameter of 270 mm are clamped by screws between a blank-holder and a die. A fixed volume of water is pressed under the blank by the displacement of an actuator (Zwick 8803, Zwick Roell, Ulm, Germany). The blank bends over the die radius of 8 mm and then bulges inside the die cavity of diameter 184 mm. A pressure sensor (SensorTechnics, BTE6000 / PTU6000 series, Berlin, Germany) gives the fluid pressure and the strain field is measured by digital image correlation in an area around the center point corresponding to the maximum displacement. The strain state on the surface is recorded during the test; an average of the strain components ϵ_{xx} and ϵ_{yy} over a small area around the specimen center is performed, with $\vec{x}$ and $\vec{y}$ parallel respectively to the rolling and transverse directions. The center of the circular specimen is pencil-marked before the bulge test and checked after bulging and unloading, in order to assess precisely the maximum height of the sample, also output from Digital Image Correlation (DIC, Aramis 4M, GOM mbH, Braunschweig, Germany) measures. DIC data is calculated as a post-treatment and cannot be used to control the test. Therefore, the end criterion of the bulge test is defined on the pressure.

2.2.2. Heat Treatment

In order to highlight the influence of bake hardening, some bulged specimen are then submitted to the same thermal treatment as the tensile samples. Figure 3 shows the specimen in the oven; two specimen are heat-treated at the same time. Furthermore, some bulged specimen are not subjected to the thermal heat-treatment.

Figure 3. Thermal treatment of bulged specimen.

2.2.3. Dent Tests

A dedicated device is designed to characterise the dent resistance using a universal tensile machine (INSTRON 5566A, Norwood, MA, USA). Figure 4 shows the different components, i.e., a supporting plate, four clamping devices and a hemispherical punch of radius 12.5 mm connected to the load cell of maximum capacity 1 kN. The supporting plate lies on a precision positioning table to put the sample such as its pole is located right beneath the punch. The displacement of the punch and the applied load are recorded during the test. The stiffness of the device is evaluated by performing tests using a very rigid block as a sample. The corresponding displacement, which is related to gaps and elastic deformation of the tools, is removed from the total displacement, to calculate the real local displacement—or corrected displacement—*d* imposed to the bulged specimen. This local value is used to plot the results. Tests are conducted up to a maximum displacement of 2.4 mm whatever the time and deformation history. The samples are then unloaded.

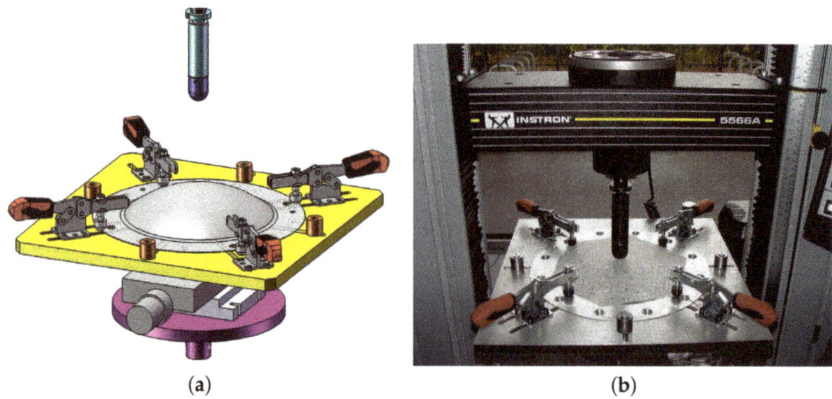

(a)   (b)

Figure 4. Device designed to characterise the dent resistance. (**a**) Schematic representation of dent test; (**b**) Experimental setup and sample.

Moreover, specific tests are performed with several loadings-unloadings, for samples without or with a thermal treatment after the pre-strain. DIC is used also to visualise the dent depth, as shown in Figure 5. The pole after bulging corresponds to the point in the middle and the permanent dent in clearly seen.

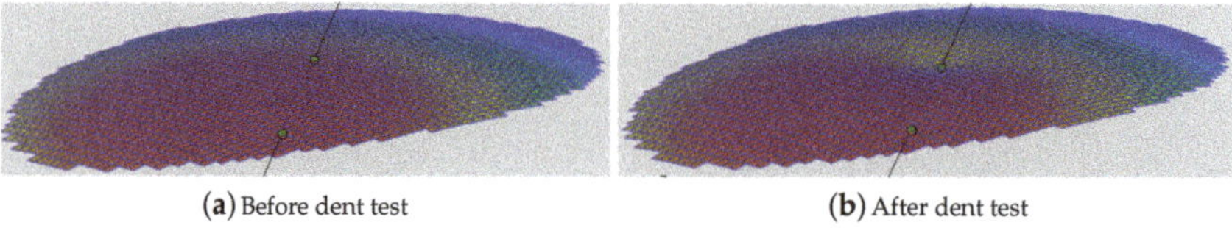

(**a**) Before dent test   (**b**) After dent test

Figure 5. Visualisation of the dent depth with digital image correlation (DIC): pole geometry after bulging (**a**) and visualisation of the dent depth (**b**). This images correspond to the smallest pre-strain, with the permanent dent depth around 1 mm.

3. Experimental Results

3.1. Mechanical Behavior in Tension

The influence of heating on virgin samples (without tensile pre-strain) is also investigated and, as expected (though not presented here), the yield stress is not affected but only the intensity of the Piobert-Lüders phenomenon, with an increase of the length of the stress stagnation. Heating results in diffusion of carbon atoms and therefore, to a slight increase of the plateau magnitude.

Figure 6 shows the Cauchy stress-logarithmic strain curves, for both the pre-strain and the subsequent tension. For the lowest pre-strain of 0.02, an initial stress increase of 46 MPa is recorded, which represents 17% of the stress reached at the end of the first tension, before the heat-treatment. The stress then sharply decreases and then, increases again and the stress-strain curve tends towards the one of the virgin sample. The magnitude of the BH effect decreases slightly with the pre-strain amount, down to 39 MPa at 0.04 pre-strain (13%) and 38 MPa at 0.08 pre-strain (11%). Similarly to previous results, the BH magnitude tends to decrease when the pre-strain increases, both intrinsically and as the ratio of BH over the yield stress of the virgin sample at the same pre-strain value. A rather similar evolution is noted for the higher pre-strains, though strain needed to reach back the curve of the virgin sample is longer and premature rupture takes place very clearly. The BH magnitude is similar to results obtained for a similar material [7], as well as DP steel [10], but lower than the value reported for E180BH [13]. Such a difference may arise from different concentrations of solute atoms in the materials.

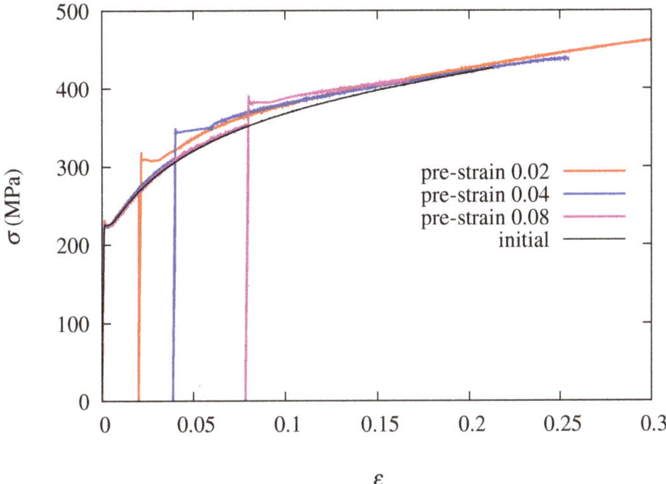

Figure 6. Tensile pre-strain followed by heat treatment and tensile deformation. Bake hardening (BH) effect is evidenced by the stress jump after the second loading in tension. The curves of the subsequent tension, after baking, are shifted horizontally by the amplitude of the pre-strain.

3.2. Dent Resistance of Bulged Specimen

The evolution of the pressure with the pole height h is plotted in Figure 7a. It can be seen that a pressure around 1.5 MPa corresponds to an average height of 13 mm, and respectively 28 mm and 29.5 mm for pressures of 3.5 MPa and 3.6 MPa. The strain at the pole is also calculated with DIC. Figure 7b shows the evolution of the fluid pressure with the strain component ϵ_{xx} at the pole. Two main pre-strain values are achieved, after unloading: around 0.016 and 0.07, though the dispersion is higher for the higher pre-strain, values ranging from 0.065 up to 0.077. A total of 14 samples are deformed, both to check the reproducibility and to have enough samples for the subsequent dent test, without and with a thermal treatment. However, as the pre-strain is not controlled, each test is specific and no average pre-strain is used in the following.

Figure 8a shows the load evolution with the corrected punch displacement, for a pre-strain $\epsilon_{xx} = 0.02$. The curves exhibit a first rather linear part followed by an elastic-plastic rounded transition; the unloading is linear and the permanent displacement is slightly below 1 mm. The load evolution for two tests is plotted, to highlight the very good reproducibility. Such an evolution is close to results presented in the literature for mild steel [6].

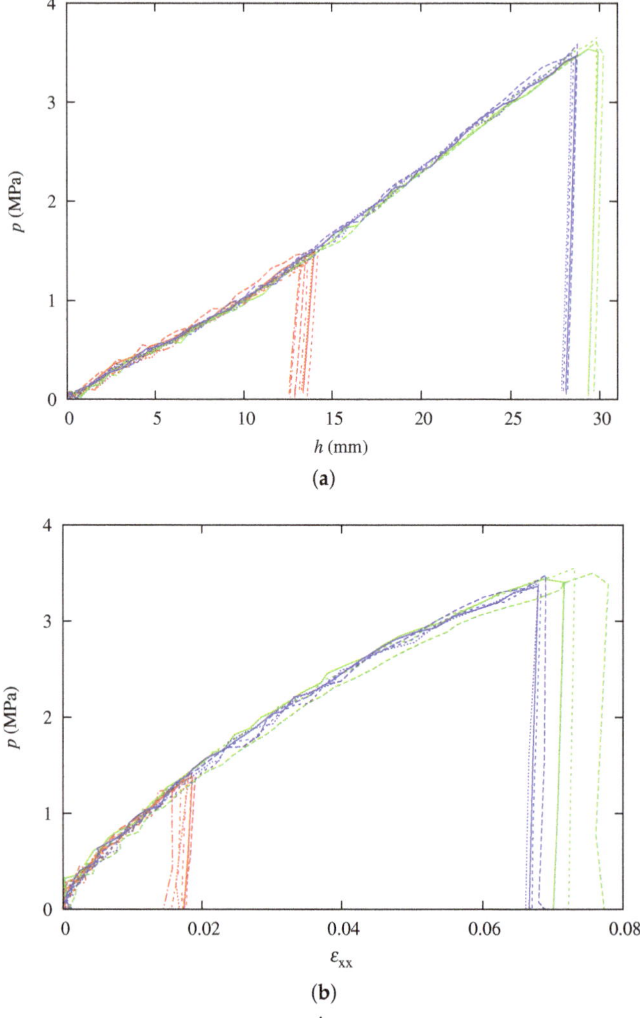

Figure 7. Bulge test. Each colour and line type stands for a single test. The tests can be gathered around the maximum pole height: 13 mm in red, 28 mm in blue and 29.5 mm in green. (**a**) Evolution of the fluid pressure as a function of the maximum height, as measured on the outer surface; (**b**) evolution of the fluid pressure as a function of the strain component ϵ_{xx}

The influence of the thermal treatment is around 40 N at the maximum displacement, representing a contribution of 10% of the maximum load. Though the initial slope is the same as for the non-thermally treated sample, the unloading slope is higher, leading to a residual displacement of the same order of magnitude, around 1 mm. The reproducibility is also very good.

The load evolution is similar for the higher pre-strain $\epsilon_{xx} = 0.07$ (Figure 8b), though the unloading becomes more significantly non-linear. The residual displacement is also the same for both configurations (i.e., without and with a thermal treatment) and is around 1.3 mm.

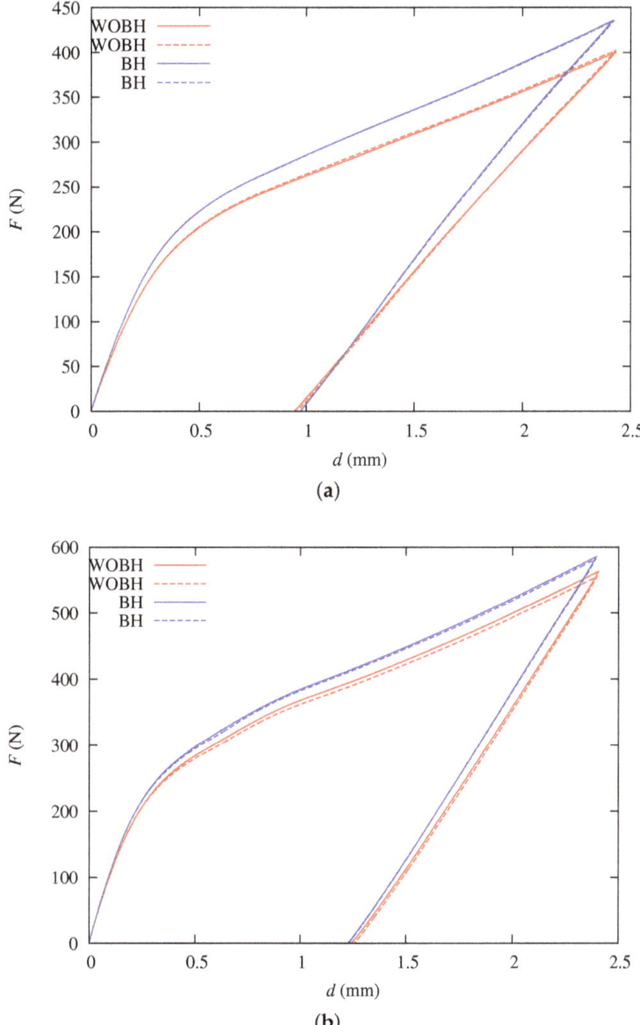

Figure 8. Load versus corrected displacement (d) during dent test with (BH) or without bake hardening (WOBH), after a pre-strain in hydraulic bulging. Continuous and dashed lines represent two tests having the same pre-strain magnitude. (**a**) Pre-strain $\epsilon_{xx} \approx 0.02$; (**b**) Pre-strain $\epsilon_{xx} \approx 0.07$.

Concerning the loading-unloading-reloading during the dent tests, the same influence can be shown in Figure 9, i.e., the load level is increased by the bake hardening effect, which implies that the dent resistance is increased. Indeed, for a given applied load, the residual displacement is lower for bake-hardened samples. The hysteresis loop at unloading changes when the displacement increases. Indeed, Figure 9a clearly shows that a concave unloading is recorded up to applied displacements of 2.4 mm and then a convex one is noted for a higher displacement of 3.5 mm, for the pre-strain $\epsilon_{xx} = 0.02$. The same trend is noticed for the higher pre-strain $\epsilon_{xx} = 0.07$ (Figure 9b).

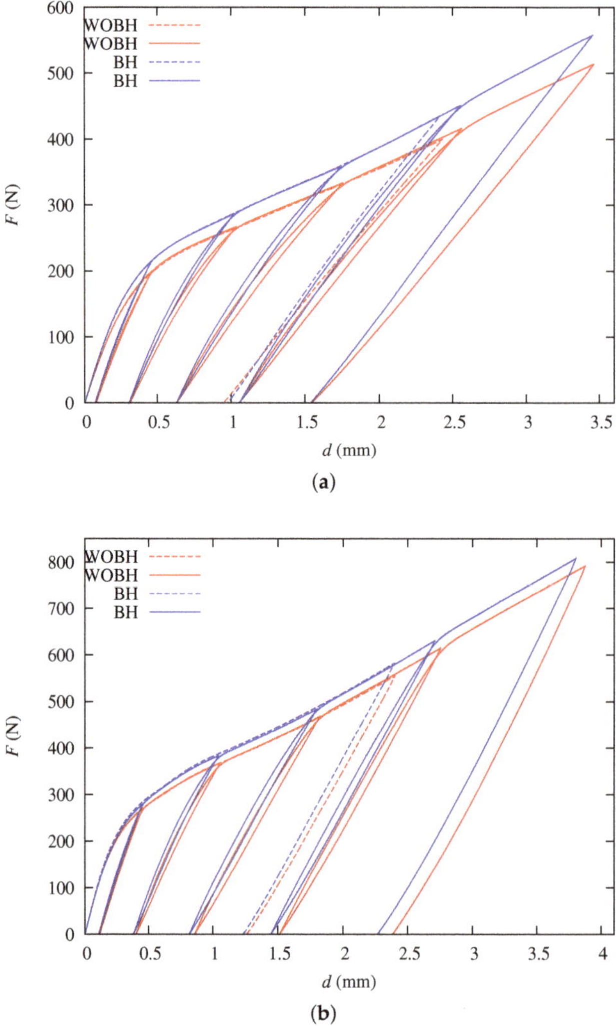

Figure 9. Dent test with (BH) or without bake hardening (WOBH), after a pre-strain in hydraulic bulging. Several loading-unloading-reloading sequences are applied during the tests. Continuous and dashed lines represent two tests having the same pre-strain magnitude. (**a**) Pre-strain $\epsilon_{xx} = 0.02$; (**b**) Pre-strain $\epsilon_{xx} = 0.07$.

4. Numerical Simulation of the Multi-Step Process

A numerical model to predict the influence of bake hardening on the dent resistance is proposed, based on a phenomenological description. Constitutive equations, material parameter identification and validation on dent tests of bulged specimen are presented in the following sections.

4.1. BH Modeling and Material Parameter Identification

The main idea of the model proposed in this study is to take into account both the flow stress increase after bake hardening, $\Delta\sigma$, and the convergence toward the initial flow stress. Moreover, the stress increase should be dependent on the pre-strain amount, represented by the equivalent plastic strain after the pre-strain $\bar{\epsilon}^*$. Certainly, the baking temperature T_{bake} and the duration of the baking should also be taken into account but are not considered in a first step in this study, for they kept the same values for all the experiments.

The constitutive model is written within the large transformation framework and isotropic hardening is considered. The yield function $\mathcal{F}$ is given by Equation (1):

$$\mathcal{F} = \bar{\sigma} - \sigma_Y - f\Delta\sigma \tag{1}$$

where $\bar{\sigma}$ stands for the equivalent stress according to von Mises yield criterion. The parameter f is set equal to zero for a test without bake hardening whereas it is set equal to 1 in case of bake hardening. The evolution of the yield contour with equivalent plastic strain $\bar{\epsilon}$ is described with a Swift equation (Equation (2)):

$$\sigma_Y = K\left(\bar{\epsilon} + \epsilon_0\right)^n \quad \text{and} \quad \epsilon_0 = (\sigma_0/K)^{1/n} \tag{2}$$

where K, n and σ_0 are material parameters. Finally, the overstress $\Delta\sigma$ related to bake hardening is modelled as given by Equation (3):

$$\Delta\sigma = C_1\left(2\exp\left(-\gamma_1\left(\bar{\epsilon} - \bar{\epsilon}^*\right)\right) - \exp\left(-\gamma_1\bar{\epsilon}\right)\right) \tag{3}$$

where C_1 and γ_1 are material parameters. The overstress has an initial value higher than the stress without bake hardening, that depends on C1 and γ_1, at the same strain level and then decreases down to the stress level without heat treatment. Indeed, when considering the reloading after the pre-strain and the heat treatment, with $\bar{\epsilon} = \bar{\epsilon}^*$, the yield stress is given by $\sigma_Y + \Delta\sigma$, with:

$$\Delta\sigma = C_1\left(2 - \exp\left(-\gamma_1\bar{\epsilon}^*\right)\right) \tag{4}$$

The first term of the right-hand side of Equation (4) is a constant equal to $2C_1$ and the second term is negative and decreases in absolute value when $\bar{\epsilon}^*$ increases. Therefore, it leads to an increasing overstress when the pre-strain increases.

Inverse identification of the material parameters is carried out using the in-house Matlab toolbox SMAT developed at Xi'an Jiaotong University [24]. A cost function $\mathcal{L}(A)$ is defined and minimized iteratively in the least square sense by Equation (5).

$$\mathcal{L}(A) = \sum_{n=1}^{N} \mathcal{L}_n(A) = \sum_{n=1}^{N}\left[\frac{1}{M_n}\sum_{i=1}^{M_n} D_n\left|\sigma(A, t_i) - \sigma^*(t_i)\right|^2\right] \tag{5}$$

where N is the number of tests in the database, A is the set of material parameters. For the variable n, numbers of 1, 2 and 3 correspond to the cases of tensile pre-strains of 0.02, 0.04 and 0.08 before thermal treatment. M_n is the number of experimental points of the n-th test, $\sigma(A, t_i) - \sigma^*(t_i)$ is the gap between experimental value for the stress $\sigma(A, t_i)$ and simulated stress $\sigma^*(t_i)$ at time t_i, and D_n is a weighting coefficient for the n-th test. Material parameters thus identified are given in Table 2.

The constitutive equations presented above are implemented in the finite element code Abaqus via a user subroutine, for plane stress states $\sigma_{i3} = 0, i = 1, 2, 3$. The equivalent plastic strain is stored at each integration point during the pre-strain, and in particular at the end of the first step. Then, upon reloading during the subsequent step after bake hardening, the parameter $\bar{\epsilon}^*$ is set equal to the last stored value of the equivalent plastic strain during the first step.

Table 2. Material parameters identified from tensile tests for E220BH steel.

E (MPa)	v	σ_0 (MPa)	K (MPa)	n	C_1 (MPa)	γ_1
219,700	0.29	235.9	607.7	0.23	21.5	22.2

Comparison of experimental and numerical results for the tensile tests are shown in Figure 10. The numerical model uses only one finite element for the tensile sample, assuming a homogeneous strain field. Several steps are performed: firstly, a pre-strain in tension followed by unloading, using the initial mechanical behavior exhibited in Figure 10a ($f = 0$). Then, for the following steps, $f = 1$ and the overstress $\Delta\sigma$ depends on the pre-strain amount. It can be seen that the model captures well the maximum stress after the thermal treatment as well as the convergence towards the initial flow stress. However, no effort was dedicated to the work-hardening stagnation, associated to a flow localisation. Indeed, in this case, the tensile test can no longer be considered as homogeneous. It would be necessary to model the whole structure in order to reproduce also the localisation in bands. Though it is an interesting challenge, this modelling is out of the scope of the present article.

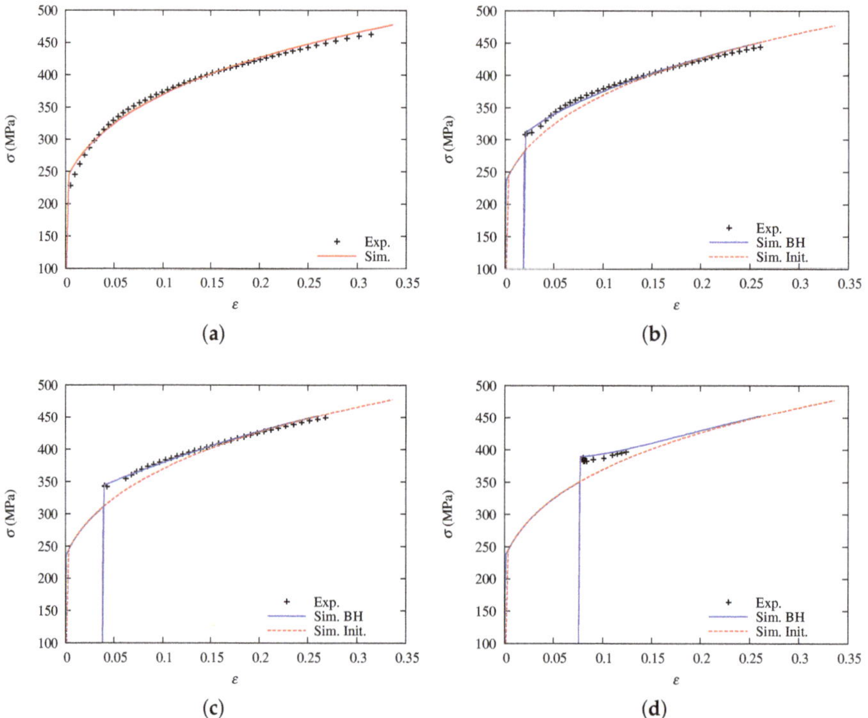

Figure 10. Modeling of the mechanical behavior. (**a**) Initial; (**b**) Pre-strain 0.02; (**c**) Pre-strain 0.04; (**d**) Pre-strain 0.08.

Moreover, although some authors note a decrease of the BH magnitude when the pre-strain increases [13], it seems that predominantly the BH magnitude evolves similarly to the pre-strain, e.g., it increases when the pre-strain increases [7,10,16]. The dependence of the proposed model to the pre-strain magnitude follows this trend, as given by Equation (3). Figure 11 compares the magnitude of the bake hardening effect (BH) calculated from experimental and numerical data. It can be seen that from the experimental data, BH does not evolve significantly over the strain range, though it tends slightly to decrease when the pre-strain increases. This slight discrepancy may come from the fact that material parameters are identified by an inverse procedure over all the tests, and optimisation leads to a compromise, whereas BH magnitude is evaluated only upon the highest stress reached upon reloading.

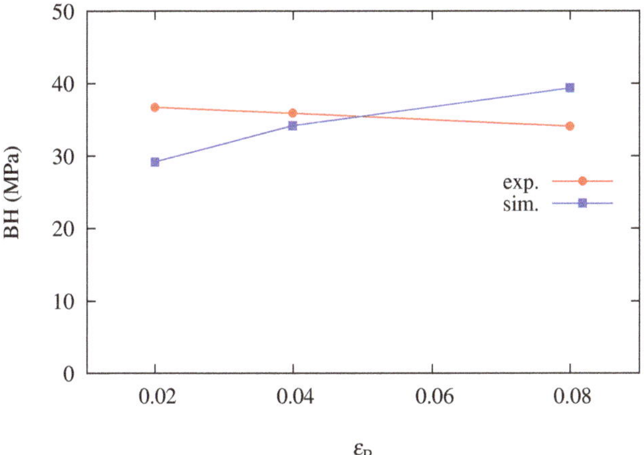

Figure 11. Comparison of the BH magnitude calculated from experimental and numerical data.

These steps are used out of identification purposes. Therefore, a next step is the validation of the model, using a different experimental database.

4.2. Numerical Simulation of the Dent Test

The aim is now to develop a numerical model for the multi-step process used in this study with the software Abaqus. Though the material is assumed isotropic, the initial circular blank (radius 135 mm) is modelled with one fourth of the geometry. Symmetry conditions are applied on the two straight edges. The tools are defined as analytical surfaces and are considered rigid. Three-node triangular shell elements are used to mesh the blank (total number of 9271). The mesh size is of the order of 4 mm on the outer diameter and in the area that remains flat under the die. It decreases down to 1.5 mm and is constant in the circular ring for radii in between 40 mm and 100 mm. Then, it further decreases down to 0.6 mm close to the pole. Numerical simulations of both the bulge and the dent tests were performed using axisymmetric elements, 4-node (regular mesh) and 3-node (random mesh) shell elements. The load-displacement curve during dent test was not sensitive to the type of element. Therefore, 3-node shell element, with a random mesh, is chosen in this study. The mesh sensitivity is analysed with regard to both the bulge and the dent tests, to obtain a solution rather independent of the mesh size.

For the hydraulic bulging, as no sliding of the blank between the die and the blank-holder is evidenced experimentally, the blank is clamped on the perimeter located at the outer diameter of the blank; this clamping replaces the blank-holder force applied experimentally and the blank-holder is therefore not modelled. Only the die is introduced in the model, to give the right shape to the blank related to the radius of curvature. The fluid pressure is applied on a gauge area of diameter 100 mm of the blank surface opposite to the die. The die is removed in the dent step whereas the punch is activated in the dent step.

The calculation is subdivided into 4 steps: (i) bulging under the applied pressure, (ii) unloading of the blank by removing the applied pressure and the clamping, (iii) dent at the pole by a vertical movement of the hemispherical punch, of magnitude 2.4 mm as in the experiments, the blank being only pinned along the vertical direction on the intermediate radius and finally (iv) removal of the punch and final springback of the blank.

Concerning the mechanical model, the parameter f is set equal to zero for the 2 first steps. Then, it remains equal to zero when no thermal treatment is performed or is switched to 1, in case of occurrence of a thermal treatment.

4.3. Sensitivity of Dent Resistance to BH Effect

Concerning the numerical simulation, Figure 12 shows the numerical shape for the two pre-strain after bulging and after the dent test. Isovalues of the vertical displacement, i.e., normal to the sheet plane, are displayed. For a pre-strain $\epsilon_{xx} = 0.02$, the maximum height after removing the internal pressure, corresponding to the shape at the end of the bulge test, a value of 10.4 mm is calculated (respectively 22 mm for a pre-strain of $\epsilon_{xx} = 0.07$). At the end of the dent test, when the punch is removed, the maximum height decreases, by 2.6 mm (respectively, 2.9 mm).

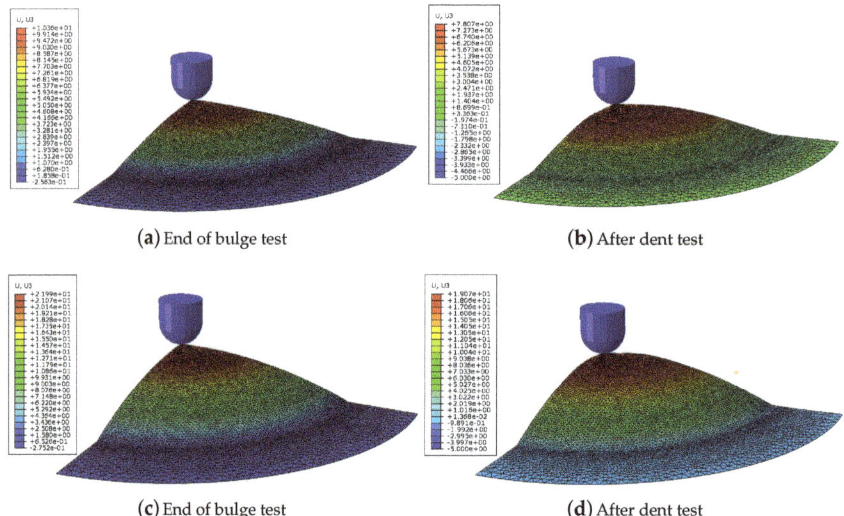

(**a**) End of bulge test

(**b**) After dent test

(**c**) End of bulge test

(**d**) After dent test

Figure 12. Numerical deformed geometries. Pre-strain intensity $\epsilon_{xx} = 0.02$ for (**a**,**b**), and $\epsilon_{xx} = 0.07$ for (**c**,**d**) respectively.

Figure 13 shows a comparison, for the two pre-strain values, of the load evolution during the dent test. The oscillations on the numerical load, clearly visible for the highest pre-strain, seems to come from a local structural instability. For the lowest pre-strain value, it can be seen that the predictions overestimate the loading part, whatever the state of the material, i.e., without or with bake hardening. The gap at the maximum displacement, with or without BH, is around 6.5%. However, the trend is different for the highest pre-strain, and the predicted values lie below the experimental ones, though they converge to exactly the same value at the maximum displacement. The maximum gap, calculated at the displacement of 1 mm, is around 6.2%. However, the unloading numerical slope is higher than the experimental value. This effect is also observed during loading-unloading tensile tests and a decrease of the unloading slope in the constitutive equations, via a dependency of the Young's modulus with the equivalent plastic strain, should be taken into account to represent this effect [25–27]. Globally, the model takes account of the bake hardening effect and further efforts should be given to the numerical simulation of the whole process.

As a whole, the numerical simulation of the dent test with bake hardening gives a close representation to the experimental data. As it is a two-step process, the quality of the end results depend on the prediction of the two stages, i.e., the bulge test and the dent test. The bake hardening magnitude is identified in tension, as the hardening characteristics and used in the numerical simulation of 3D tests, like bulge test, using von Mises yield criterion. Neglecting the anisotropy at this stage could lead to some difficulties to represent correctly the behaviour of the material under a biaxial stress state. Then, the stress state during the dent test is also fully 3D. However, Figure 13 highlights that the proposed approach captures rather well the influence of bake hardening on the dent resistance, all the more that the pre-strain is higher.

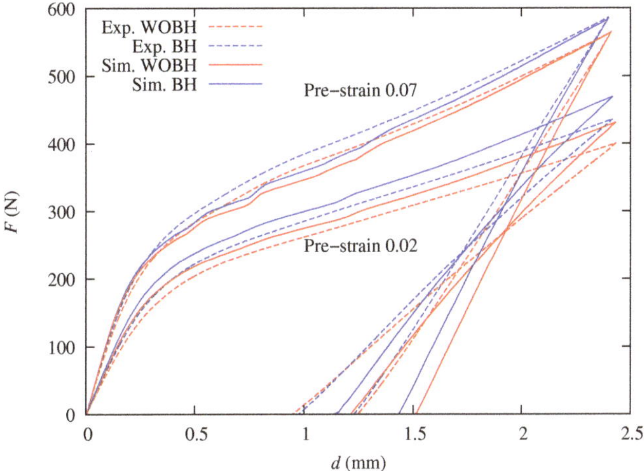

Figure 13. Evolution of load versus displacement during dent test: experiments (dashed lines) and numerical predictions (solid lines).

5. Conclusions

This study deals with the influence of bake hardening on the experimental and numerical prediction of dent resistance. In the first part, the bake hardening effect for E220BH steel, for a given time and temperature of the thermal treatment, is studied in uniaxial tension. In a second step, specimen of the same material are pre-strained in hydraulic bulging, to shape them into a curved geometry. Then, the static dent resistance is characterised with a dedicated device designed for this purpose. An original phenomenological model is proposed, which takes into account the influence of the pre-strain of the overstress upon reloading. The material parameters are then identified from tensile data using an inverse methodology and three pre-strain magnitudes are considered. A numerical model is developed, which uses as input the new constitutive equations implemented in a user subroutine. The bulge and dent tests are predicted numerically. Results for the dent test show that the bake hardening influences the load reached for an imposed displacement but that the residual displacement in nearly the same for virgin and bake-hardened specimen. The numerical model gives a good prediction of the loading part but tends to overestimate the residual displacement.

Author Contributions: Conceptualization, S.T. and P.Y.M.; Investigation, J.T. and A.J.; Methodology, S.T. and S.L.Z.; Supervision, S.T. and P.Y.M.; Validation, S.T., S.L.Z., J.T. and A.J.; Writing—original draft, S.T.; Writing—review and editing, S.T. and S.L.Z.

Funding: This research was funded by the French "Fond unique interministériel" as part of the collaborative project EMOA (2007–2011).

Acknowledgments: S. Zang would like to thank University of South Brittany for the financial support received for his stay in IRDL in 2014.

Conflicts of Interest: The authors declare no conflict of interest. The founding sponsors had no role in the design of the study; in the collection, analyses, or interpretation of data; in the writing of the manuscript, and in the decision to publish the results.

References

1. Asnafi, N. On strength, stiffness and dent resistance of car body panels. *J. Mater. Process. Technol.* **1995**, *49*, 13–31. [CrossRef]
2. Ekstrand, G.; Asnafi, N. On testing of the stiffness and the dent resistance of autobody panels. *Mater. Des.* **1998**, *19*, 145–156. [CrossRef]
3. Holmberg, S.; Thilderkvist, P. Influence of material properties and stamping conditions on the stiffness and static dent resistance of automotive panels. *Mater. Des.* **2002**, *23*, 681–691. [CrossRef]
4. Holmberg, S.; Nejabat, B. Numerical assessment of stiffness and dent properties of automotive exterior panels. *Mater. Des.* **2004**, *25*, 361–368. [CrossRef]
5. Shen, H.; Li, S.; Chen, Q. Numerical analysis of panels' dent resistance considering the Bauschinger effect. *Mater. Des.* **2010**, *31*, 870–876. [CrossRef]
6. Lee, J.Y.; Lee, M.G.; Barlat, F.; Chung, K.H.; Kim, D.J. Effect of nonlinear multi-axial elasticity and anisotropic plasticity on quasi-static dent properties of automotive steel sheets. *Int. J. Solids Struct.* **2016**, *87*, 254–266.
7. Kvackaj, T.; Mamuzic, I. Development of bake hardening effect by plastic deformation and annealing conditions. *Metallurgija* **2006**, *45*, 51–55.
8. De, A.K.; Vandeputte, S.; De Cooman, B.C. Static strain aging behavior of ultra low carbon bake hardening steel. *Metall. Mater. Trans. A* **1999**, *41*, 831–837. [CrossRef]
9. Kuang, C.F.; Li, J.; Zhang, S.G.; Wang, J.; Liu, H.F.; Volinsky, A.A. Effects of quenching and tempering on the microstructure and bake hardening behavior of ferrite and dual phase steels. *Mater. Sci. Eng. A* **2014**, *613*, 178–183. [CrossRef]
10. Ormsuptave, N.; Uthaisangsuk, V. Modeling of bake-hardening effect for fine grain bainite-aided dual phase steel. *Mater. Des.* **2017**, *118*, 314–329. [CrossRef]
11. Ramazani, A.; Bruehl, S.; Gerber, T.; Bleck, W.; Prahl, U. Quantification of bake hardening effect in DP600 and TRIP700 steels. *Mater. Des.* **2014**, *57*, 479–486. [CrossRef]
12. Sha, M.H.; Shi, G.D.; Wang, Y.; Qiao, J. Paint-bake response of AZ80 and AZ31 Mg alloys. *Trans. Nonferrous Met. Soc. China* **2010**, *20*, 571–575. [CrossRef]
13. Berbenni, S.; Favier, V.; Lemoine, X.; Berveiller, M. A micromechanical approach to model the bake hardening effect for low carbon steels. *Scr. Mater.* **2004**, *51*, 303–308. [CrossRef]
14. Vasilyev, A.A.; Lee, H.C.; Kuzmin, N.L. Nature of strain aging stages in bake hardening steel for automotive application. *Mater. Sci. Eng. A* **2008**, *485*, 282–289. [CrossRef]
15. Woo, C.J. Effect of Prestrain on Aging and Bake Hardening of Cold-Rolled, Continuously Annealed Steel Sheets. *Metall. Mater. Trans. A* **1998**, *29*, 463–467.
16. Kilic, S.; Ozturk, F.; Sigirtmac, T.; Tekin, G. Effects of Pre-strain and Temperature on Bake Hardening of TWIP900cr Steel. *Int. J. Iron Steel Res.* **2015**, *22*, 361–365. [CrossRef]
17. Ballarin, V.; Soler, M.; Perlade, A.; Lemoine, X.; Forest, S. Mechanisms and Modeling of Bake-Hardening Steels: Part I. *Metall. Mater. Trans. A* **2009**, *40*, 1367–1374. [CrossRef]
18. Ballarin, V.; Perlade, A.; Lemoine, X.; Bouaziz, O.; Forest, S. Mechanisms and Modeling of Bake-Hardening Steels: Part II. Complex Loading Paths. *Metall. Mater. Trans. A* **2009**, *40*, 1375–1382. [CrossRef]
19. Dehghani, K.; Shafiei, A. Predicting the bake hardenability of steels using neural network modeling. *Mater. Lett.* **2008**, *62*, 173–178.
20. Dehghani, K.; Nekahi, A. Artificial neural network to predict the effect of thermomechanical treatments on bake hardenability of low carbon steels. *Mater. Des.* **2010**, *31*, 2224–2229.
21. Nekahi, A.; Dehghani, K. Modeling the thermomechanical effects on baking behavior of low carbon steels using response surface methodology. *Mater. Des.* **2010**, *31*, 3845–3851. [CrossRef]
22. Thuillier, S.; Manach, P.Y. Comparison of the work-hardening of metallic sheets using tensile and shear strain paths. *Int. J. Plast.* **2009**, *25*, 733–751. [CrossRef]
23. Zang, S.L.; Thuillier, S.; Le Port, A.; Manach, P.Y. Prediction of anisotropy and hardening for metallic sheets in tension, simple shear and biaxial tension. *Int. J. Mech. Sci.* **2011**, *53*, 338–347. [CrossRef]

24. Zang, S.; Lee, M.G.; Sun, L.; Kim, J.H. Measurement of the bauschinger behavior of sheet metals by three-point bending springback test with pre-strained strips. *Int. J. Plast.* **2014**, *59*, 84–107. [CrossRef]
25. Morestin, F.; Boivin, M. On the necessity of taking into account the variation in the Young modulus with plastic strain in elastic-plastic software. *Nucl. Eng. Des.* **1996**, *162*, 107–116. [CrossRef]
26. Jung, J.; Jun, S.; Lee, H.S.; Kim, B.M.; Lee, M.G.; Kim, J.K. Anisotropic Hardening Behaviour and Springback of Advanced High-Strength Steels. *Metals* **2017**, *7*, 480. [CrossRef]
27. Seo, K.Y.; Kim, J.H.; Lee, H.S.; Kim, J.H.; Kim, B.M. Effect of Constitutive Equations on Springback Prediction Accuracy in the TRIP1180 Cold Stamping. *Metals* **2018**, *8*, 18. [CrossRef]

 © 2018 by the authors. Licensee MDPI, Basel, Switzerland. This article is an open access article distributed under the terms and conditions of the Creative Commons Attribution (CC BY) license (http://creativecommons.org/licenses/by/4.0/).

 *metals*

Article

Numerical Study on the Formability of Metallic Bipolar Plates for Proton Exchange Membrane (PEM) Fuel Cells

Diogo M. Neto [1,*], Marta C. Oliveira [1], José L. Alves [2] and Luís F. Menezes [1]

[1] CEMMPRE, Department of Mechanical Engineering, University of Coimbra, Rua Luís Reis Santos, Pinhal de Marrocos, 3030-788 Coimbra, Portugal

[2] CMEMS, Microelectromechanical Systems Research Unit, University of Minho, Campus de Azurém, 4800-058 Guimarães, Portugal

* Correspondence: diogo.neto@dem.uc.pt; Tel.: +351-239790700

Received: 10 July 2019; Accepted: 21 July 2019; Published: 23 July 2019

Abstract: Thin stamped bipolar plates (BPPs) are viewed as promising alternatives to traditional graphite BPPs in proton exchange membrane fuel cells. Metallic BPPs provide good thermal/electrical conductivity and exhibit high mechanical strength, to support the loads within the stack. However, BPPs manufactured by stamping processes are prone to defects. In this study, the effect of the tool's geometry on the thin sheet formability is investigated through finite element simulation. Despite the broad variety of flow field designs, most of BPPs comprise two representative zones. Hence, in order to reduce the computational cost, the finite element analysis is restricted to these two zones, where the deformation induced by the stamping tools is investigated. The channel/rib width, the punch/die fillet radii, and the channel depth are the parameters studied. The analysis is conducted for a stainless steel SS304 with a thickness of 0.15 mm. The results show that the maximum value of thinning occurs always in the U-bend channel section, specifically in the fillet radius of the die closest to the axis of revolution.

Keywords: numerical simulation; stamping; formability; metallic bipolar plate; fuel cells

---

1. Introduction

In the last years, fuel cell technology has received increasing attention due to the growing concerns about the depletion of fossil fuels and climate changes [1]. The proton exchange membrane (PEM) fuel cells emerged as one promising candidate to replace internal combustion engines in the automotive industry, producing electricity from the electrochemical reaction between hydrogen and oxygen [2]. They are characterised by: (i) low operation temperatures (<100 °C); (ii) quick start-up; (iii) high power density; (iv) high efficiency; and (v) low greenhouse gas emissions [3]. The main drawbacks of the fuel cells are the high manufacturing cost and the low durability, which prevent their widespread commercialization [4]. For transportation applications, the 2015 US Department of Energy (DOE) targets for the fuel cell cost and lifetime are $30/kW and 5000 h, respectively, which are in line with the automotive internal combustion engine systems [5]. The key components of a PEM fuel cell are the bipolar plates (BPPs) and the membrane electrode assembly (MEA). The last comprises the PEM, the gas diffusion layer (GDL), and a catalyst layer, as presented schematically in Figure 1.

The BPPs are key elements of a PEM fuel cell, comprising about 60–80% of the stack weight and up to 30–50% of the stack manufacturing cost [6]. They are multifunctional components responsible for: (i) supplying a uniform distribution of the reactant gases (H_2 and O_2) over the electrodes via flow channels; (ii) removing the heat and reaction products (water) from the cell assembly; (iii) connecting electrically the cathode of one cell to the anode of the adjacent cell (that is why they are called BPPs);

and (iv) providing structural support for the thin and mechanically weak MEA. Therefore, an ideal material for BPPs should comprise the following properties: high electrical conductivity, low gas permeability, high corrosion resistance in acidic environments, high mechanical strength, and low cost [7]. The earlier BPPs were fabricated from high-density graphite, which is chemically stable (excellent corrosion resistance) and possesses high thermal/electrical conductivity [8]. Nevertheless, the graphite plates are brittle, exhibit low mechanical strength, and present high manufacturing cost, resulting from the need to mill the flow field channels [3]. Accordingly, several studies have been performed in order to develop more suitable and cost-effective materials for the fabrication of BPPs, such as metals and composites [9]. Further, the adoption of metallic materials allows for the application of other manufacturing techniques, including stamping [10], hydroforming [11], rubber pad forming [12], micro-electrical discharge machining (μEDM) [13], electrochemical micro-machining [14], and vacuum die casting [15].

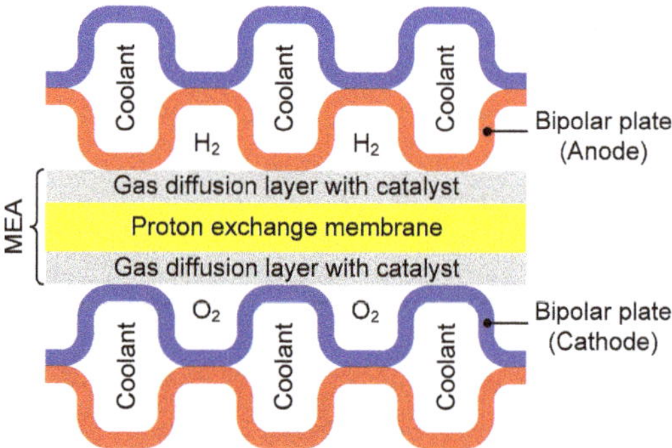

Figure 1. Schematic representation of a proton exchange membrane fuel cell composed by the bipolar plates (BPPs) and the membrane electrode assembly (not to scale).

Among the metals candidates for BPPs, stainless steels, Ni-based alloys, Ti-based alloys, and Al-based alloys have been considered for PEM fuel cells [6]. The stainless steel is presently a consensual material for BPPs, due to its relatively high strength, high chemical stability, high electrical conductivity, low gas permeability, and much lower manufacturing cost in comparison with graphite [7]. The main drawbacks of metals are the high density and the weak corrosion resistance [16]. Regarding the high density, it can be alleviated by using ultra-thin sheets (51 μm of thickness) [17], which requires the adopting of different forming methods to produce the BPPs [18]. The corrosion of the BPPs leads to a release of metal ions, which contaminate the PEM [19,20]. In addition, a passive film of oxides is generated on the BPP surface during the fuel cell operation, which increases the interfacial contact resistance between the BPPs and the GDL [21]. Both previously mentioned conditions significantly reduce the stack performance and lifetime [22]. Thus, several studies have been carried out in the last years to improve the corrosion resistance (eliminate the passive film) using coatings [23]. Some corrosion-resistant-treated metal BPPs show a 12% saving in hydrogen consumption and higher efficiency in relation to graphite [24].

The metallic BPPs manufactured by forming processes received raised interest in the last years, namely the stainless steels, since they are promising candidates to replace the graphite at low cost for massive production of fuel cells [7]. However, different forming defects can occur in the forming of ultra-thin BPPs, such as springback, wrinkles, thinning, and fracture [10]. Accordingly, several strategies have been proposed to improve the formability and reduce the forming defects. In order to obtain deeper channels on the BPPs, dynamic loads (sine and square waves) are applied in the stamping process of the 0.1 mm-thick austenitic stainless steel SS304 [25]. The forming depth of the BPP increases up to 10%, in comparison with the static load, when the number of cycles is over five. On the other hand, Park et al. [26] explored the use of solution heat treatment to improve the formability of two stainless steels (SS304 and SS316). Since the ductility increases after applying the heat treatment, the channel depth achieved in the stamping process doubles for the heat-treated sheet. Recently, Bong et al. [27] proposed the adoption of a multi-stage forming approach to improve the formability of ultra-thin ferritic stainless steel.

Since the flow field configuration of BPPs is usually defined by a channel pattern (see Section 2), most of the numerical studies of the stamping process reported in the literature consider plane strain conditions and they are restricted to the analysis of a single channel. Nevertheless, the accurate modelling of the forming conditions requires the study of different regions, which comprise distinct strain paths. Therefore, the purpose of this study is to assess the formability of BPPs manufactured by stamping, considering two representative zones of the BPPs (straight and the U-bend channel sections), which are analysed in detail using finite element simulation. The developed numerical model aims to provide a reference for optimizing stamping process of BPPs. Section 2 contains a brief presentation of the flow field configurations currently used in PEM fuel cells, in order to highlight their common geometrical features. The proposed finite element model adopted in the analysis of the stamping process is presented in Section 3, namely the model adopted to describe the mechanical behaviour of the stainless steel and the process conditions. Section 4 comprises the results for the straight channel section, where the effect of the cross-section geometry on the deformation is analysed considering plane strain conditions. The results regarding the U-bend channel section are presented in Section 5, highlighting the influence of tools geometry on the formability of stamped BPPs, as well as the importance of the adopted boundary conditions. The main conclusions of this study are discussed in Section 6.

2. Flow Field Configurations

The main functions of the flow field in a BPP is to distribute evenly the reactant gases (H_2 and O_2) over the respective GDL and remove the water produced during the reaction. Since the performance of the PEM fuel cell is strongly affected by the flow field design, several numerical models have been developed to analyse the coupled transport process and electrochemical reaction in PEM fuel cells [28]. Typically, the flow field configurations can be divided into five types: parallel, interdigitated, pin-type, spiral, and serpentine, which are schematically illustrated in Figure 2. The serpentine flow field, either containing single (Figure 2e) or multiple channels (Figure 2f), is the most commonly used design in commercial fuel cells [29]. The effect of the gas flow fields design on the fuel cell performance was investigated experimentally by Dhahad et al. [30]. On the other hand, the modelling of an optimum flow field design was presented by Kahraman and Orhan [31], including a parametric study with respect to different design and performance parameters in a flow field plate. In addition to the conventional flow field patterns, some nature-inspired flow field designs have been studied recently [32]. The results show that the bio-inspired interdigitated designs improve the fuel cell performance by about 20–25%, in comparison with the conventional designs [33]. However, the manufacturing complexity leads to significant costs because typically these BPPs are made from graphite composite and produced by a selective laser sintering process.

In addition to the gas flow field configuration, several studies have been carried out to optimise the channel geometries (dimensions and shape) in order to achieve better fuel cell's performance [34]. Since originally the channels of the BPPs were milled on graphite plates, most of the studies are focused on the rectangular channel geometry [35]. The influence of the channel cross-section aspect ratio (height/width) on the performance of a PEM fuel cell was numerically investigated by Manso et al. [36]. They concluded that fuel cell models with high channel cross-section aspect ratios present better performance, due to the increasing mass transport. On the other hand, the effect of the channel/rib width ratio on PEM fuel cell performance was assessed by Shimpalee and Van Zee [34], using computational fluid dynamics simulation. The results show that for automotive applications the performance is higher for a wider channel with narrower rib spacing, indicating that the heat removal from the MEA is less important than the water management.

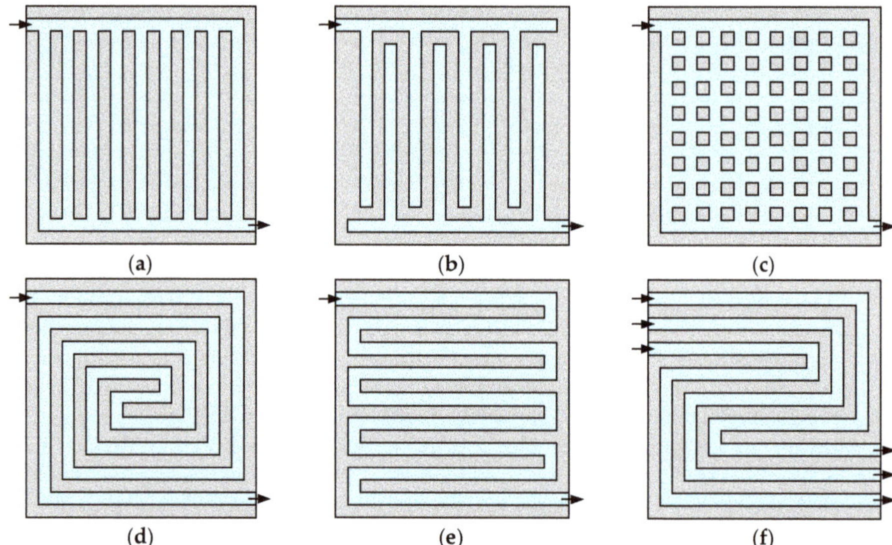

Figure 2. Scheme of typical flow field configurations: (**a**) straight parallel; (**b**) interdigitated; (**c**) pin-type; (**d**) spiral; (**e**) single-channel serpentine; (**f**) multiple-channel (triple) serpentine.

The adoption of graphite BPPs or thick metallic BPPs fabricated by milling allows obtaining independent flow field configurations on each side of the plate. Nevertheless, for thin metallic BPPs manufactured by a stamping process, the flow field configuration of one side of the plate is always determined by the configuration of the opposite side, because both sides are formed simultaneously. Indeed, it is difficult to obtain two continuous channels on both sides of a single BPP to provide reactant and cooling flow field, respectively [37]. Thus, typically the reactant flow channel is continuous, with exception to the interdigitated channel (Figure 2b). Moreover, the cross-section of the flow channels in BPPs manufactured by a stamping process is shaped like a trapezoidal with fillets. Hence, Xu et al. [38] recently established a model to calculate the influence of the tapered channel geometry on the pipe resistance and flow distribution.

3. Finite Element Model

In order to analyse the stamping process used to manufacture metallic BPPs, the influence of the tools geometry on the formability was assessed using finite element simulation. The numerical simulations of the stamping process were carried out with the in-house static implicit finite element code DD3IMP [39], specifically developed to simulate sheet metal forming processes [40,41]. Its main characteristic is the use of a fully implicit algorithm of Newton–Raphson type to solve, within a single iterative loop, the non-linearities related with the frictional contact problem and the elastoplastic behaviour of the deformable body. All simulations were performed on a computer machine equipped with an Intel® Core™ i7-4770K Quad-Core processor (3.5 GHz frequency) and the Windows® 10 (64-bit platform) operating system.

3.1. Material Properties

The metal sheet considered in this study is of austenitic stainless steel SS304 with a thickness of 0.15 mm, which is commonly used in the sheet metal forming of BPPs [25]. The mechanical behaviour of this ultra-thin sheet of stainless steel was experimentally evaluated by Pham et al. [42]. The stress–strain curve recorded in the experimental uniaxial tensile test (initial strain rate of 10^{-3} s^{-1}) is used in the present study to define the material parameters in the numerical model. Hence, the elastic behaviour of the stainless steel is considered isotropic and constant, which is described by the Hooke's law with an Young's modulus of 206.2 GPa and a Poisson ratio of 0.30 [42]. Regarding the plastic response, the work hardening behaviour is described by the phenomenological Swift hardening law, where the flow stress, Y, is given by:

$$Y = K(\varepsilon_0 + \bar{\varepsilon}^{\mathrm{P}})^n \quad \text{with} \quad \varepsilon_0 = \left(\frac{Y_0}{K}\right)^{1/n}, \tag{1}$$

where $\bar{\varepsilon}^{\mathrm{P}}$ denotes the equivalent plastic strain, while K, ε_0, and n are the material parameters.

The parameters of the Swift hardening law were identified using the stress–strain curve of the experimental tensile test, presented in Figure 3. The identification procedure is based on the minimization of a cost function, which evaluates the difference between numerical and experimental stress values, using least squares estimation. The obtained material parameters for the isotropic hardening law are listed in Table 1, which are identical to the ones adopted in the numerical model used by Hu et al. [10]. The comparison between the experimental and numerical stress–strain curves is presented in Figure 3, highlighting the accurate description of the work hardening behaviour by the Swift law.

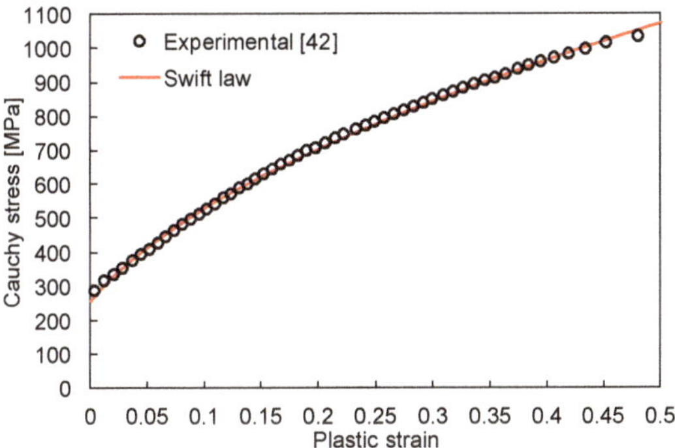

Figure 3. Comparison between experimental and numerical stress–strain curves from the uniaxial tensile test in the rolling direction (stainless steel SS304).

Table 1. Material parameters used in the isotropic Swift hardening law to describe the SS304 stainless steel.

Y_0 (MPa)	K (MPa)	n
255.02	1481.84	0.508

The plastic anisotropy coefficients of the stainless steel SS304 were experimentally calculated by Pham et al. [43] in the rolling and transverse directions. They obtained similar values for both directions, close to 0.9, which are in agreement with the values presented by Raj [44] for the SS304 with a thickness of 0.5 mm. However, in the present study, the behaviour of the metal sheet is assumed isotropic and modelled by the von Mises yield function, in order to avoid the influence of the orientation of the sheet. In fact, the experimental study conducted by Park et al. [26] shows that the influence of the sheet orientation (channel parallel or perpendicular to the rolling direction of the sheet) on the maximum channel depth is insignificant.

3.2. Stamping Process

The high productivity and the low cost of mass production are the main advantages of the stamping process for BPPs manufacturing. The desired geometry of the ultra-thin BPP (flow field and channel geometry) is obtained by plastic deformation induced by the forming tools operation (displacement control). Typically, the forming tools are assumed rigid in the numerical simulation, while the deformation of the metallic sheet is described by an elastoplastic material model (strain rate-insensitive), as described in the previous section. Hence, in the present study, the tools surface is discretized with Nagata patches [45,46]. The blank sheet is discretized with linear hexahedral finite elements using a selective reduced integration technique [47] to avoid volumetric locking. In order to accurately capture the through-thickness gradients (stress and strain), all simulations use six finite elements in the thickness direction. Despite the large computation time associated with the solid finite elements, they are required for accurate predictions when the ratio between the tool fillet radius and the sheet thickness is lower than five [48]. The friction between the blank and the forming tools is modelled through the classical Coulomb's law. The value of the constant friction coefficient adopted in the finite element model is selected as $\mu = 0.1$, which is within the range of values commonly used in the numerical simulation of BPPs stamping [10,27].

Due to the large geometric complexity of the ultra-thin BPPs (see the example shown in Figure 4 [49]), the finite element simulation of the stamping process is commonly carried out under plane strain conditions, i.e., simplified two-dimensional (2D) finite element models are used to study locally a few numbers of channels [50]. In fact, the number of finite elements required to describe accurately the entire BPP is very high, since the minimum fillet radius of the channel is considerably smaller in comparison with the BPP size. Therefore, in the present work, specific areas of the BPP are studied by finite element simulation, namely the straight and the U-bend channel sections indicated in Figure 4. Despite the variety of flow field configurations (see Figure 2), these two channel sections are usually present in the BPPs, allowing evaluating different deformation mechanisms existent in the BPPs stamping process.

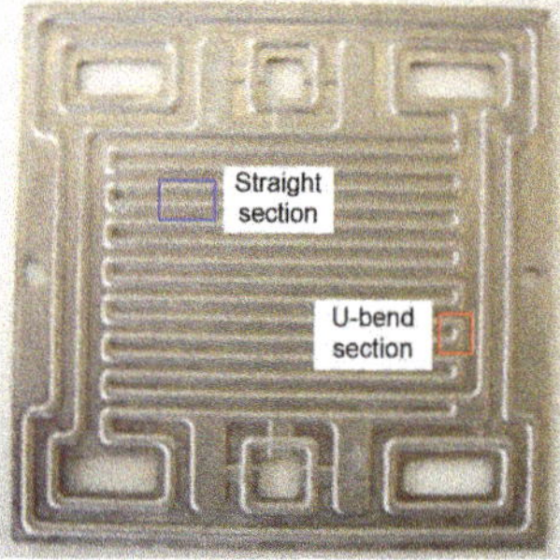

Figure 4. Example of a bipolar plate manufactured by forming, indicating both the straight and the U-bend channel sections: a BPP with an interdigitated flow field for reactant gases and a serpentine flow field for the coolant fluid on the opposite side [49].

4. Straight Channel Section

Since the main region of a flow channel is straight (see Figure 4), the effect of the cross-section geometry on the deformation is numerically analysed using plane strain conditions. The geometry of the trapezoidal channel obtained by sheet metal forming (see Figure 5a) is usually defined by five key parameters: (i) channel depth (h); (ii) channel width (w_1); (iii) rib width (s); (iv) draft angle (θ); and (v) fillet radii (r and R). These dimensions are dictated mainly by the geometry of the forming tools (punch and die) and the stamping process conditions. Hence, the influence of the tool dimensions (see Figure 5a) on the formability of stamped BPPs was assessed in the present study through the thinning and the equivalent plastic strain predicted by numerical simulation. The fillets in the trapezoidal shape of the channel cross-section are defined by the fillet radius of the punch and die (see Figure 5b). For a predefined channel height, the angle between the walls of the channel (draft angle θ) is directly related to the ratio between the channel width w_1 and the dimension w_2, as shown in Figure 5. Since the die position is fixed in the numerical model, the channel depth is simply dictated by the prescribed vertical displacement of the punch.

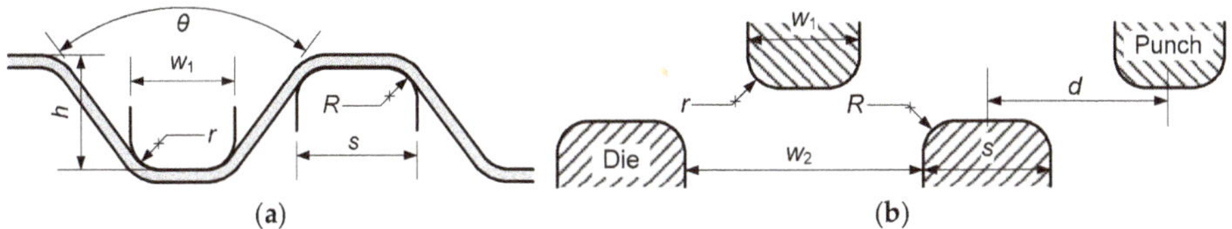

Figure 5. Stamping process used to manufacture BPPs: (**a**) cross-section of the obtained trapezoidal channel with key dimensions; (**b**) cross-section of the tools (punch and die including the main dimensions).

Regarding the stainless steel BPPs manufactured by a stamping process, the review carried out by Peng et al. [7] summarizes the range of values currently used for each key parameter that defines the flow channel (cross-section). The initial thickness of the stainless steel sheets used in the forming process of BPPs ranges between 0.051 mm [18] and 0.15 mm [10]. Based on the previous studies, the values of channel width range from 0.75 mm [18] to 1.5 mm [10]. On the other hand, since the channel cross-section aspect ratio is usually lower than 0.5, the channel depth ranges between 0.20 mm [18] and 0.60 mm [51]. In fact, the channel depth reached by forming (before fracture) is always inferior to 5 times the initial thickness. In order to simplify the sensitivity analysis performed in this study, some simplifications are introduced in the geometry of the forming tool. Hence, the rib width is selected to be equal to the channel width ($w_1 = s$) and the fillet radius of the punch and die are assumed identical in the 2D plane strain finite element model, i.e., $r = R$.

4.1. Multiple Channels

Although the numerical simulation of the forming process is usually restricted to a single channel [50], BPPs are composed by a set of parallel channels (see Figure 4). The proximity of the channel to the blank edges can change the local process conditions in terms of restraining force. The border effect on the channel geometry was experimentally studied by Mahabunphachai et al. [18] for stamped BPPs. They show that for a BPP with 26 parallel channels, the height of each one varies according to its position, presenting variations of about 25% (higher values of channel height closer to the blank edges). Therefore, in order to quantify the influence of the adopted boundary conditions on the numerical results, a stamped BPP composed by 25 parallel channels is analysed under plane strain conditions. Due to symmetry conditions, only half BPP width is simulated (see Figure 6), which is discretized with $3000 \times 6 = 18,000$ finite elements. In order to account for the border effect, the rib width is extended (10 mm) and a blank-holder is added to allow clamping the blank during the forming, as shown in Figure 6a. The dimensions of the forming tools (punch and die) are presented in Table 2.

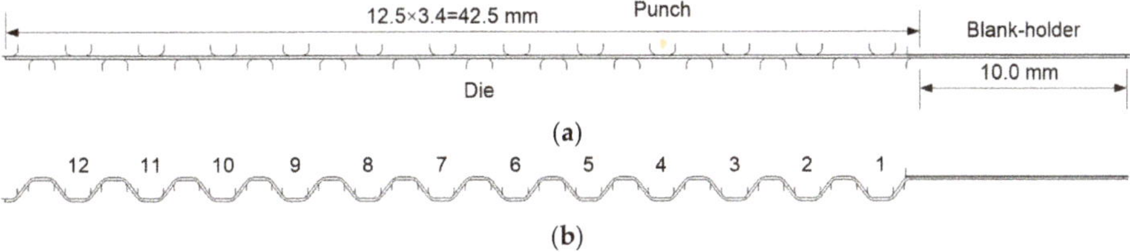

Figure 6. Finite element model (plane strain conditions and half width) of a stamped BPP composed by 25 parallel channels: (**a**) initial position; (**b**) final configuration.

Table 2. Reference values for the main dimensions of the forming tools (punch and die) adopted in the analysis of the straight channel section.

w_1	w_2	S	r	R	$d = (w_2 + s)/2$
1.2 mm	2.2 mm	1.2 mm	0.3 mm	0.3 mm	1.7 mm

The predicted final geometry of the BPP is presented in Figure 6b assuming a channel depth of $h = 1.0$ mm (punch displacement), highlighting the trapezoidal shape of the channel cross-section in stamped BPPs. In order to cover a wide range of experimental clamping conditions, applied during the stamping process, three distinct boundary conditions are adopted in the numerical model. They are defined by: (i) an unconstrained free edge of the flange with the blank-holder placed over the flange with a fixed gap (initial thickness of the blank); (ii) an unconstrained free edge of the flange and application of a constant force on the blank-holder (initial contact pressure of 10 MPa); (iii) a free edge of the flange constrained in the x-direction with the blank-holder placed over the flange with a fixed gap (initial thickness of the blank). They are denoted by an unconstrained free edge, a fixed free edge, and a clamped flange, respectively.

The predicted cross-section geometry of the two channels closest to the flange is presented in Figure 7, comparing the three boundary conditions applied to the flange. The shape of channel #1 (see Figure 6) is strongly affected by the boundary conditions adopted, as highlighted in Figure 7. Since the flange is completely free to slide ($\approx$2.29 mm length) when the free edge is unconstrained while the blank-holder presents a fixed gap, the bottom of channel #1 is curved, while the fillet radii of the trapezoidal shape are larger. This induces differences in the height of the channels and consequent dimensional errors of the metallic BPP, which creates variations in the GDL assembly pressure distribution [52]. On the other hand, when the free edge of the flange is fixed in the x-direction, the geometry of all channels is identical, as shown in Figure 7. The application of a constant blank-holder force on the flange allows for a controlled sliding of the flange ($\approx$1.16 mm length), providing a channel shape sited between the other two conditions.

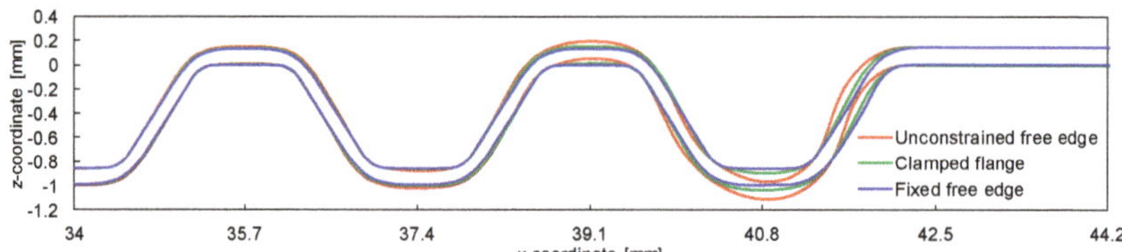

Figure 7. Cross-section of the BPP (2 channels closest to the flange) using three different boundary conditions in the flange region and considering plane strain conditions.

The predicted thickness distribution on the BPP along the x-direction is presented in Figure 8, comparing the three different boundary conditions applied on the flange. The effect of the boundary conditions on the final thickness is negligible in the central region of the BPP, i.e., the three finite element models provide identical results for the channels away from the boundary. On the other hand, the predicted thickness distribution of the channels adjacent to the flange is strongly influenced by the boundary conditions adopted, as shown in Figure 8. The thinning is lower in this region of the BPP due to the sliding of the flange, particularly when the free edge of the flange is unconstrained. In contrast, the minimum value of thickness arises in the central region of the BPP, specifically in the fillet radii of the flow channels.

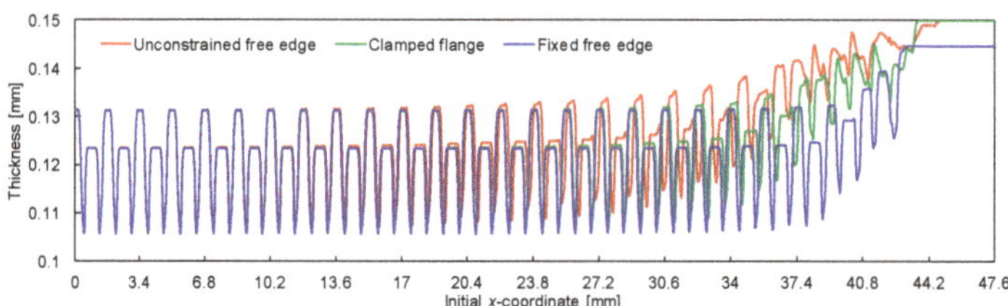

Figure 8. Thickness distribution on the BPP along the x-direction using three different boundary conditions in the flange region and considering plane strain conditions.

Since both the cross-section geometry of the flow channels (Figure 7) and the thickness distribution (Figure 8) present cyclic symmetry along the BPP width direction, particularly for channels away from the border, most of the studies are focused on a single channel [50]. Moreover, the maximum value of thinning occurs in the central region of the BPP, which dictates the formability analysis of the stamping process. Therefore, the analysis of a single channel is representative of the deformation occurring in the straight channel section of the BBPs. Hence, the sensitivity analysis carried out in the present study considers only a single channel in the finite element model, allowing reducing significantly the computational cost.

4.2. Single Channel

Due to symmetry conditions, only half-channel width is simulated under plane strain conditions, as shown in Figure 9. The displacement of the nodes located at the mid-width of the channel/rib is constrained in the x-direction, providing the cyclic symmetry conditions observed in the thickness distribution (Figure 8). Since the symmetry conditions are applied in both extremities of the channel, only the punch and the die are required in the numerical simulation (see Figure 9). Regardless of the tool dimensions tested during the sensitivity analysis, the half-distance between two consecutive channels $d = 1.7$ mm is fixed in all simulations (see Figure 5). The blank is discretized with $100 \times 6 = 600$ finite elements.

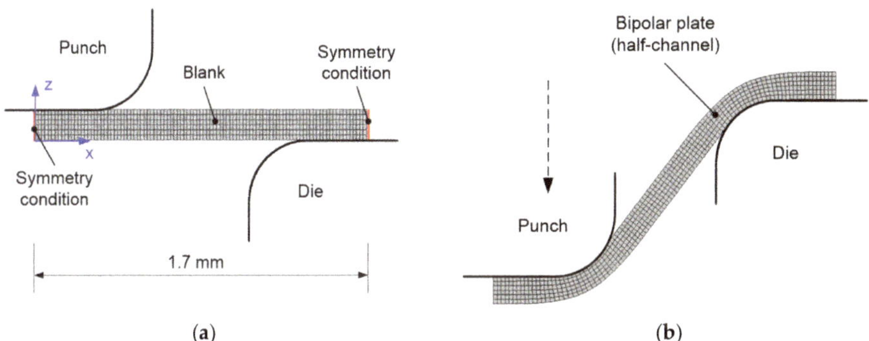

Figure 9. Finite element model of the stamping process (half-channel) assuming plane strain conditions: (**a**) initial configuration; (**b**) final configuration.

The reference values for the dimensions of the forming tools (punch and die) are presented in Table 2. The die is fixed, while the punch presents a prescribed displacement in the vertical direction, providing a maximum channel depth of $h = 1.0$ mm (see Figure 5b). In order to study the influence of the tools geometry on the formability, the channel/rib width, the punch/die radius, and the channel depth are selected as variable parameters in the present numerical model.

Effect of Tool Dimensions

The channel/rib width is dictated by the punch/die width (assumed identical: $w_1 = s$), while the fillet radius of the trapezoidal shape of the channel is defined by the fillet radius of the punch/die (assumed identical: $r = R$), as shown in Figure 5. The predicted geometry of the channel cross-section is presented in Figure 10, comparing three values of punch/die width, namely $w_1 = s = 1.0$ mm, $w_1 = s = 1.2$ mm, and $w_1 = s = 1.4$ mm, and three values of punch/die fillet radius, namely $r = R = 0.2$ mm, $r = R = 0.3$ mm, and $r = R = 0.4$ mm. The maximum value of plastic strain (fillet zones) increases when the channel/rib width rises and when the fillet radius of the punch/die decreases. Both conditions lead to a decrease of the draft angle, as shown in Figure 10. The maximum value of equivalent plastic strain increases by approximately 40% from the narrow to the broad channel and increases by about 50% from the largest to the lower fillet radius.

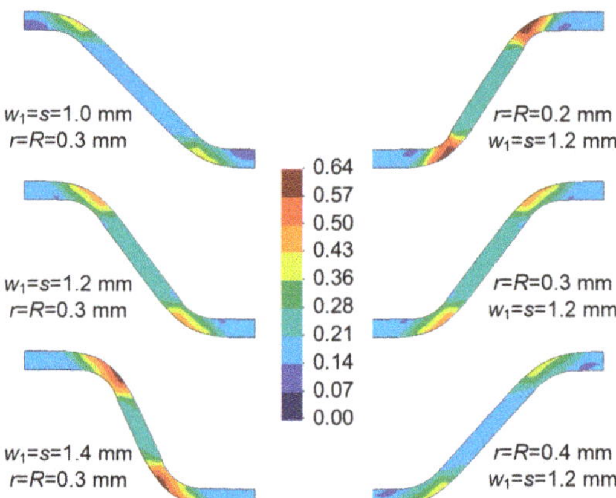

Figure 10. Equivalent plastic strain distribution plotted on the deformed configuration of the channel for three values of channel/rib width (left side) and for three values of punch/die fillet radius (right side) and $h = 1.0$ mm.

The predicted thickness distribution is presented in Figure 11 (half-channel), for the three different values of channel/rib width and the three values of punch/die fillet radius ($h = 1.0$ mm). The thickness is roughly constant in the wall, presenting a value slightly lower than the one measured in the rib/bottom of the channel. The minimum value of thickness occurs in the fillet zones of the flow channel, which is in agreement with the equivalent plastic strain distribution shown in Figure 10. Using the lower fillet radius, the minimum thickness is approximately 0.092 mm, while the adoption of the highest fillet radius leads to a minimum thickness of about 0.114 mm. Increasing the channel/rib width leads to a reduction of the minimum thickness, as shown in Figure 11. The comparison of the thickness distribution presented in Figure 11 with the thickness distribution in a BPP composed by 25 parallel channels (Figure 8) shows that the boundary conditions adopted in the single channel model allow representing accurately the process conditions.

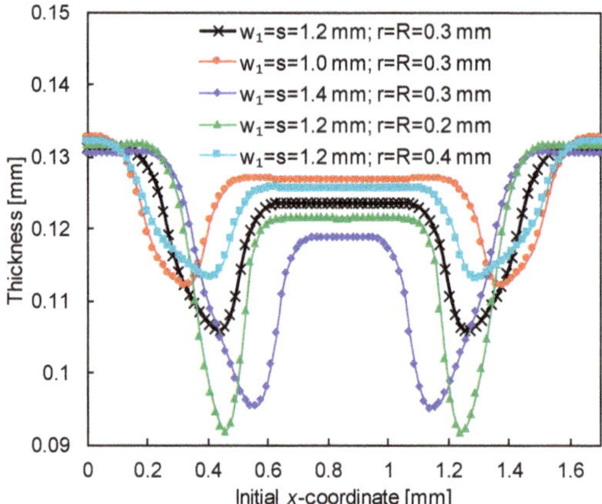

Figure 11. Thickness distribution in half-channel for three values of channel/rib width and three values of punch/die fillet radius ($h = 1.0$ mm).

The evolution of the maximum thinning in the channel is presented in Figure 12 for the three different values of channel/rib width and the three values of punch/die fillet radius. The increase of the maximum thinning is approximately linear for a punch displacement larger than 0.5 mm. However, considering either the lower fillet radius ($r = R = 0.2$ mm) or the broader channel ($w_1 = s = 1.4$ mm), the thinning rate increases gradually after 0.8 mm of punch displacement (see Figure 12). This behaviour can indicate the occurrence of necking in the fillet radius of the flow channel, which is in agreement with the larger thickness strain predicted in this region (see Figure 11). For this channel geometry and considering the depth of 1.0 mm, the predicted maximum thinning is nearly 40%. Therefore, the formability of stamped BPPs is strongly affected by both the channel/rib width and the punch/die fillet radius.

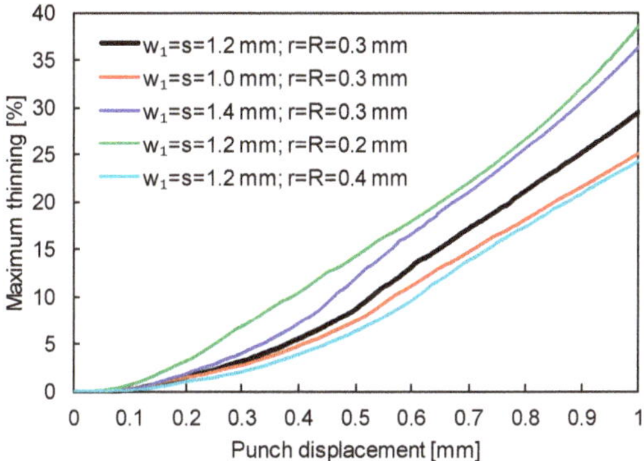

Figure 12. Evolution of the maximum thinning in the channel with the punch displacement for three values of channel/rib width and three values of punch/die fillet radius (plane strain conditions).

Since the draft angle of the flow channel is dictated by the channel width w_1, the dimension w_2, and the channel depth (see Figure 5a), its value decreases with the punch displacement rising. The evolution of the draft angle with the punch displacement is presented in Figure 13, for the three different values of channel/rib width and the three values of punch/die fillet radius. For all channel geometries analysed, the decrease of the draft angle is approximately linear up to a channel depth of 0.5 mm However, the draft angle is always smaller for wider channel/rib and for smaller fillet radii of the punch/die, as shown in Figures 10 and 13. In fact, considering a channel depth of 1.0 mm and $r = R = 0.3$ mm, the draft angles are about 51° and 91° in the broader and narrower channel, respectively.

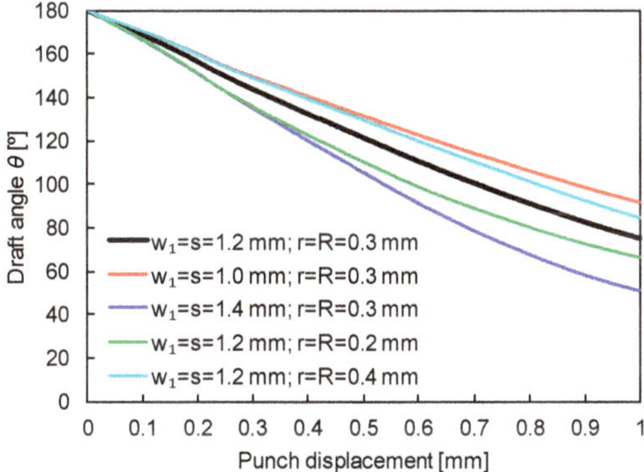

Figure 13. Draft angle evolution obtained for three values of channel/rib width and three values of punch/die fillet radius (plane strain conditions).

The punch force required for the stamping of a single channel, assuming plane strain conditions, is presented in Figure 14, for three different values of channel/rib width and three different values of punch/die fillet radius. Since the thinning is significantly higher in wider channels with smaller fillet radii (see Figure 12), the required punch force is also considerably higher. Indeed, the punch force required for forming the wider channel is at least 40% higher than the one necessary for forming the narrowest channel, as shown in Figure 14. Besides, the slowest increase of the punch force after 0.7 mm of punch displacement indicates the onset of necking in the fillet radius of the flow channel (see Figure 11).

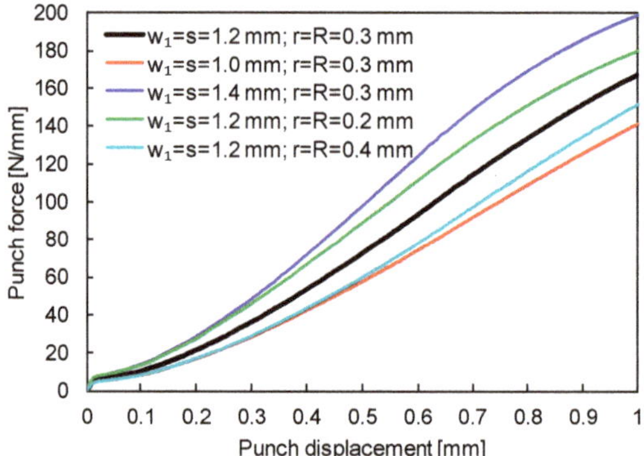

Figure 14. Punch force evolution obtained for three values of channel/rib width and three values of punch/die fillet radius (plane strain conditions).

The last analysis considers different values for the channel and rib widths, i.e., $w_1 \neq s$, but the summation of them is always 2.4 mm in order to keep $d = 1.7$ mm (see Table 2). Assuming $r = R = 0.3$ mm and $h = 1.0$ mm, the final geometry of the channel cross-section is presented in Figure 15, comparing three different configurations. The predicted equivalent plastic strain distribution is identical in all channel configurations, because the draft angle is the same. Indeed, the flat zone in the bottom of the channel is reduced while the flat zone in the rib top is enlarged in the same proportion, as shown in Figure 15. Accordingly, this leads only to a shift in the thickness distribution, preserving the maximum thinning value during the punch force evolution.

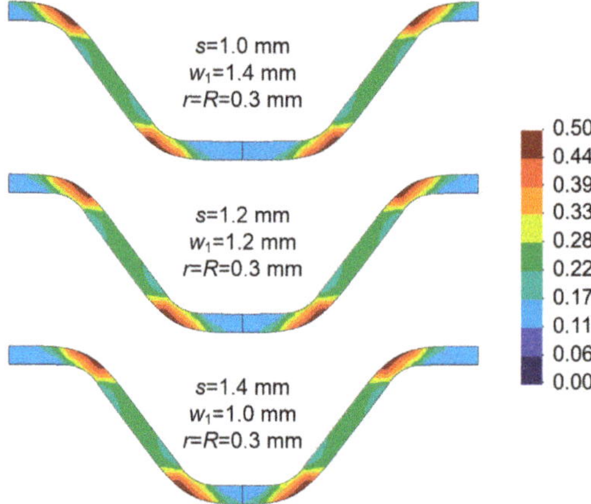

Figure 15. Equivalent plastic strain distribution plotted on the deformed configuration of the channel for three different values of channel width and rib width (plane strain conditions) and $h = 1.0$ mm.

5. U-bend Channel Section

In addition to the straight channel section, the BPP manufactured by forming is also composed by U-bend channel sections, as indicated in Figure 4. Indeed, the straight channel sections are connected between them by U-bend channel sections, leading to different flow field configurations (see Figure 2). Nevertheless, the modelling of the U-bend channel section requires the development of three-dimensional (3D) finite element models. This section contains the numerical analysis of this zone of the BPP, namely the effect of the tools geometry on the BPP deformed configuration.

5.1. Boundary Conditions

Besides the geometrical dimensions that describe the cross-section geometry of the channel, the U-bend section requires the definition of other dimensions. Anyway, revolving the cross-section geometry is the simplest way to create a U-bend channel section. Figure 16 presents the finite element model of the forming tools obtained by revolving the cross-section geometry (half-model), previously used as a reference in Section 4. The square blank (3.4 mm width) is discretized with $150 \times 150 \times 6 = 135{,}000$ finite elements. Besides, symmetry conditions are applied in each of the four blank edges, in order to take into account the effect of the neighbouring channels, as illustrated in Figure 16.

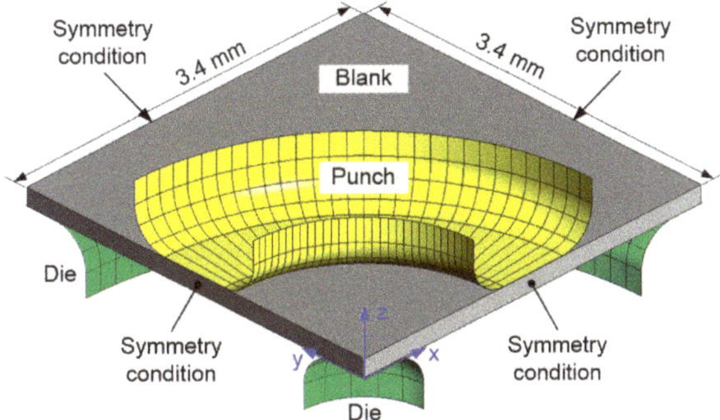

Figure 16. Finite element model adopted to simulate the stamping of the U-bend channel section (revolved tool geometry).

Considering the dimensions of the cross-section geometry listed in Table 2, the thickness distribution in the flow channel at $y = 0$ (see Figure 16) is presented in Figure 17. Figure 17 also shows the thickness distribution for a straight channel with the same geometrical dimensions, assuming plane strain conditions. Note that the 3D model considered has a total length of 3.4 mm, instead of the 1.7 mm (half) used in the previous section (see Figure 9). The thickness is substantially different from the one obtained with the finite element model previously presented in Section 4.2 (plane strain conditions). Since the thinning is considerably higher in the U-bend channel section than in the straight section, the channel depth (vertical displacement of the punch) is limited to 0.7 mm in this analysis. The thinning is significantly larger in the fillet radius of the channel close to the axis of revolution (near axisymmetric boundary conditions), as shown in Figure 17. On the other hand, the thicknesses predicted by both models are similar in the punch fillet radii. Besides, the computational time is higher than 10 h, when assuming 3D conditions, while when considering plane strain conditions, it is less than half-minute.

In order to assess the effect of the adopted boundary conditions on the deformation behaviour, the straight channel section is included in the revolved tool geometry. The finite element model of the forming tools is presented in Figure 18, using 5.0 mm as the length of the straight channel section. The blank is rectangular with dimensions $3.4 \times 8.4 \text{ mm}^2$ and discretized with $150 \times 167 \times 6 = 150{,}300$ finite elements. The symmetry conditions are applied on all edges of the blank (see Figure 18), as shown in the previous model.

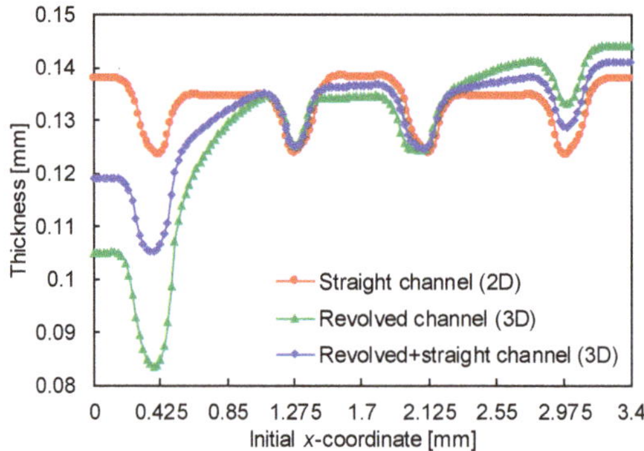

Figure 17. Thickness distribution in a single channel at $y = 0$ for the three different models. The parameters are $w_1 = s = 1.2$ mm, $r = R = 0.3$ mm, and $h = 0.7$ mm.

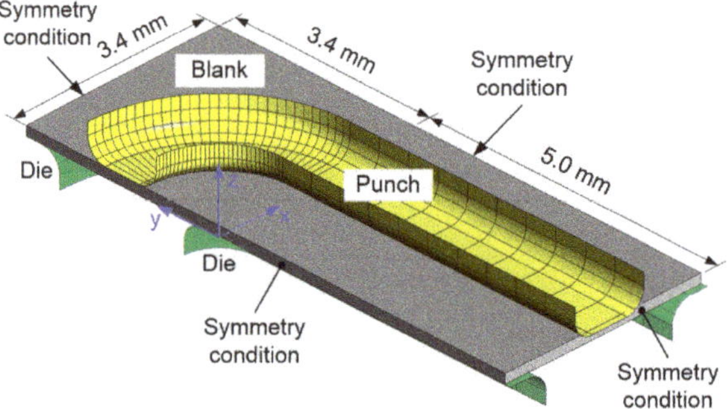

Figure 18. Finite element model adopted to simulate the stamping of the U-bend channel section (revolved tool geometry with a straight section).

Using the finite element model presented in Figure 18 (labelled revolved + straight channel) with the dimensions listed in in Table 2, the predicted thickness distribution in the flow channel at $y = 0$ is presented in Figure 17. Since this numerical model is composed by the revolved tool geometry (Figure 16) and the extruded cross-section geometry (Figure 9), the predicted thickness is in-between the values obtained with the two previous models, as shown in Figure 17. Since the U-bend channel section is always connected to the straight channel (see Figure 4), the blank deformation is better predicted using this model, namely in the transition between the U-bend and the straight sections ($y = 0$). In fact, the maximum thinning occurs always in the region of the U-bend channel [49].

5.2. Bent Geometry

Considering the numerical model shown in Figure 18, the equivalent plastic strain distribution is presented in Figure 19a. The maximum value of equivalent plastic strain arises in the fillet radius of the die close to the axis of revolution. On the other hand, considering the fillet radius furthest away from the axis of revolution, the equivalent plastic strain increases from the U-bend to the straight channel section (see Figure 19a). This tool geometry leads to a cross-section profile (trapezoidal) which remains the same along the flow channel.

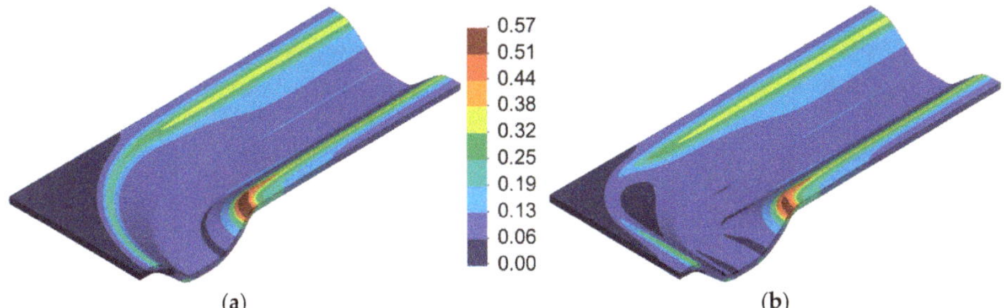

Figure 19. Equivalent plastic strain distribution plotted on the deformed configuration of the channel: (**a**) revolved tool geometry in the U-bend channel section; (**b**) rounded bend in the U-bend channel section.

Since the lowest values of equivalent plastic strain (almost zero) are located in the exterior rib of the U-bend channel section (Figure 19a), the tool geometry (punch and die) was modified according to Figure 20. The revolved radius in the exterior rib of the die is 1.2 mm, while the revolution radius of the interior rib is kept identical (0.6 mm). The cross-section profile of the punch geometry is constant, while the clearance between the punch and die, measured along the diagonal direction, is equally distributed on both sides (see Figure 20). Thus, the U-bend channel section is composed of straight sections of the punch and die.

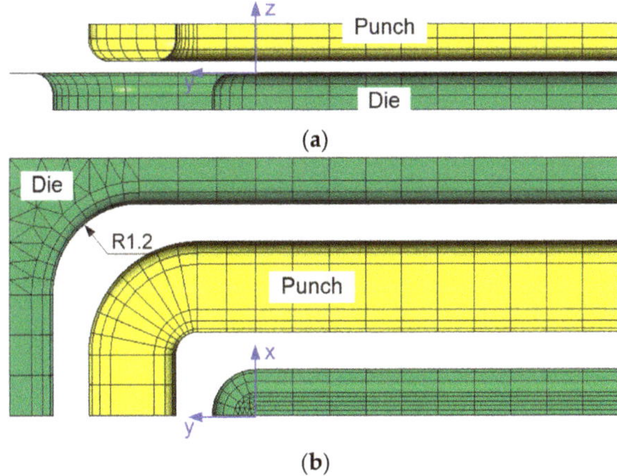

Figure 20. Finite element model of the tools adopted in the stamping of the U-bend channel section (rounded bend): (**a**) lateral view; (**b**) top view.

Considering the tool geometry presented in Figure 20, the equivalent plastic strain distribution plotted on the deformed configuration of the stamped channel is illustrated in Figure 19b. The area with larger values of equivalent plastic strain is reduced in comparison with the one obtained with the model given in Figure 18. Therefore, the formability of the stamped BPP is improved using the rounded bend tool geometry (Figure 20). Nevertheless, the cross-section profile of the flow channel is not constant along the entire U-bend section. The draft angle is larger in the diagonal direction (see Figure 19b), which results from the large clearance between the punch and the die in this zone of the channel, as highlighted in Figure 20b.

The final thickness distribution at $x = 0$ is presented in Figure 21, comparing the finite element models presented in Figures 18 and 20. According to the axis system indicated in Figure 20, the negative values of y-coordinate correspond to the straight channel section. The minimum value of thickness occurs in the symmetry plane ($x = 0$) of the U-bend channel section (compare Figures 17 and 21), specifically in the fillet radius of the die. However, the maximum thinning predicted with the tool geometry defined in Figure 20 (rounded bend) is lower than the one obtained using the model present in Figure 18. Besides, the tool geometry with rounded bend provides a flow channel with more uniform thickness, i.e., the final thickness ranges between 0.143 mm and 0.101 mm in this section, as shown in Figure 21. This improves the contact pressure distribution in the assembly due to the low dimensional error in terms of channel height of the metallic BPP [52].

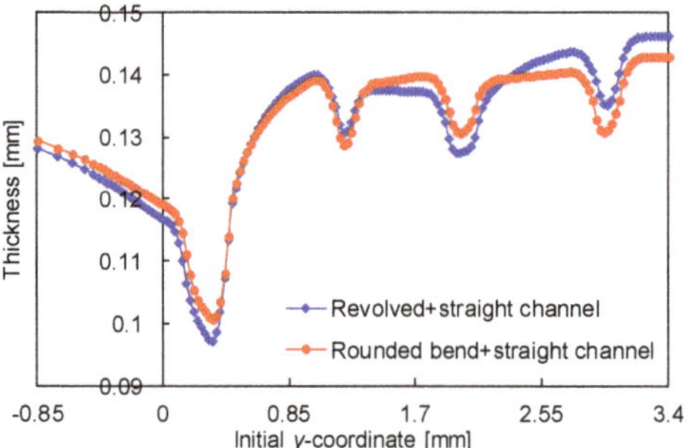

Figure 21. Thickness distribution in a single channel at $x = 0$, for the two different models. The parameters are $w_1 = s = 1.2$ mm, $r = R = 0.3$ mm, and $h = 0.7$ mm.

5.3. Effect of Tool Dimensions

The rib width adjacent to the axis of revolution dictates the maximum value for the revolution radius of the cross-section geometry (see die in Figure 20b). Taking into account the results obtained assuming plane strain conditions (see Figures 11 and 12), the thinning decreases with the reduction of the channel/rib width. Therefore, a new model was analysed, for which the rib width is increased to $s = 1.4$ mm in comparison with the tool geometry illustrated in Figure 20, while the other dimensions are kept constant (see in Table 2). In another model, the fillet radius R of the die is increased to $R = 0.4$ mm because, according to the straight channel results, the thinning decreases by increasing of the punch/die fillet radii (see Figures 11 and 12). Finally, a model was built that takes into account both the rib width and the fillet radius of the die, which are increased to 1.6 mm and 0.5 mm, respectively.

The predicted thickness distribution in the stamped flow channel at $x = 0$ is presented in Figure 22, comparing the three die geometries previously described. The punch dimensions considered in the simulation are $w_1 = 1.2$ mm and $r = 0.3$ mm, and the channel depth is $h = 0.7$ mm. Considering the U-bend channel section, the minimum value of thickness arises always in the fillet radius of the die closest to the axis of revolution, as shown in Figure 22. Indeed, the change of the die geometry, namely the rib width and the fillet radius, affects predominantly the thinning in this zone of the channel. The increase of both parameters leads to a reduction of the thinning. Considering the largest values for the rib width and fillet radius ($s = 1.6$ mm and $R = 0.5$ mm), the minimum thickness is about 0.11 mm, i.e., the maximum thinning is approximately 27% for $h = 0.7$ mm.

The equivalent plastic strain distribution is presented in Figure 23, comparing the three die geometries adopted in the finite element simulation. Although the distribution seems similar for all conditions, the maximum value of equivalent plastic strain is reduced from about 0.53 to approximately 0.43, which occurs always in the elliptic surface of the U-bend channel section (see Figure 23). Therefore, the increase of both the rib width and the fillet radius leads to a significant reduction of the equivalent plastic strain in this zone of the channel. Despite the modification of the die geometry, the flow channel cross-section areas are identical for all conditions. The obtained draft angle is approximately 95° in the trapezoidal shape of the channel cross-section.

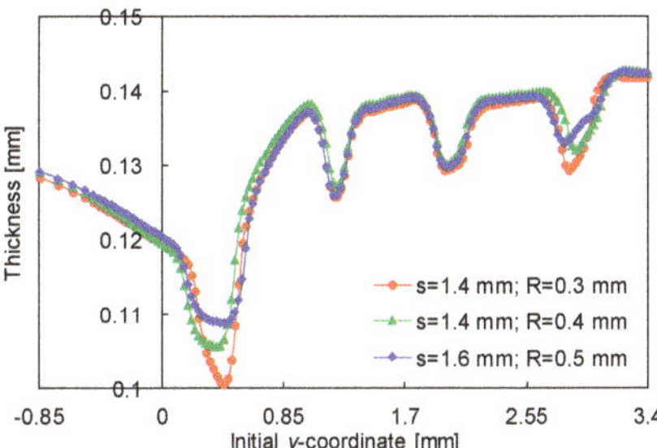

Figure 22. Thickness distribution in the flow channel at $x = 0$, comparing three different geometries of the die. The parameters are $w_1 = 1.2$ mm, $r = 0.3$ mm, and $h = 0.7$ mm.

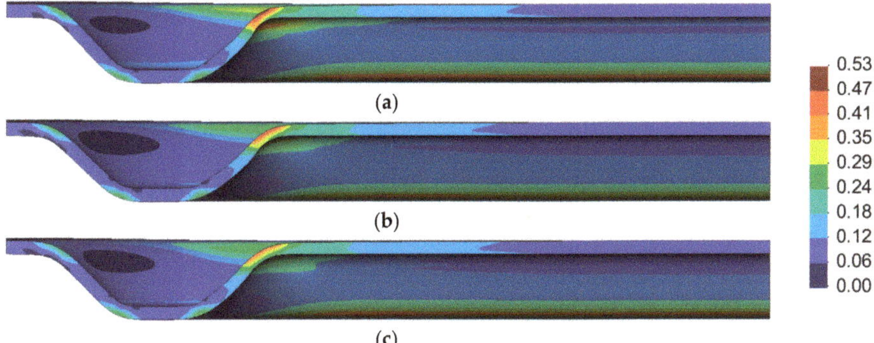

Figure 23. Equivalent plastic strain distribution plotted on the deformed configuration of the channel for different die dimensions: (**a**) $s = 1.4$ mm and $R = 0.3$ mm; (**b**) $s = 1.4$ mm and $R = 0.4$ mm; (**c**) $s = 1.6$ mm and $R = 0.5$ mm. The other parameters are $w_1 = 1.2$ mm, $r = 0.3$ mm, and $h = 0.7$ mm.

The draft angle decreases when the punch width increases (Figure 13), due to the reduction of the gap between the punch and the die (see Figure 9b). The results obtained assuming plane strain conditions indicate that the channel width and the punch fillet radius have opposite effects on the thinning (see Figure 11). Thus, both the channel width and the fillet radius of the punch are increased. Considering the punch dimensions of $w_1 = 1.4$ mm and $r = 0.4$ mm, the equivalent plastic strain distribution plotted on the deformed configuration of the channel is presented in Figure 24. The maximum value of equivalent plastic strain arises on the upper surface of the flow channel, specifically in the fillet radius of the die closest to the axis of revolution. On the other hand, the maximum value of equivalent plastic strain on the lower surface of the flow channel is located in the punch fillet radii (straight section), as shown in Figure 24b. In fact, regarding the straight channel section, the largest values of equivalent plastic strain occur always in the fillet radii, namely in the die fillet (upper surface) and in the punch fillet (lower surface).

The minor–major strain plot is presented in Figure 25, comparing the predictions obtained for the upper and lower surface of the flow channel. The strain paths of the points located on the upper surface are different from the one predicted for the lower surface of the channel, which is in accordance with the equivalent plastic strain gradient through the thickness (Figure 24). The deformation mode conditions in the BPP are predominantly between plane strain and equi-biaxial stretching, as shown in Figure 25. Nevertheless, the upper surface of the flow channel presents several points with nearly equi-biaxial stretching (large strain), which corresponds to the fillet radius of the die closest to the axis of revolution (see Figure 24a). On the other hand, the lower surface of the flow channel is mainly under plane strain conditions. Indeed, the largest value of major strain arises in that surface (see Figure 25), namely in the straight channel section.

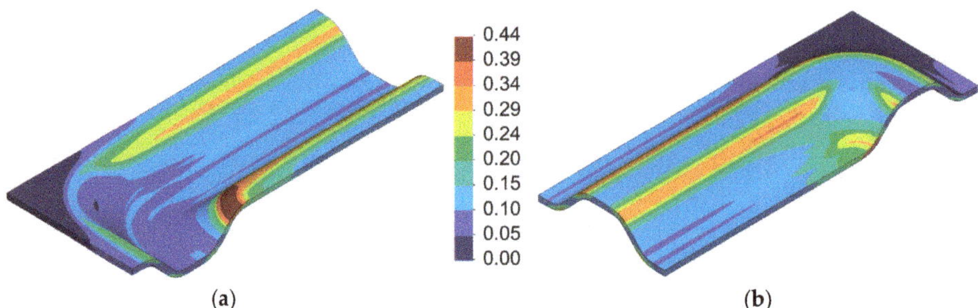

(**a**)   (**b**)

Figure 24. Equivalent plastic strain distribution plotted on the deformed configuration of the channel parameters considering $w_1 = 1.4$ mm, $s = 1.6$ mm, $r = 0.4$ mm, $R = 0.5$ mm, and $h = 0.7$ mm: (**a**) upper surface of the flow channel; (**b**) lower surface of the flow channel.

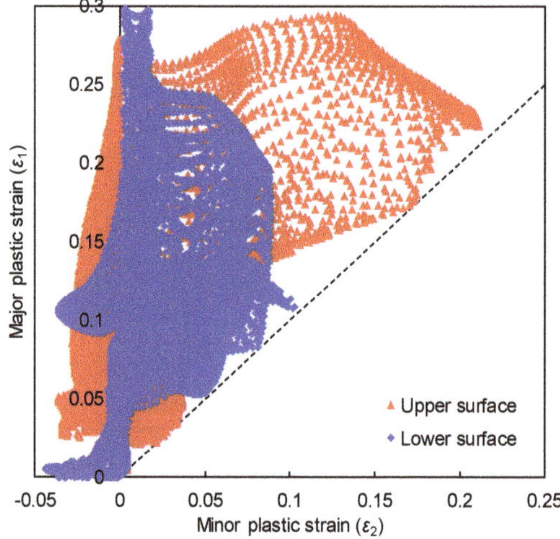

Figure 25. Minor–major strain plots for both surfaces of the channel (upper and lower) considering $w_1 = 1.4$ mm, $s = 1.6$ mm, $r = 0.4$ mm, $R = 0.5$ mm, and $h = 0.7$ mm.

The predicted thickness distribution is presented in Figure 26 for three different cross-sections of the stamped flow channel. The minimum value of thickness occurs in the fillet radius of the die (closest to the axis of revolution), specifically in the cross-section at $x = 0$. On the other hand, the predicted thickness distribution in the cross-section at $y = -5$ mm is identical to the one obtained considering plane strain conditions (straight channel section). Therefore, the formability analysis of this stamping process requires the study of the U-bend channel section, where the thinning is larger, as highlighted in Figure 26. Indeed, the maximum thinning (arising in the U-bend channel section) decreases by increasing the rib width, whereas assuming plane strain conditions (straight channel section) the increase of the rib width leads to increase of the thinning (see Figure 12).

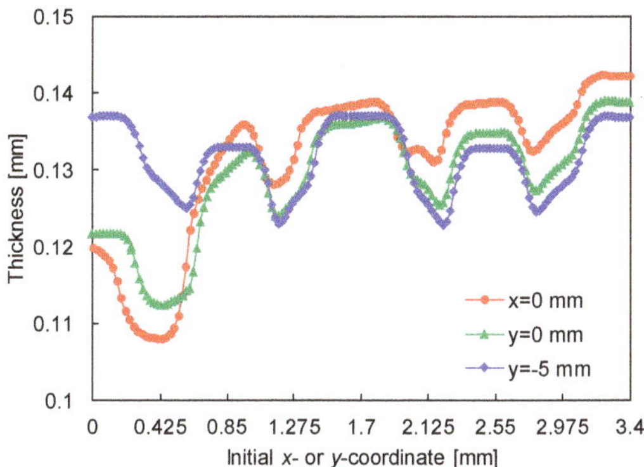

Figure 26. Thickness distribution in three different localizations of the flow channel considering $w_1 = 1.4$ mm, $s = 1.6$ mm, $r = 0.4$ mm, $R = 0.5$ mm, and $h = 0.7$ mm.

6. Conclusions

This study presents the finite element analysis of the stamping process used in the manufacturing of metallic BPPs for PEM fuel cells. The effects of the geometrical dimensions of the forming tools on the formability and final thickness were discussed considering the stainless steel SS304 with a thickness of 0.15 mm. In order to reduce the computational cost of the simulation, only two representative zones of the BPP were studied, namely the straight and the U-bend channel sections.

Considering the finite element model composed by several parallel flow channels, the predicted geometry of each stamped channel varies according to its relative position, which results from the border effect. Both the height and the cross-section of the flow channel closest to the free edge are substantially different from the one adjacent to the symmetry condition. Nevertheless, the geometry of most flow channels was accurately predicted by the numerical model that considers half-channel width under plane strain conditions. Regarding the influence of the tools (punch and die) geometry on the formability, the amount of thinning decreases with the reduction of the channel/rib width and increase of the punch/die fillet radii. On the other hand, the predicted thinning increases as the channel depth increases.

The deformation mode changes gradually from plane strain in the straight channel section to biaxial strain in the U-bend section. Therefore, the accurate analysis of this zone requires the inclusion of a portion of the straight channel section in addition to the revolved cross-section geometry. In fact, the thinning is considerably overestimated when the model comprises only the revolved cross-section geometry. The maximum value of thinning occurs always in the U-bend channel section, namely in the fillet radius of the die closest to the axis of revolution. This zone presents the highest value of equivalent plastic strain, which increases by reducing both the rib width and the fillet radius.

The validation of the presented numerical model with experimental results will allow improving the reliability of the numerical results. The conclusions obtained from the numerical results should be attested using an experimental design of experiences.

Author Contributions: D.M.N. carried out the numerical simulations; J.L.A. developed software; D.M.N., M.C.O., and L.F.M. performed the formal analysis; D.M.N. wrote the original manuscript; all the authors participated in the writing, review and editing process.

Funding: This research was funded by the Portuguese Foundation for Science and Technology (FCT) under projects PTDC/EMS-TEC/0702/2014 (POCI-01-0145-FEDER-016779) and PTDC/EMS-TEC/6400/2014 (POCI-01-0145-FEDER-016876) by UE/FEDER through the program COMPETE 2020. The support under the project MATIS (CENTRO-01-0145-FEDER-000014) is also acknowledged.

Conflicts of Interest: The authors declare no conflicts of interest. The funders had no role in the design of the study; in the collection, analyses, or interpretation of data; in the writing of the manuscript, or in the decision to publish the results.

References

1. Dunn, S. Hydrogen futures: Toward a sustainable energy system. *Int. J. Hydrogen Energy* **2002**, *27*, 235–264. [CrossRef]

2. Wang, Y.; Chen, K.S.; Mishler, J.; Cho, S.C.; Adroher, X.C. A review of polymer electrolyte membrane fuel cells: Technology, applications, and needs on fundamental research. *Appl. Energy* **2011**, *88*, 981–1007. [CrossRef]

3. Karimi, S.; Fraser, N.; Roberts, B.; Foulkes, F.R. A Review of Metallic Bipolar Plates for Proton Exchange Membrane Fuel Cells: Materials and Fabrication Methods. *Adv. Mater. Sci. Eng.* **2012**, *2012*, 1–22. [CrossRef]

4. Wang, J. Barriers of scaling-up fuel cells: Cost, durability and reliability. *Energy* **2015**, *80*, 509–521. [CrossRef]

5. DOE. *Hydrogen and Fuel Cell Activities, Progress, and Plans*; DOE: New York, NY, USA, 2009.

6. Hermann, A.; Chaudhuri, T.; Spagnol, P. Bipolar plates for PEM fuel cells: A review. *Int. J. Hydrogen Energy* **2005**, *30*, 1297–1302. [CrossRef]

7. Peng, L.; Yi, P.; Lai, X. Design and manufacturing of stainless steel bipolar plates for proton exchange membrane fuel cells. *Int. J. Hydrogen Energy* **2014**, *39*, 21127–21153. [CrossRef]

8. Barbir, F. *PEM Fuel Cells: Theory and Practice*; Elsevier Academic: Cambridge, MA, USA, 2005.

9. Taherian, R. A review of composite and metallic bipolar plates in proton exchange membrane fuel cell: Materials, fabrication, and material selection. *J. Power Sources* **2014**, *265*, 370–390. [CrossRef]

10. Hu, Q.; Zhang, D.; Fu, H.; Huang, K. Investigation of stamping process of metallic bipolar plates in PEM fuel cell—Numerical simulation and experiments. *Int. J. Hydrogen Energy* **2014**, *39*, 13770–13776. [CrossRef]

11. Peng, L.; Lai, X.; Liu, D.; Hu, P.; Ni, J. Flow channel shape optimum design for hydroformed metal bipolar plate in PEM fuel cell. *J. Power Sources* **2008**, *178*, 223–230. [CrossRef]

12. Liu, Y.; Hua, L. Fabrication of metallic bipolar plate for proton exchange membrane fuel cells by rubber pad forming. *J. Power Sources* **2010**, *195*, 3529–3535. [CrossRef]

13. Hung, J.C.; Chang, D.H.; Chuang, Y. The fabrication of high-aspect-ratio micro-flow channels on metallic bipolar plates using die-sinking micro-electrical discharge machining. *J. Power Sources* **2012**, *198*, 158–163. [CrossRef]

14. Lee, S.J.; Lee, C.Y.; Yang, K.T.; Kuan, F.H.; Lai, P.H. Simulation and fabrication of micro-scaled flow channels for metallic bipolar plates by the electrochemical micro-machining process. *J. Power Sources* **2008**, *185*, 1115–1121. [CrossRef]

15. Jin, C.; Jang, C.; Kang, C.; Jin, C.K.; Jang, C.H.; Kang, C.G. Vacuum Die Casting Process and Simulation for Manufacturing 0.8 mm-Thick Aluminum Plate with Four Maze Shapes. *Metals* **2015**, *5*, 192–205. [CrossRef]

16. Antunes, R.A.; Oliveira, M.C.L.; Ett, G.; Ett, V. Corrosion of metal bipolar plates for PEM fuel cells: A review. *Int. J. Hydrogen Energy* **2010**, *35*, 3632–3647. [CrossRef]

17. Koç, M.; Mahabunphachai, S. Feasibility investigations on a novel micro-manufacturing process for fabrication of fuel cell bipolar plates: Internal pressure-assisted embossing of micro-channels with in-die mechanical bonding. *J. Power Sources* **2007**, *172*, 725–733. [CrossRef]

18. Mahabunphachai, S.; Cora, Ö.N.; Koç, M. Effect of manufacturing processes on formability and surface topography of proton exchange membrane fuel cell metallic bipolar plates. *J. Power Sources* **2010**, *195*, 5269–5277. [CrossRef]

19. Kim, A.R.; Vinothkannan, M.; Yoo, D.J. Sulfonated fluorinated multi-block copolymer hybrid containing sulfonated (poly ether ether ketone) and graphene oxide: A ternary hybrid membrane architecture for electrolyte applications in proton exchange membrane fuel cells. *J. Energy Chem.* **2018**, *27*, 1247–1260. [CrossRef]

20. Kim, A.R.; Park, C.J.; Vinothkannan, M.; Yoo, D.J. Sulfonated poly ether sulfone/heteropoly acid composite membranes as electrolytes for the improved power generation of proton exchange membrane fuel cells. *Compos. Part B Eng.* **2018**, *155*, 272–281. [CrossRef]

21. André, J.; Antoni, L.; Petit, J.P.; De Vito, E.; Montani, A. Electrical contact resistance between stainless steel bipolar plate and carbon felt in PEFC: A comprehensive study. *Int. J. Hydrogen Energy* **2009**, *34*, 3125–3133. [CrossRef]

22. Ge, J.; Higier, A.; Liu, H. Effect of gas diffusion layer compression on PEM fuel cell performance. *J. Power Sources* **2006**, *159*, 922–927. [CrossRef]

23. Wlodarczyk, R.; Zasada, D.; Morel, S.; Kacprzak, A. A comparison of nickel coated and uncoated sintered stainless steel used as bipolar plates in low-temperature fuel cells. *Int. J. Hydrogen Energy* **2016**, *41*, 17644–17651. [CrossRef]

24. Hung, Y.; EL-Khatib, K.M.; Tawfik, H. Corrosion-resistant lightweight metallic bipolar plates for PEM fuel cells. *J. Appl. Electrochem.* **2005**, *35*, 445–447. [CrossRef]

25. Jin, C.K.; Koo, J.Y.; Kang, C.G. Fabrication of stainless steel bipolar plates for fuel cells using dynamic loads for the stamping process and performance evaluation of a single cell. *Int. J. Hydrogen Energy* **2014**, *39*, 21461–21469. [CrossRef]

26. Park, W.T.; Jin, C.K.; Kang, C.G. Improving channel depth of stainless steel bipolar plate in fuel cell using process parameters of stamping. *Int. J. Adv. Manuf. Technol.* **2016**, *87*, 1677–1684. [CrossRef]

27. Bong, H.J.; Lee, J.; Kim, J.H.; Barlat, F.; Lee, M.G. Two-stage forming approach for manufacturing ferritic stainless steel bipolar plates in PEM fuel cell: Experiments and numerical simulations. *Int. J. Hydrogen Energy* **2017**, *42*, 6965–6977. [CrossRef]

28. Afshari, E.; Mosharaf-Dehkordi, M.; Rajabian, H. An investigation of the PEM fuel cells performance with partially restricted cathode flow channels and metal foam as a flow distributor. *Energy* **2017**, *118*, 705–715. [CrossRef]

29. Wang, J. Theory and practice of flow field designs for fuel cell scaling-up: A critical review. *Appl. Energy* **2015**, *157*, 640–663. [CrossRef]

30. Dhahad, H.A.; Alawee, W.H.; Hassan, A.K. Experimental study of the effect of flow field design to PEM fuel cells performance. *Renew. Energy Focus* **2019**, *30*, 71–77. [CrossRef]

31. Kahraman, H.; Orhan, M.F. Flow field bipolar plates in a proton exchange membrane fuel cell: Analysis & modeling. *Energy Convers. Manag.* **2017**, *133*, 363–384.

32. Arvay, A.; French, J.; Wang, J.C.; Peng, X.H.; Kannan, A.M. Nature inspired flow field designs for proton exchange membrane fuel cell. *Int. J. Hydrogen Energy* **2013**, *38*, 3717–3726. [CrossRef]

33. Guo, N.; Leu, M.C.; Koylu, U.O. Bio-inspired flow field designs for polymer electrolyte membrane fuel cells. *Int. J. Hydrogen Energy* **2014**, *39*, 21185–21195. [CrossRef]

34. Shimpalee, S.; Van Zee, J.W. Numerical studies on rib & channel dimension of flow-field on PEMFC performance. *Int. J. Hydrogen Energy* **2007**, *32*, 842–856.

35. Hontañón, E.; Escudero, M.J.; Bautista, C.; García-Ybarra, P.L.; Daza, L. Optimisation of flow-field in polymer electrolyte membrane fuel cells using computational fluid dynamics techniques. *J. Power Sources* **2000**, *86*, 363–368. [CrossRef]

36. Manso, A.P.; Marzo, F.F.; Mujika, M.G.; Barranco, J.; Lorenzo, A. Numerical analysis of the influence of the channel cross-section aspect ratio on the performance of a PEM fuel cell with serpentine flow field design. *Int. J. Hydrogen Energy* **2011**, *36*, 6795–6808. [CrossRef]

37. Hu, P.; Peng, L.; Zhang, W.; Lai, X. Optimization design of slotted-interdigitated channel for stamped thin metal bipolar plate in proton exchange membrane fuel cell. *J. Power Sources* **2009**, *187*, 407–414. [CrossRef]

38. Xu, Y.; Peng, L.; Yi, P.; Lai, X. Analysis of the flow distribution for thin stamped bipolar plates with tapered channel shape. *Int. J. Hydrogen Energy* **2016**, *41*, 5084–5095. [CrossRef]

39. Menezes, L.F.; Teodosiu, C. Three-dimensional numerical simulation of the deep-drawing process using solid finite elements. *J. Mater. Process. Technol.* **2000**, *97*, 100–106. [CrossRef]

40. Oliveira, M.C.; Alves, J.L.; Menezes, L.F. Algorithms and Strategies for Treatment of Large Deformation Frictional Contact in the Numerical Simulation of Deep Drawing Process. *Arch. Comput. Methods Eng.* **2008**, *15*, 113–162. [CrossRef]

41. Menezes, L.F.; Neto, D.M.; Oliveira, M.C.; Alves, J.L. Improving Computational Performance through HPC Techniques: Case study using DD3IMP in-house code. *AIP Conf. Proc.* **2011**, *1353*, 1220–1225.

42. Pham, C.H.; Thuillier, S.; Manach, P.Y. Prediction of flow stress and surface roughness of stainless steel sheets considering an inhomogeneous microstructure. *Mater. Sci. Eng. A* **2016**, *678*, 377–388. [CrossRef]

43. Pham, C.H.; Thuillier, S.; Manach, P.Y. Mechanical Properties Involved in the Micro-forming of Ultra-thin Stainless Steel Sheets. *Metall. Mater. Trans. A* **2015**, *46*, 3502–3515. [CrossRef]

44. Raj, A.K. *Formability: Metastable Austenitic Stainless Steels*; Lulu Press: Morrisville, NC, USA, 2015.

45. Neto, D.M.; Oliveira, M.C.; Menezes, L.F.; Alves, J.L. Applying Nagata patches to smooth discretized surfaces used in 3D frictional contact problems. *Comput. Methods Appl. Mech. Eng.* **2014**, *271*, 296–320. [CrossRef]

46. Neto, D.M.; Oliveira, M.C.; Menezes, L.F.; Alves, J.L. Nagata patch interpolation using surface normal vectors evaluated from the IGES file. *Finite Elem. Anal. Des.* **2013**, *72*, 35–46. [CrossRef]
47. Hughes, T.J.R. Generalization of selective integration procedures to anisotropic and nonlinear media. *Int. J. Numer. Methods Eng.* **1980**, *15*, 1413–1418. [CrossRef]
48. Li, K.P.; Carden, W.P.; Wagoner, R.H. Simulation of springback. *Int. J. Mech. Sci.* **2002**, *44*, 103–122. [CrossRef]
49. Peng, L.; Liu, D.; Hu, P.; Lai, X.; Ni, J. Fabrication of Metallic Bipolar Plates for Proton Exchange Membrane Fuel Cell by Flexible Forming Process-Numerical Simulations and Experiments. *J. Fuel Cell Sci. Technol.* **2010**, *7*, 031009. [CrossRef]
50. Xu, S.; Li, K.; Wei, Y.; Jiang, W. Numerical investigation of formed residual stresses and the thickness of stainless steel bipolar plate in PEMFC. *Int. J. Hydrogen Energy* **2016**, *41*, 6855–6863. [CrossRef]
51. Koo, J.Y.; Jeon, Y.P.; Kang, C.G. Effect of stamping load variation on deformation behaviour of stainless steel thin plate with microchannel. *Proc. Inst. Mech. Eng. Part B J. Eng. Manuf.* **2013**, *227*, 1121–1128. [CrossRef]
52. Liu, D.; Peng, L.; Lai, X. Effect of dimensional error of metallic bipolar plate on the GDL pressure distribution in the PEM fuel cell. *Int. J. Hydrogen Energy* **2009**, *34*, 990–997. [CrossRef]

© 2019 by the authors. Licensee MDPI, Basel, Switzerland. This article is an open access article distributed under the terms and conditions of the Creative Commons Attribution (CC BY) license (http://creativecommons.org/licenses/by/4.0/).

Article

Experimental and Numerical Studies of Sheet Metal Forming with Damage Using Gas Detonation Process

Sandeep P. Patil [1,*]**, Kaushik G. Prajapati** [1]**, Vahid Jenkouk** [1]**, Herbert Olivier** [2] **and Bernd Markert** [1]

[1] Institute of General Mechanics, RWTH Aachen University, Templergraben 64, 52062 Aachen, Germany; kaushik.prajapati@rwth-aachen.de (K.G.P.); jenkouk@iam.rwth-aachen.de (V.J.); markert@iam.rwth-aachen.de (B.M.)

[2] Shock Wave Laboratory, RWTH Aachen University, Schurzelter Str. 35, 52074 Aachen-Laurensberg, Germany; olivier@swl.rwth-aachen.de

* Correspondence: patil@iam.rwth-aachen.de; Tel.: +49-241-80-90036

Received: 9 November 2017; Accepted: 6 December 2017; Published: 10 December 2017

Abstract: Gas detonation forming is a high-speed forming method, which has the potential to form complex geometries, including sharp angles and undercuts, in a very short process time. Despite many efforts being made to develop detonation forming, many important aspects remain unclear and have not been studied experimentally, nor numerically in detail, e.g., the ability to produce sharp corners, the effect of peak load on deformation and damage location and its propagation in the workpiece. In the present work, DC04 steel cups were formed using gas detonation forming, and finite element method (FEM) simulations of the cup forming process were performed. The simulations on 3D computational models were carried out with explicit dynamic analysis using the Johnson–Cook material model. The results obtained in the simulations were in good agreement with the experimental observations, e.g., deformed shape and thickness distribution. Moreover, the proposed computational model was capable of predicting the damage initiation and evolution correctly, which was mainly due to the high-pressure magnitude or an initial offset of the workpiece in the experiments.

Keywords: gas detonation forming; finite element method; Johnson–Cook material model; damage

---

1. Introduction

Sheet metal forming basically consists of stretching and bending a thin sheet into the desired shape. The produced parts can be stiff and have good strength-to-weight ratios; therefore, these products are widely used for automobiles, domestic appliances, aircraft and food and drink cans. A large number of techniques is used to make sheet metal parts. In recent years, many aspects of sheet metal forming processes have been widely studied using electromagnetic forming, especially with regard to the behavior of materials under a high strain rate, the possible future applications and numerical modeling of the process, with several works dedicated to these topics [1–7]. Moreover, a detailed review of numerical simulations in sheet metal forming and potential developments is presented by Tekkaya [8].

DC04 steel has a good ductility level, which facilitates the production of complicated component shapes where required and even allowing deep pressing processes to be carried out. Here, cup formation of DC04 steel sheets was studied using the gas detonation forming technique. It is a highly dynamic manufacturing method, which involves the release of the stored energy in a very short interval of time. There are various high-speed forming processes, which are classified based on the type of energy transfer. This can be done by active media, accelerated mass or by active energy. Here, the high kinetic energy of a fluid medium is exploited, and it is used to collide the sheet-metal workpiece in the form of a shock front, which is produced as a result of the detonation of a mixture of gases like oxyhydrogen [9,10]. This forming process has many well-known advantages, namely high degree of formability, capability to form complex geometries, including undercuts for high strain rate-dependent materials, and fine embossing without relief angle. The process consists of a clean combustion, having the advantages of easy automation and fewer safety regulations. The overall process and tooling costs are significantly reduced due to simplified tooling requirements compared to electromagnetic forming [11].

In previous works, Yasar and Yasar et al., conducted both experimental and numerical investigations of aluminum cup drawing using gas detonation. In the first work, 2D and 3D simulations were performed using both explicit and implicit dynamic analysis. The thickness and the final shape were compared to the experiments. Based on the term of deformed shape, the spring back predictions by explicit and implicit methods were discussed. In the second study, gas detonation forming experiments were performed using a mixture of acetylene and oxygen with an equal volume ratio. They also did the explicit dynamic simulations using ANSYS/LS-DYNA. The strain, thickness and volume of cup formation were compared [12,13]. Mokadem developed a dynamic forming limit diagram for this process [14]. Wijayathunga and Webb also developed a finite element model to simulate the experimental tests for the impulsive deep drawing of a brass square cup with the presence of a soft lead plug [15]. In the implementation of the finite element simulation, the effects of the medium impedance, wave reflection and refraction were considered to be negligible in order to improve the simplicity of the modeling procedure [15]. Mousavi et al., studied free underwater explosive forming of aluminum circular plates experimentally and analytically, using a central explosive charge on 2024 aluminum sheets [16]. In this study, numerical simulation results concluded that the friction coefficient and blank holder force must be sufficient and optimized in order to prevent uneven drawing and wrinkling [16].

Khalegi et al., worked on gas detonation forming of clamped circular mild steel with three conical dies having apex angles of 60°, 90° and 120°. They studied the influence of the initial ratio of the oxyhydrogen mixture and also the effect of three different initial pressures of 3, 4 and 5 bar. Moreover, FEM simulations were performed, and the results of thickness strain, hoop strain, thickness variation and deformed geometry were compared with the experiments [17]. Hashem Babaei et al., conducted experiments on clamped circular plates of mild steel, using impulse loading from the detonation of the oxygen and acetylene mixture at various volume ratios and different initial pressures. They developed an analytical and empirical model for their experiments to demonstrate the effect of the mechanical properties of the plate and gas, the impulse of applied load, plate geometry, the velocity of sound in different gases and the strain-rate sensitivity on the large deformation of circular plates in high rate energy forming [18,19]. Mirzababaie Mostofi et al., investigated the effect of the detonation of different oxyacetylene mixtures on the dynamic response of aluminum alloy and mild steel plates with different thicknesses. They examined the ductile transverse deformation of the clamped rectangular plates. Theoretical analysis was conducted, according to an upper bound solution and energy method, with theoretical models assuming a zero-order Bessel function of the first kind in the x and y directions for a transverse displacement profile to predict permanent deflections. To account for material strain rate sensitivity, a Cowper–Symonds model has been used and was compared to Jones' theoretical model [20]. In other works, they suggested new dimensionless numbers based on the dimensionless governing equations and using a new mathematical method, namely the singular value decomposition method. Their empirical model was validated against the experiments. The study revealed that the empirical model using the Cowper–Symonds constitutive equation predicted the ratio of midpoint deflection to the thickness more accurately than Jones' theoretical equation [21].

These studies are important to shed light on the gas detonation forming process. However, some of the important aspects of the experiments, as well as the simulations of sheet metal forming by this technique have not yet been studied in detail, e.g., the observation of sharp edges in the deformation process, the influence of the peak load, the reproduction of the sharp corners in the numerical analysis and, more importantly, damage. In the gas detonation forming process, fracture occurs in the sheet metal by ductile damage due to the development of micro-cracks associated with large straining or due to plastic instabilities associated with the sheet materials' micro-structure and boundary conditions.

Therefore, one of the main objectives of this work is to predict when and where the cracks can appear in the workpiece during the forming.

The present work investigates the gas detonation forming of DC04 steel cups. The 3D explicit dynamic finite element analyses are carried out using the LS-DYNA explicit solver [22]. The material description considered for this study is the Johnson–Cook plasticity material model. The deformed geometry of the cup and thickness distributions are compared with the experimentally-obtained values. The relative differences found between the experimental and simulation results are discussed. Finally, the fractured specimens in the experiments are studied using adapted damage parameters for the numerical simulations.

2. Methods and Setup

2.1. Experimental Setup

Figure 1 depicts the apparatus, and Figure 2 shows a schematic representation of the experimental setup of the gas detonation forming process. It consists of four major parts, i.e., detonation tube, die holder, die and sheet metal or workpiece. The detonation tube of 700 mm in length is clamped to the die holder, forming a tight seal between them. A small hole is drilled through both the die holder and die, which is connected to a vacuum pump. This is done to prevent the formation of an air cushion between the sheet metal and the die, enabling the sheet metal to perfectly sit into the die. Figure 3 shows the inner dimensions of the die. The diameter of the circular metal blank was 54 mm with a thickness of 1 mm. The inner diameter of the die was 30 mm. The detonation tube contains two piezo-electric sensors oriented coaxially and connected to the gas space by radial bores. The types of sensors used were Kistler 603B (closer to workpiece) and Kistler 601H.

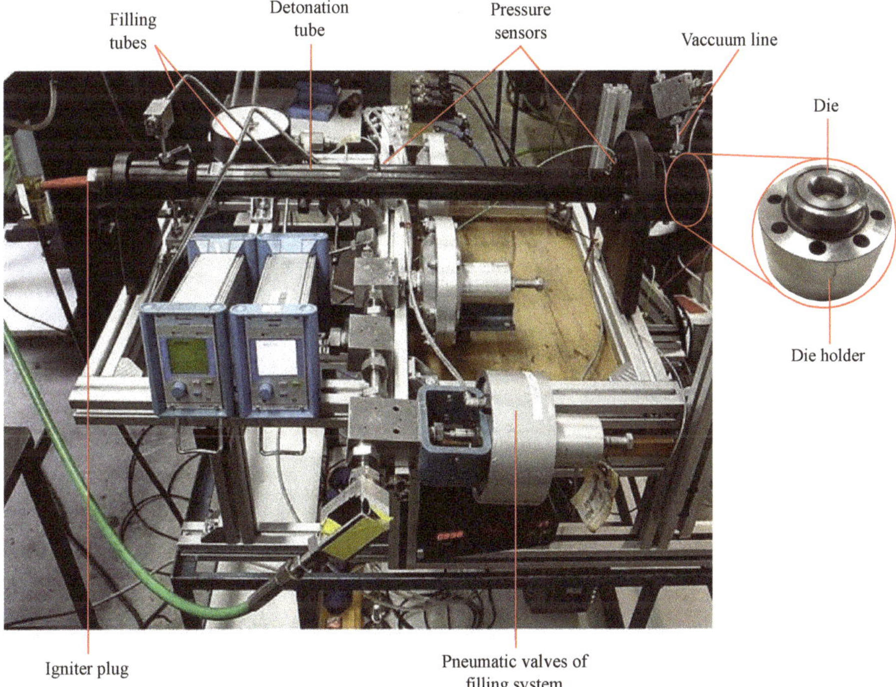

Figure 1. Gas detonation forming apparatus with peripheral equipment.

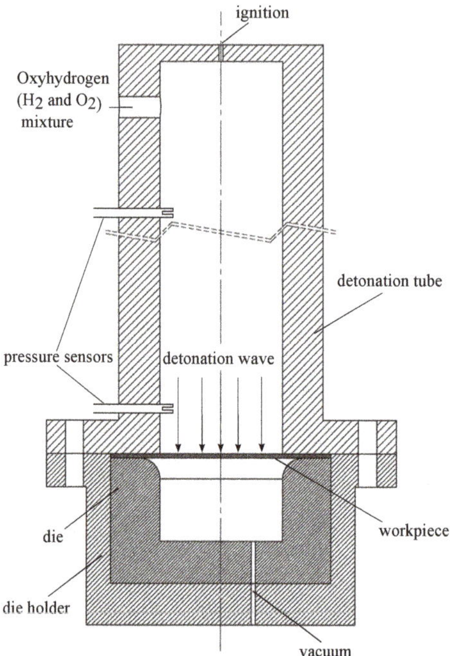

Figure 2. Schematic representation of the experimental setup of the gas detonation forming process.

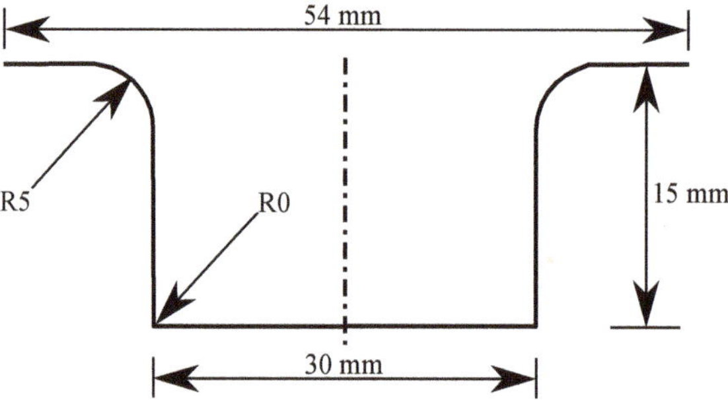

Figure 3. Internal dimensions of the die.

In the gas detonation process, the detonation tube is filled with oxyhydrogen. The gas mixture is compressed in the tube to the initial pressure. The mixture is ignited at the other end of the tube, causing the detonation wave to travel in the tube at constant supersonic velocity. A detonation wave is a joint complex of shock waves and reaction zones, implying shock waves that are strong enough to induce an immediate chemical reaction. The shock compression of the gas is sufficient to cause an instantaneous reaction of the oxyhydrogen mixture, which quickly leads to a chemical equilibrium. The released heat sustains the wave. The wave speed is approximately 3000 m/s; the thickness of wave is less than 1 mm; and the pressure directly behind it is about 20-times the initial pressure [23]. For this case study, the initial pressure in the tube was kept at 30 bar (3 MPa). The maximum pressure acting on the metal sheet was observed to be approximately 1500 bar (150 MPa). This maximum pressure loading, which occurs just at the beginning of the interaction of the detonation wave with the workpiece, is caused by the reflection of the detonation wave at the workpiece. This wave reflection is so fast that during this very short interval of time, no or only a very small deformation of the metal sheet occurs [23]. For the prevention of overheating of the workpiece from hot detonated gas, moist filter paper was placed on the blank as thermal insulation.

The averaged pressure record of the sensor (close to the workpiece) is shown in Figure 4, which was obtained during the forming of DC04 specimens. The measured signals from the two sensors have been smoothed with a half-width of 10 μs to eliminate extreme oscillations. The second pressure rise is caused by the reflection at the end wall. The forming of the cup leads to a faster and further pressure drop following the detonation wave because of the increasing volume. All experiments were conducted at the Shock Wave Laboratory, RWTHAachen University.

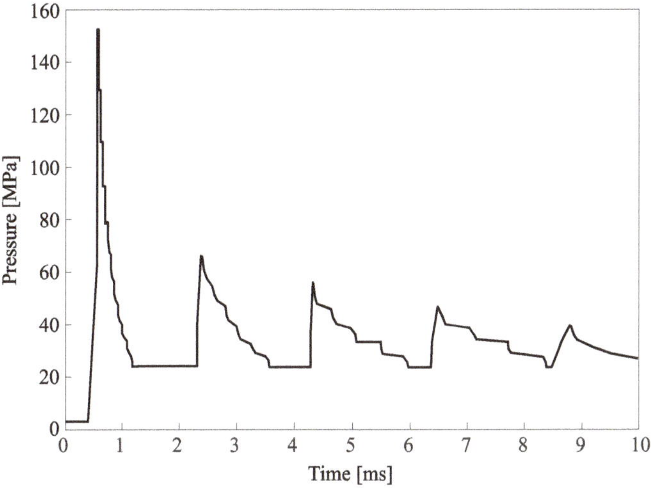

Figure 4. Averaged detonation pressure records from experiments.

2.2. Numerical Modeling

Gas detonation is a transient dynamic process involving shock waves transferring energy into the workpiece. The process simulation is based on the solution of dynamic equilibrium equations [24]. Hence, the simulations were done using an explicit time integration in the LS-DYNA solver (version: ls971 R7.1.1) [22]. Since we are interested in the deformation process of the workpiece and not in the shock wave propagation in the detonation tube, the problem was simplified by directly applying the detonation pressure as a load in the finite element (FE) models.

Figure 5 shows one-quarter section view of the 3D FE model. Due to the axial symmetry of the problem, only a quarter of the whole system with symmetric boundary conditions was considered to reduce the computational time. The model includes the die, the top plate (bottom part of the detonation tube) and the sheet metal workpiece. The one-quarter FE model was used to study the deformation of the workpiece into the cup with no misalignment. However, to study damage in the workpiece during forming, we considered the complete (full) model because the symmetry boundary conditions would highly influence the prediction of the damage areas. Moreover, offsets or improper alignment of the workpiece in experiments can be very well studied using the full model in order to capture workpiece formability in all directions.

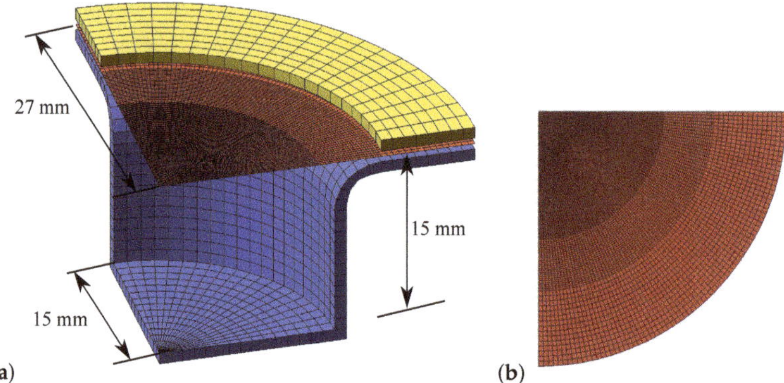

Figure 5. (**a**) One-quarter section view of the 3D finite element model; (**b**) FE meshed workpiece.

The die and the top plate were modeled using solid elements and rigid material (MAT_020), because of their high stiffness compared to the blank and as they are not the active components during the forming process. Belytschko–Tsay shell elements with five integration points [22] were used to create the meshed workpiece, which resulted in a total of 11,076 elements. The die and the holder were considered to be fixed, and the contact between them was defined using the surface-to-surface segment-based contact formulation [22], assuming planer segments. The pressure load was applied only on the free surface of the blank. According to the EN 10130-2006 standards, DC04 steel contains carbon, manganese, phosphorus and sulfur at 0.08% , 0.04% , 0.03% and 0.03% by weight, respectively. The mechanical properties of the blank DC04 are given in Table 1.

Table 1. Mechanical properties of DC04.

Property	Value
Young's modulus (GPa)	180
Poisson's ratio	0.3
Density (kg/m^3)	7870
Tensile strength (MPa)	210

The material model used for the workpiece was the Johnson–Cook phenomenological material model (MAT_015) [25], which is probably the most used and available in most of the commercial finite element commercial codes. This material model reproduces several important material responses observed in the forming, impact and penetration of metals. In this model, the three key plastic material responses are considered strain hardening, strain rate sensitivity and thermal softening. These three effects are combined in a multiplicative manner, such that the Johnson–Cook constitutive stress reads:

$$\sigma_y = \left(A + B\bar{\varepsilon}^{p^n} \right)\left(1 + C \ln \frac{\dot{\bar{\varepsilon}}^p}{\dot{\varepsilon}_0}\right)\left(1 - \left[\frac{T - T_{room}}{T_{melt} - T_{room}}\right]^m\right), \tag{1}$$

where $\bar{\varepsilon}^p$ is the effective plastic strain, T_{room} the ambient temperature, T_{melt} the melting point or solidus temperature, T the effective temperature, A the yield stress, B the hardening modulus, n the strain exponent, m the temperature exponent and C the strain rate factor. Furthermore, $\dot{\varepsilon}_0$ represents the strain rate for the quasi-static reference loading $\dot{\varepsilon}_0 = 5.6 \times 10^{-4}$ s^{-1} [26].

However, in this work, using the proposed strain rate by Verleysen et al. [26], we observed that there were no sharp corners at the bottom of the deformed cup; the final diameter of the cup did not match with the experimental value; and also, the formability for different loading profiles was not similar to that of the experiments. All the above-mentioned numerical issues were due to the highly dynamic process. Schwer et al., introduced optional strain rate forms calibrated to laboratory data for A36 steel [27]. Comparing the calibrated model response to quasi-static A36 steel data, they illustrated the role of the $\dot{\varepsilon}_0$ parameter in the Johnson–Cook material model. This is not simply a parameter for making the effective plastic strain-rate non-dimensional, as is often incorrectly cited, but this parameter must be specified as the effective plastic strain rate of the quasi-static testing. Therefore, in order to get the experimental shapes, the value of $\dot{\varepsilon}_0$ was increased by two times, and the obtained results were close to the experimental ones. Hence, the different trial simulations were performed with increasing values of $\dot{\varepsilon}_0$, and the results were compared with the experiments. Therefore, we propose a strain rate of $\dot{\varepsilon}_0 = 7.3 \times 10^{-3}\,\text{s}^{-1}$ for the gas detonation process.

The first bracketed term of the right-hand side of Equation (1) describes the isothermal static material behavior, i.e., the strain hardening of the yield stress. Consequently, the parameters A, B and n are determined using static tensile tests. The second term expresses the strain rate hardening with the parameter C. The last term represents a softening of the yield stress due to local thermal effects. In the experiments, a moistened filter paper was placed on the workpiece, in order to prevent the heating by contact with the hot detonated gas. Hence, the material surface remains unaffected despite the gas temperature. Therefore, in the Johnson–Cook material model, the thermal softening effect was not considered. The required material parameters for the simulations are given in Table 2 [26].

Table 2. Values for the Johnson–Cook material model parameters [26].

Property	Value
Yield stress, A (MPa)	162
Strength coefficient, B (MPa)	598
Deformation hardening, n	0.6
Strain rate coefficient, C	2.623
Deformation sensitivity, m	0.009

The ductile rupture of materials is described by three phases, namely void nucleation, growth and coalescence [28,29]. The void growth depends not only on the equivalent plastic strain, but also on triaxiality, which is defined as the ratio of the mean stress to the von Mises effective stress. Therefore, the damage behavior of a material depends strongly on the load type and on the geometry. In addition, the damage behavior is influenced by the strain rate.

To simulate the damage in the workpiece, damage parameters were included in the Johnson–Cook material model. Damage in the material tries to take path dependency into account by accruing the incremental effective plastic strain as the forming process proceeds [30]. In this material model, the failure strain is a function of the effective stress, the strain rate and the temperature. The equation of the fracture strain is given as:

$$\varepsilon^f = [D_1 + D_2\exp(D_3\sigma^*)][1 + D_4\ln\dot{\varepsilon}^*][1 + D_5 T^*],\tag{2}$$

where D_1 to D_5 are five constants. σ^* is the ratio of pressure divided by effective stress:

$$\sigma^* = \frac{P}{\sigma_{eff}}.\tag{3}$$

where P is the average of the normal stresses and σ_{eff} is the von Mises equivalent stress. $\dot{\varepsilon}^*$ is normalized effective plastic strain, given by:

$$\dot{\varepsilon}^* = \frac{\dot{\bar{\varepsilon}}^p}{\dot{\varepsilon}_0}.\tag{4}$$

and T^* is the homologous temperature:

$$T^* = \frac{T - T_{room}}{T_{melt} - T_{room}}.$$ (5)

The expression in the first set of brackets in Equation (2) represents that the strain to fracture decreases as the average normal stresses, P, increase. The second set of brackets represents the effect of the strain rate, and that in the third set of brackets represents the effect of temperature [30]. In this numerical simulation work, only D_1 to D_4 are considered, since we are assuming the temperature of the workpiece to be constant during the forming process.

The damage to an element is defined as:

$$D = \sum \frac{\Delta \bar{\varepsilon}^p}{\varepsilon^f}$$ (6)

where $\Delta \bar{\varepsilon}^p$ is the increment of the equivalent plastic strain, which would occur during the integration cycle, and ε^f is the equivalent strain to fracture under the current condition of pressure, equivalent stress, strain rate and temperature. Fracture occurs when the damage parameter D reaches the value of one, and the corresponding failed elements are deleted.

3. Results and Discussion

3.1. Cup Formation

A number of gas detonation experiments of cup formation were carried out using a DC04 steel sheet of 1 mm in thickness and 54 mm in diameter. The shock wave acting on the blank was approximately 1500 bar (150 MPa). The blank sits perfectly into the die seat, with a depth of 15 mm. In our previous work [31], as well as Yasar [12], it is clear that when the applied load is triangular, i.e., load increases with a lower slop than that of the experiments, the spring-back effects are observed. In the present work, load was instantaneously (high slop of pressure loading profile) like that in the experiments, and hence, the spring-back effect was not observed. Since the process takes place in a very short period of time and at a very high-pressure, sharp corners were observed at the bottom of the cup. There were no observable wrinkles on the flange or on the skirt of the cup.

Figure 6 shows the qualitative comparison between the deformed shape of cups in the experiment and the numerical simulation. In the numerical simulations, the one-quarter section was considered with symmetry boundary conditions. The experimental pressure profile was the input loading curve for the simulations.

Figure 6. (**a**) Detonation formed cup in the shock tube; (**b**) Cup formation predicted by numerical simulations; (**c**) One-quarter section view of the FE model after the application of load.

Our numerical simulation studies produced remarkably similar results compared to the experiments. The numerical simulation shows no wrinkles on the flange or skirt of the formed cup (Figure 6). Furthermore, very sharp corners were observed at the bottom of the cup.

Figure 7 shows a comparison of the final diameters of the deformed cups in the experiment and simulation. The mean value of the final diameter of the workpiece was 44 mm, which was obtained from eight samples in the experiment, and 44.6 mm was the diameter of the cup in the numerical simulation. In the experiments, it was unclear what caused the flange diameter to be 44 mm, i.e., friction or the inertia effects of the blank.

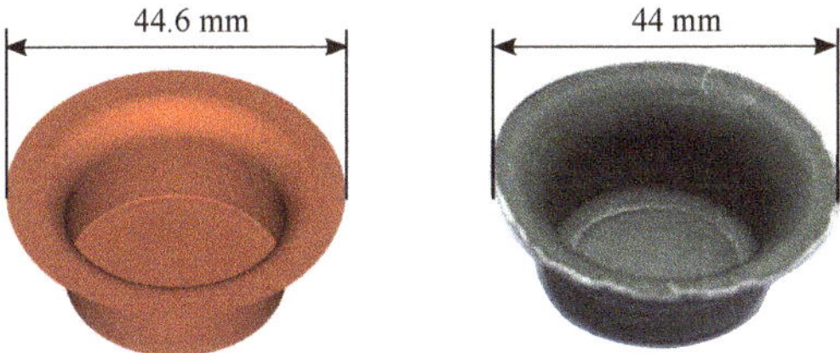

Figure 7. Diameter comparison of numerically- and experimentally-formed DC04 cups.

Therefore, different static and dynamic friction coefficients were considered between the workpiece and the top plate, as well as between the workpiece and the die. The detailed analysis is discussed in the Supplementary Materials. Initially, a static coefficient of friction of 0.6 and a dynamic coefficient of friction of 0.7 were considered between the workpiece and the top plate, as well as between the workpiece and die [32]. However, the outer diameter was nearly 53.9 mm, and also, the bottom corners were not sharp. In case the friction parameters were zeros, the outer diameter was nearly matched to the experimental diameter of 44.6 mm. Moreover, the bottom corners were sharp like the experimentally-formed cup. Therefore, we concluded that the final flange diameter is the result of the inertia effects.

Figure 8 depicts the shape of the deformed blank with respect to the loading time. The analysis of the blank shape with respect to the time highlights the fact that the whole deformation process takes place within the first approximately 60 µs, and the reflective waves do not play a major role in the formation of the cup.

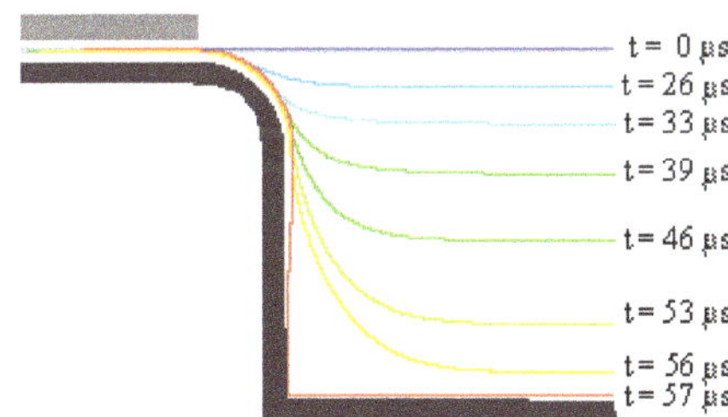

Figure 8. Cup shape formation during the simulation of the gas detonation forming process.

Figure 9 depicts the displacement of the center point of the blank over the time. From the graph, it is clear that the workpiece took some time initially to deform, and then, there is instantaneous deformation. There is no kink or decrease in the displacement of the center point of the workpiece; therefore, there was no spring back effect observed in the simulations.

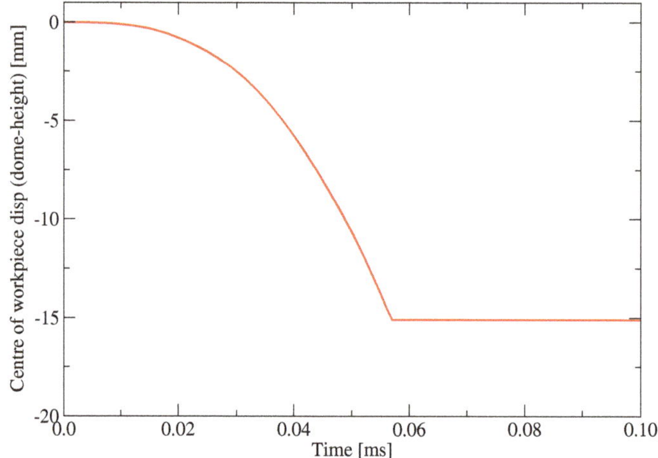

Figure 9. Displacement of the center point during cup formation in the simulations.

Thickness Variation

In our previous work [31,33], we concluded that the loading rate, i.e., the time instant at which the peak pressure acts on the blank, has a significant influence on the thickness distribution and the radial strain of the blank. The work highlighted the fact that the obtained results from the numerical model were very sensitive to the loading rate. Therefore, it is necessary to get an accurate experimental loading curve using well-calibrated measuring devices and to consider it in the numerical simulations.

The parameters of the Johnson–Cook material model can significantly affect the deformation behavior and the thickness distribution. One of the aims of this work was to correctly predict the thickness distribution in the numerical simulations along the base and the wall of the deformed cup. Figure 10 shows the thickness distribution along the initial radius of the workpiece. The obtained results from the numerical simulation were in good agreement with the experiments. The model was clearly able to predict a local minimum in the thickness value close to the 90° bend, which was towards the center of the cup (approximately 10 mm in radius). The minimum thickness obtained was nearly 0.6 mm. However, the area that stays between the die and top plate, which does not go into the cavity, experiences pure radial pulling. Moreover, due to the circumferential stresses, the thickness was increased [12,34].

In the literature, the thickness distribution has been studied using finite element models; however, experimental loading rates were not considered, and smooth variations of the thickness were observed [12,13]. In this work, an experimental loading rate was considered, and the simulations were competent to predict the experimental thickness variation pattern.

Along the wall of the cup, lower strain was observed suggesting preservation of wall thickness. On the outer area of the workpiece, a strain of approximately 0.1 was predicted due to material concentration, resulting in increased thickness. However, close to the center (approximately 5 mm radius), the radial strain was constant, where the thickness distribution was also nearly constant.

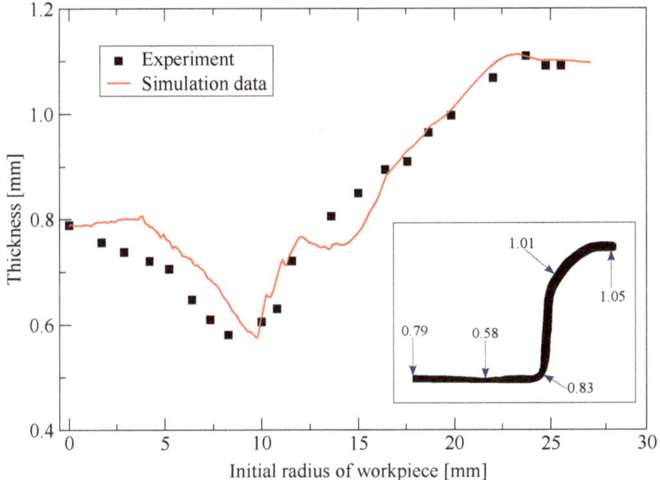

Figure 10. Thickness variation in the deformed cup with respect to the initial radius of the workpiece. Inset: half cut section view of the formed cup in the experiment (all the marked thickness values are in mm).

3.2. Damage in the Cup

Damage has a significant effect on the mechanical properties of a metal during deformation. The internal defects in the material act as nucleation sites and induce damage. The evolution of damage is essentially related to the dominant deformation mechanism. This mechanism depends on the deformation temperature, effective stress, strain rate, material micro-structure and chemical composition. In this work, the Johnson–Cook material model was considered; therefore, the focus was on the damage occurring in the forming process due to the effective stress and strain rate.

In 1985, the damage parameters of 4340 steel were investigated by Johnson and Cook [30]. Furthermore, the Johnson–Cook material model parameters were studied for Ti-6Al-4V and 7075-T6 aluminum alloy by Wang and Shi [35] and Zhang et al. [36], respectively. Recently, Buchkremer et al. [37] focused on the damage parameters of the Johnson–Cook material model of AISI 1045 steel. However, to the best of our knowledge, the damage parameters of DC04 steel have never been investigated for such a highly dynamic process. We have a number of fractured cups for different pressures, as well as misalignment of the workpiece from the experiments. The goal was to reproduce the experimental fracture patterns using the Johnson–Cook material model. Initially, the numerical simulations were performed using the proposed damage parameters of 4340 steel [30]. Then, we changed all the damage parameters in such way that the changed parameters can reproduce the experimental results. The damage parameters used in the Johnson–Cook material model are shown in Table 3.

Table 3. Damage parameters used in the Johnson–Cook material model.

Property	Value
D_1	0.02
D_2	3.9
D_3	−4.6
D_4	0.002

3.2.1. Pressure Magnitude

Figure 11 shows the variation in the pressure load curves acting on the blank to study its influence on the cup formation. As is clear from the Johnson–Cook damage material model, the fracture strain depends on the effective stress and strain rate. Therefore, in this work, the pressure load profile has been scaled, and we studied the resultant shape of the deformed cups with damage evolution.

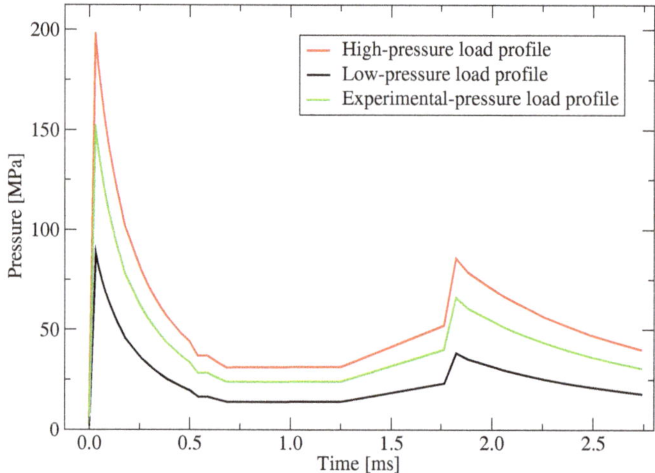

Figure 11. Pressure load curves acting on the workpiece in the experiments and simulations.

In the case of the low-pressure load profile, the workpiece was unable to deform completely. Figure 12 depicts the final shape of the workpiece when only 60% of the optimum experimental load, i.e., approximately 90 MPa peak pressure was applied. As mentioned earlier, the detonation process is highly inertia dependent, and low peak pressure was insufficient to introduce the required energy into the system. Hence, a dome-shaped output has been observed in the experiments, as well as the numerical simulations.

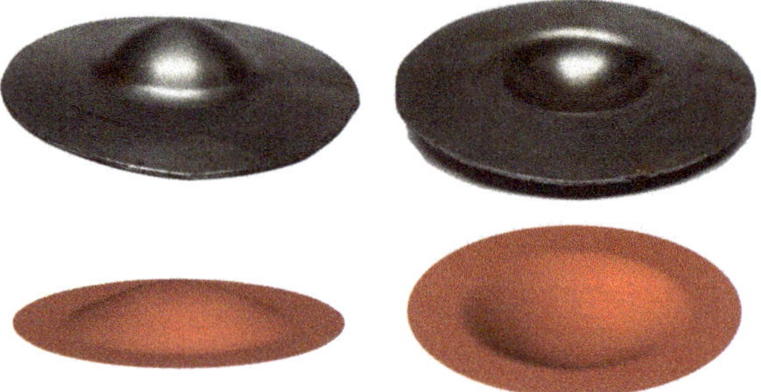

Figure 12. Comparison of the final shape obtained in the experiments (**top**) and numerical simulations (**bottom**) due to low-pressure load (60% of the optimum experimental load).

Figure 13 depicts the minor damage along the corner at the bottom. This kind of fracture was observed when pressure loading was increased up to 110% of the optimum experimental load. In the numerical simulations, a minor fracture was observed at the corners.

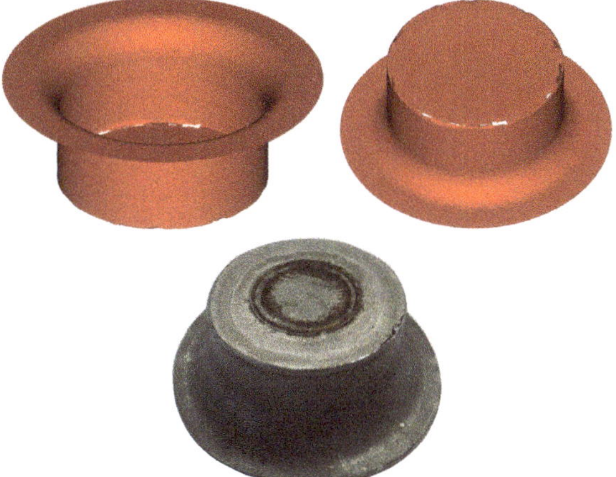

Figure 13. Damage occurring due to scaled load (110% of the optimum experimental load). Minor damage was observed along the corner. Comparison of the snapshots obtained in the numerical simulations (**top**) and experiments (**bottom**).

Figure 14 compares the simulation results at high pressure with those of the experiments. In the case of the high-pressure load, i.e., 130% of experimental optimum pressure load, the workpiece fails along the 90° bend; as a result, we observed a through hole in the workpiece. Similar observations were made in the simulations.

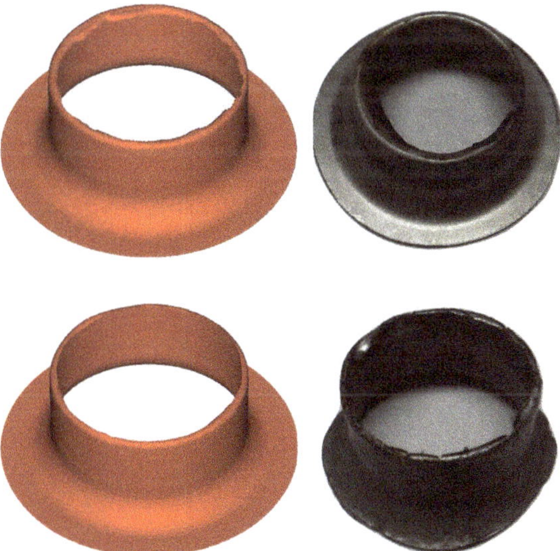

Figure 14. Damage caused by the high-pressure load profile (130% of experimental load). Snapshots of the fully damaged samples: numerical simulations (**left**) and experiments (**right**).

Figure 15 depicts the damage parameter distribution just before the fracture starts. In the vicinity of the bend, there was the highest stress concentration, as well as the maximum change in the effective plastic strain observed. Ultimately, the highest (close to 1.0) value of the damage parameter D was observed in this region. Therefore, the cup failed along the corner at the bottom.

Figure 15. Distribution of damage parameter D in the high-pressure load profile simulation (the snapshot was taken just before $D = 1$).

3.2.2. Effect of Offset

A workpiece is properly placed between the die and the top plate, i.e., the center of the workpiece matches with the center of the die, as well as the top plate. This alignment is very important in order to apply the pressure load at the central part of the workpiece, and it deforms equivalently in all radial directions to form a perfect cup.

Offsetting of a workpiece, i.e., misalignment while placing of the workpiece in between the die and the top plate, can greatly affect the final shape. In this work, the offset influence on the final shape of the cup for the optimum experimental pressure load was investigated. For this purpose, we considered 3 mm of center offset of the workpiece in the experiments, as well as the numerical simulations. Figure 16 depicts the fracture occurring in the workpiece due to an initial misalignment of 3 mm between the blank and the die. A tearing effect was observed along the side where the material was less. This was because the amount of material available was less, and the strain value was high in this region. Subsequently, due to high energy in the process, the material failed along the skirt, as well as in the corners of the die, where the high local strain was concentrated.

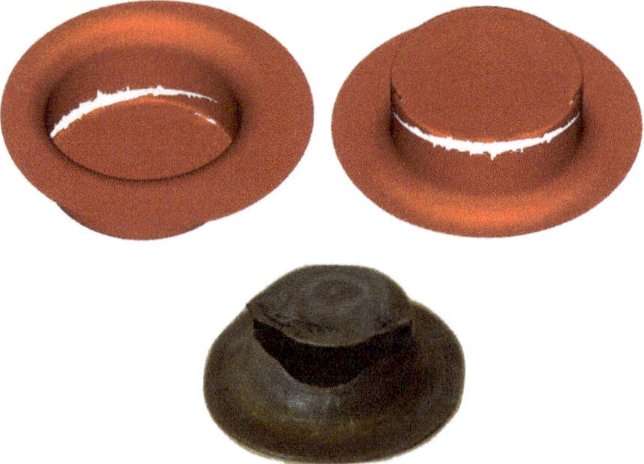

Figure 16. Initial misalignment of 3 mm between the blank and the die. Fractured workpieces in numerical simulations (**top**) and specimen in the experiment (**bottom**).

Figure 17 shows the distribution of the damage parameter D in the simulation of a misaligned workpiece before damage. As previously mentioned, failure occurs when this parameter reaches 1.0. For the offset or misaligned workpiece simulation, it was observed that D increased rapidly in the region of high strain, where less material was available. Hence, the change in equivalent plastic strain was higher, implying that in Equation (6), the ratio would reach 1.0 more quickly, compared to the other regions. Therefore, failure occurred in this region.

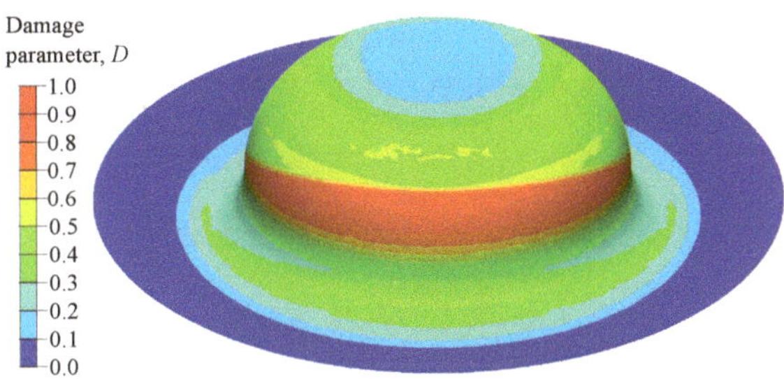

Figure 17. Distribution of damage parameter D in the simulation of the misaligned workpiece (the snapshot was taken just before the damage initiation; as the simulation continues, we obtained the final shape of the cup as shown in Figure 16).

4. Conclusions and Future Work

The experimental investigations of forming by gas detonation have shown the ability to produce sharp corners at the bottom of the formed DC04 steel cup without observable wrinkles on the flange or skirt. Furthermore, experiments conclude that the magnitude of the peak load has a high influence on the deformation.

Numerical simulations of the dynamic forming process were carried out with the Johnson–Cook plasticity model, which can mimic metal behavior on a wide range of strains, strain rates and temperatures. This plasticity model is the best choice to predict deformation during the forming process due to its moderate complexity and well-established methods to predict the material constants. Furthermore, damage parameters were included in this material model in order to study fracture behavior.

The proposed computational model was able to predict experimental results accurately, e.g., the shape of the cup and thickness distribution along the radius of the cup. Moreover, the model was capable of predicting damage initiation and evolution areas in the workpiece, which was mainly due to the high peak pressure magnitude and the initial misalignment of the same between the die and the top plate.

Further improvements can be made to the model by systematically performing a number of experiments of differently-shaped geometries. Furthermore, in the numerical study, temperature effects can be included and compared to the experiments. Moreover, the Johnson–Cook damage parameters can be studied in detail using more experiments and different sheet materials. This will help to approach more accurate results for a specific forming process.

Supplementary Materials: The supplementary materials are available online at www.mdpi.com/2075-4701/7/12/556/s1.

Acknowledgments: We thank Carlos Alberto Hernandez Padilla (Institute of General Mechanics, RWTH Aachen University) for fruitful discussions on the experimental data.

Author Contributions: Sandeep P. Patil and Bernd Markert proposed the idea; Sandeep P. Patil, Kaushik G. Prajapati and Bernd Markert conceived and designed the numerical simulations; Sandeep P. Patil, Kaushik G. Prajapati and Bernd Markert wrote the paper; Sandeep P. Patil, Kaushik G. Prajapati, Vahid Jenkouk and Herbert Olivier analyzed the data; Herbert Olivier performed the experiments.

Conflicts of Interest: The authors declare no conflict of interest.

References

1. Bruschi, S.; Altan, T.; Banabic, D.; Bariani, P.; Brosius, A.; Cao, J.; Ghiotti, A.; Khraisheh, M.; Merklein, M.; Tekkaya, A. Testing and modelling of material behaviour and formability in sheet metal forming. *CIRP Ann. Manuf. Techn.* **2014**, *63*, 727–749.
2. Marré, M.; Brosius, A.; Tekkaya, A. New aspects of joining by compression and expansion of tubular workpieces. *Int. J. Mater. Form.* **2008**, *1*, 1295–1298.
3. Marré, M.; Brosius, A.; Tekkaya, A.E. Joining by compression and expansion of (none-) reinforced profiles. *Adv. Mater. Res.* **2008**, *43*, 57–68.
4. Marré, M.; Ruhstorfer, M.; Tekkaya, A.; Zaeh, M. Manufacturing of lightweight frame structures by joining of (None-) reinforced profiles. *Adv. Mater. Res.* **2008**, *43*, 2573–2584.
5. Marré, M.; Ruhstorfer, M.; Tekkaya, A.; Zaeh, M. Manufacturing of lightweight frame structures by innovative joining by forming processes. *Int. J. Mater. Form.* **2009**, *2*, 307.
6. Martins, P.; Bay, N.; Tekkaya, A.; Atkins, A. Characterization of fracture loci in metal forming. *Int. J. Mech. Sci.* **2014**, *83*, 112–123.
7. Psyk, V.; Risch, D.; Kinsey, B.; Tekkaya, A.; Kleiner, M. Electromagnetic forming—A review. *J. Mater. Process. Technol.* **2011**, *211*, 787–829.
8. Tekkaya, A.E. State-of-the-art of simulation of sheet metal forming. *J. Mater. Process. Technol.* **2000**, *103*, 14–22.
9. Shchelkin, K.I.; Troshin, Y.K. *Gasdynamics of Combustion*, 1st ed.; Mono Book Corporation: Baltimore, MD, USA, 1965.
10. Honda, A.; Suzuki, M. Sheet metal forming by using gas imploding detonation. *J. Mater. Process. Technol.* **1999**, *85*, 198–203.
11. Mynors, D.J.; Zhang, B. Applications and capabilities of explosive forming. *J. Mater. Process. Technol.* **2002**, *125*, 1–25.
12. Yasar, M. Gas detonation forming process and modeling for efficient spring-back prediction. *J. Mater. Process. Technol.* **2004**, *150*, 270–279.
13. Yasar, M.; Demirci, H.I.; Kadi, I. Detonation forming of aluminium cylindrical cups experimental and theoretical modelling. *Mater. Design.* **2006**, *27*, 397–404.
14. El-Mokadem, A. Finite element modeling of sheet metal forming using shock tube. In Proceedings of the 9th International Conference Mechanical Design and Production (MDP-9), Cairo, Egypt, 8–10 January 2008.
15. Wijayathunga, V.; Webb, D. Experimental evaluation and finite element simulation of explosive forming of a square cup from a brass plate assisted by a lead plug. *J. Mater. Process. Technol.* **2006**, *172*, 139–145.
16. Mousavi, S.A.; Riahi, M.; Parast, A.H. Experimental and numerical analyses of explosive free forming. *J. Mater. Process. Technol.* **2007**, *187*, 512–516.
17. Khaleghi, M.; Aghazadeh, B.S.; Bisadi, H. Efficient oxyhydrogen mixture determination in gas Detonation forming. *Int. J. Mech. Mechatron. Eng.* **2013**, *7*, 1748–1754.
18. Babaei, H.; Mostofi, T.M.; Sadraei, S.H. Effect of gas detonation on response of circular plate-experimental and theoretical. *Struct. Eng. Mech.* **2015**, *56*, 535–548.
19. Babaei, H.; Mostofi, T.M.; Alitavoli, M.; Darvizeh, A. Empirical modelling for prediction of large deformation of clamped circular plates in gas detonation forming process. *Exp. Technol.* **2016**, *40*, 1485–1494.
20. Mirzababaie Mostofi, T.; Babaei, H.; Alitavoli, M. Experimental and theoretical study on large ductile transverse deformations of rectangular plates subjected to shock load due to gas mixture detonation. *Strain* **2017**, doi:10.1111/str.12235.
21. Mostofi, T.M.; Babaei, H.; Alitavoli, M. The influence of gas mixture detonation loads on large plastic deformation of thin quadrangular plates: Experimental investigation and empirical modelling. *Thin-Walled Struct.* **2017**, *118*, 1–11.
22. *LS-DYNA Theory Manual*; Livermore Software Technology Corporation: Livermore, CA, USA, 2006.
23. Kleiner, M.; Hermes, M.; Weber, M.; Olivier, H.; Gershteyn, G.; Bach, F.W.; Brosius, A. Tube expansion by gas detonation. *Prod. Eng.* **2007**, *1*, 9–17.
24. Nikiforakis, N.; Clarke, J. Numerical studies of the evolution of detonations. *Math. Comput. Model.* **1996**, *24*, 149–164.

25. Johnson, G.R.; Cook, W.H. A constitutive model and data for metals subjected to large strains, high strain rates and high temperatures. In Proceedings of the 7th International Symposium on Ballistics, The Hague, The Netherlands, 19–21 April 1983; Volume 21, pp. 541–547.
26. Verleysen, P.; Peirs, J.; Van Slycken, J.; Faes, K.; Duchene, L. Effect of strain rate on the forming behaviour of sheet metals. *J. Mater. Process. Technol.* **2011**, *211*, 1457–1464.
27. Schwer, L. Optional strain-rate forms for the Johnson Cook constitutive model and the role of the parameter Epsilon_0. In Proceedings of the 6th European LS-DYNA Users' Conference, Frankenthal, Germany, October 2007; pp. 11–22.
28. Schmitt, J.; Jalinier, J. Damage in sheet metal forming—I. Physical behavior. *Acta Mater.* **1982**, *30*, 1789–1798.
29. Jalinier, J.; Schmitt, J. Damage in sheet metal forming—II. Plastic instability. *Acta Mater.* **1982**, *30*, 1799–1809.
30. Johnson, G.R.; Cook, W.H. Fracture characteristics of three metals subjected to various strains, strain rates, temperatures and pressures. *Eng. Fract. Mech.* **1985**, *21*, 31–48.
31. Patil, S.P.; Popli, M.; Jenkouk, V.; Markert, B. Numerical modelling of the gas detonation process of sheet metal forming. *J. Phys. Conf. Ser.* **2016**, *734*, 032099.
32. Sullivan, J.F. *Technical Physics*, 99th ed.; Wiley: Hoboken, NJ, USA, 1988.
33. Jenkouk, V.; Patil, S.; Markert, B. Joining of tubes by gas detonation forming. *J. Phys. Conf. Ser.* **2016**, *734*, 032101.
34. Hosford, W.F.; Caddell, R.M. *Metal Forming: Mechanics and Metallurgy*; Cambridge University Press: Cambridge, UK, 2011.
35. Wang, X.; Shi, J. Validation of Johnson-Cook plasticity and damage model using impact experiment. *Int. J. Impact Eng.* **2013**, *60*, 67–75.
36. Zhang, D.N.; Shangguan, Q.Q.; Xie, C.J.; Liu, F. A modified Johnson–Cook model of dynamic tensile behaviors for 7075-T6 aluminum alloy. *J. Alloys Compd.* **2015**, *619*, 186–194.
37. Buchkremer, S.; Wu, B.; Lung, D.; Münstermann, S.; Klocke, F.; Bleck, W. FE-simulation of machining processes with a new material model. *J. Mater. Process. Technol.* **2014**, *214*, 599–611.

 © 2017 by the authors. Licensee MDPI, Basel, Switzerland. This article is an open access article distributed under the terms and conditions of the Creative Commons Attribution (CC BY) license (http://creativecommons.org/licenses/by/4.0/).

 *metals*

Article

Dynamic Uniform Deformation for Electromagnetic Uniaxial Tension

Xiaohui Cui [1,2,3,*], Zhiwu Zhang [1], Hailiang Yu [1,2,3], Yongqi Cheng [4] and Xiaoting Xiao [4]

1. Light alloy research institute, Central South University, 410083 Changsha, China; 163812020@csu.edu.cn (Z.Z.); yuhailiang@csu.edu.cn (H.Y.)
2. College of Mechanical and Electrical Engineering, Central South University, 410083 Changsha, China
3. State Key Laboratory of High Performance Complex Manufacturing, Central South University, 410083 Changsha, China
4. School of Material and Energy, Guangdong University of Technology, 510006 Guangzhou, China; y.q.cheng@163.com (Y.C.); xiaoxt@gdut.edu.cn (X.X.)
* Correspondence: cuixh622@csu.edu.cn; Tel.: +86-153-8802-8791

Received: 12 March 2019; Accepted: 2 April 2019; Published: 9 April 2019

Abstract: To compare with quasi-static uniaxial tensioning, researchers designed an electromagnetic uniaxial tension method using a runway coil. However, the requirements to obtain a uniformly deformed sample and the ways the stress changes on the sample using a runway coil have not been studied in the past. In this study, a three-dimensional (3D) sequential coupling method was developed to analyze the factors affecting on-sheet deformation inhomogeneity under electromagnetic uniaxial tension. Two main process parameters, comprising the die type and the relative position of the coil and sheet, were evaluated. Under the optimal parameters, the experiment and simulation both obtained uniformly deformed samples with different discharge conditions, and the simulation method had a high accuracy in modeling the deformation process. The stress state of the sample is approximately unidirectional tensile stress before 240 μs. After 240 μs, the three main stresses showed significant oscillations.

Keywords: magnetic-pulse forming; high-frequency oscillation; uniform deformation

---

1. Introduction

Both the automobile and aerospace industries use large-scale, advanced manufacturing methods that ideally minimize energy consumption and environmental impact in general. Low-density materials with good strength properties, such as aluminum alloys, help to achieve these goals. The main disadvantage of aluminum alloys is their poor formability relative to steel in conventional forming processes. Several studies have indicated that the formability of aluminum alloy increases dramatically when it is formed by high-speed processes.

Electromagnetic forming (EMF) is one of the most widely used high-speed forming methods, which can accelerate workpiece deformation by applying a magnetic force according to the electromagnetic induction theorem. A detailed review by Psyk et al. [1] showed that EMF has several advantages compared with conventional quasi-static forming, such as increased forming limits. Many scholars have proven that the formability of various materials can be dramatically increased under EMF, such as aluminum [2], steel [3], magnesium [4], copper [5], and titanium [6].

To obtain bidirectional tensile stress on a sheet, the flat spiral circular coils were used to make the sheet bulge. Imbert et al. [2] performed EMF experiments with AA5754 and AA6111 sheets, and found that strain states beyond the quasi-static forming limit diagram could be achieved using EMF. Li et al. [6] found that the forming limit of Ti-6Al-4V sheets increased by 24.37% when subjected to electromagnetic free bulging compared with quasi-static methods. Fang et al. [7] used electromagnetic pulse-assisted incremental drawing to obtain a large height-diameter ratio. Compared with the conventional drawing method, the height of the drawn cylinder was increased by 116% relative to the cylinder drawn by a conventional process.

To obtain uniaxial tensile stress on a sheet, different coil structures were developed. Li et al. [8] designed special specimens to obtain a state of uniaxial tension by EMF using a single-turn coil. Li et al. [9] experimentally investigated the formability of AA5052 sheets in a combined quasi-static-dynamic tensile process using a flat coil containing two square spiral coils. It was found that the formability of the aluminum alloy sheets after the combined quasi-static-dynamic tensile process dramatically increased compared with quasi-static-processed tensile tests. Xu et al. [10,11] designed a flat spiral coil, which can be named a "runway coil", to make AZ31 sheets uniaxially tensioned by EMF. Based on the experimental results, the strain state of uniaxial tension can be obtained. However, a deformed sample with incomplete symmetry can also be obtained.

Therefore, determining the factors that affect uniform uniaxial tension in a sample by EMF using a runway coil is a key problem. Further, determining how the stress changes in a sample during EMF is another question that should be answered. In this study, a three-dimensional (3D) sequential coupling method was used to analyze the factors affecting a sample's deformation uniformity and to obtain the optimum process parameters for electromagnetic uniaxial tension.

2. Experimental and Finite Element Simulation

The relative positions of the specimen, coil, and die are shown in Figure 1. A five-turn runway coil, combined with a capacitor bank, was used to obtain a unidirectional tensile stress state in the sheet material. The main parameters of the EMF equipment were the rated voltage (10 kV) and capacitance (1000 μF). The detailed experimental conditions were (1) the material (pure aluminum sheets with a high hardness; the elasticity modulus was taken as 0.68 GPa, Poisson's ratio was taken as 0.33, and the density was taken as 2.7×10^3 kg/m^3) and (2) the gauged width and length of the tensioned specimens, which were 12.5 and 50 mm, respectively. The materials making up the coil core were a glass epoxy board with a rectangular-shaped groove and a copper wire with a section area of 3×10 mm^2, which was imbedded in the coil core. The length and width of the coils were 165 and 68 mm, respectively. The sheet was placed above a rectangular die. The length, width, and height of the die were 280, 280, and 30 mm, respectively. The die had an open rectangle window with dimensions of 70×50 mm^2 and an entry radius of 10 mm. To facilitate subsequent analysis, we defined path 1 and path 2 on the sample, as shown in Figure 1b.

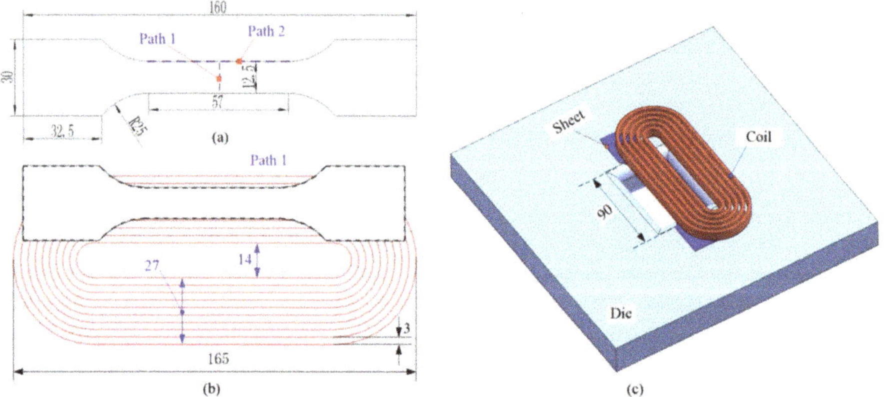

Figure 1. Schematic diagram of magnetic-pulse bending: (**a**) sample structure, (**b**) relative position of coil and sample, and (**c**) 3D structure of the forming process. Units: mm.

Figure 2 shows the finite-element model established for magnetic-pulse uniaxial tension. ANSYS/EMAG software was used to calculate the magnetic force acting on the sheet. The electromagnetic field model consists of the far air region, air region, coils, die, and sheet. The sheet and coil were meshed into eight-node elements with the SOLID97 element type. The air region was meshed into tetrahedrons with the SOLID97 element type. In the mechanical model, ANSYS/LSDYNA software was used to simulate the dynamic deformation. During the forming process, the die and holder were treated as rigid bodies. Contact conditions were considered between the die and sheet, as well as between the sheet and holder.

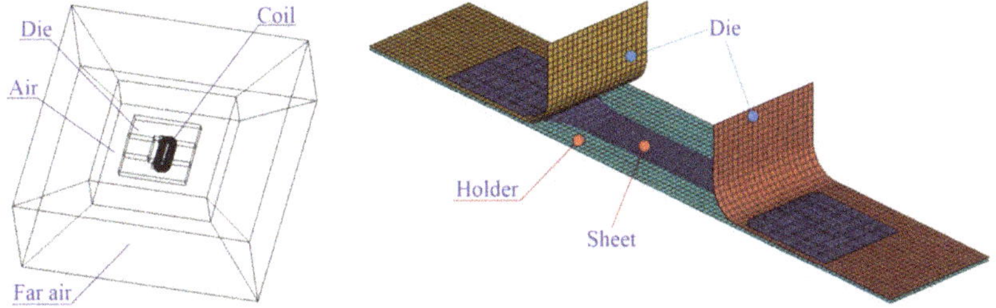

Figure 2. Finite-element models for magnetic pulse forming: (**a**) electromagnetic field model and (**b**) structure model.

For the EMF process, there are two main simulation methods for electromagnetic forming [12]: (1) the loose coupling method and (2) the sequential coupling method. If the magnetic forces are calculated based on the updated EM (electromagnetic) model, the simulation approach can be deemed to be the sequential coupling method. Thus, the sequential coupling method was adopted in this paper due to its higher accuracy. According to the electromagnetic induction theorem, the formula for the induced current and electromagnetic force on the sheet is as shown in Equations (1)–(3) [13]:

$$\frac{1}{\mu}(\nabla \times B) = \nabla \times H = J \tag{1}$$

$$f = J \times B = \frac{1}{\mu}(\nabla \times B) \times B \tag{2}$$

$$F = \frac{1}{\mu}\int (\nabla \times B) \times B \cdot dV \tag{3}$$

where μ is the permeability (H/m), B is the magnetic flux (Wb/m^2), H is the magnetic intensity (A/m), J is current density (A/m^2), f is the magnetic force per unit volume (N/m^3), V is the volume (m^3), and F is the magnetic force (N).

Pure aluminum sheets with a high hardness were used in all the experiments. One millimeter thick standard specimens were used. Table 1 shows the chemical composition of the main impurities of a pure aluminum sheet. The engineering stress and strain can be obtained by tensile tests at room temperature. The true stress and strain can be calculated by Equations (4) and (5). Thus, the true stress-strain curve of pure aluminum under quasi-static conditions can be seen in Figure 3.

$$\sigma_T = \sigma_e(1 + \varepsilon_e) \tag{4}$$

$$\varepsilon_T = \ln(1 + \varepsilon_e) \tag{5}$$

where σ_T is the true stress, σ_e is the engineering stress, ε_T is the true strain, and ε_e is the engineering strain.

Table 1. Chemical composition of the main impurities of a pure aluminum sheet.

Element	Fe	Cu	Mg	Mn	Ga	Cr	K
wt. %	0.24	0.013	0.011	0.0055	0.019	0.0099	0.01

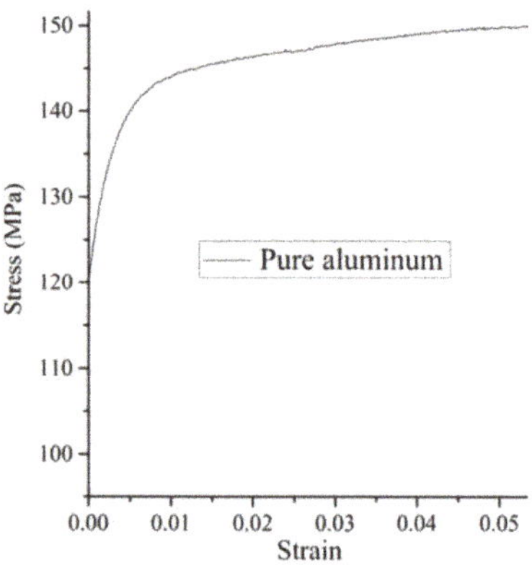

Figure 3. Stress-strain curve.

To consider the effect of a high strain rate on forming, the behavior of a viscoplastic material was modeled with a rate-dependence law (the Cowper-Symonds constitutive model) in the ANSYS/LSDYNA finite-element analysis software. Thus, the Cowper-Symonds constitutive model was used to model the high-speed forming, as shown in Equation (6):

$$\sigma = \sigma_s \left(1 + \left(\frac{\dot{\varepsilon}}{P}\right)^m\right) \tag{6}$$

where σ is the dynamic flow stress, σ_s is the quasi-static constitutive behavior of the sheet (cf. Figure 3), $\dot{\varepsilon}$ is the strain rate, $P = 6500 \text{ s}^{-1}$, and $m = 0.25$ is the specific parameter of the aluminum alloy.

3. Results

3.1. Effect of Die on Current Distribution and Magnetic Force

To prove that the effect of the die shown in Figure 1 can significantly affect the current distribution on the sample, two sets of dies were selected: (1) a conductive die made of Q235 steel and (2) a nonconductive die made of epoxy plate. Figure 4a shows the current curves through the coil, measured using a Rogowski coil. A higher current amplitude and the same pulse width were obtained at a higher discharge voltage by using a nonconductive die made of epoxy plate. Compared with the nonconductive die made of epoxy plate, the current amplitude and pulse width obtained by using the steel die were larger and smaller at 1250 V, respectively. This is because coil discharging not only produces electromagnetic induction to the sheet, but also to the steel die. As the induction current on the steel die is opposite to the current passing through the coil, the steel die will generate mutual inductance with the coil, reducing the inductance of the entire system. Thus, the current amplitude obtained using the steel die was higher than that obtained using the epoxy die. Figure 4b shows a schematic diagram of counterclockwise current loaded on the coil. To facilitate subsequent analysis, we defined the distance between the right-end face of sample and the coil's inner turn as *h*, as shown in Figure 4b.

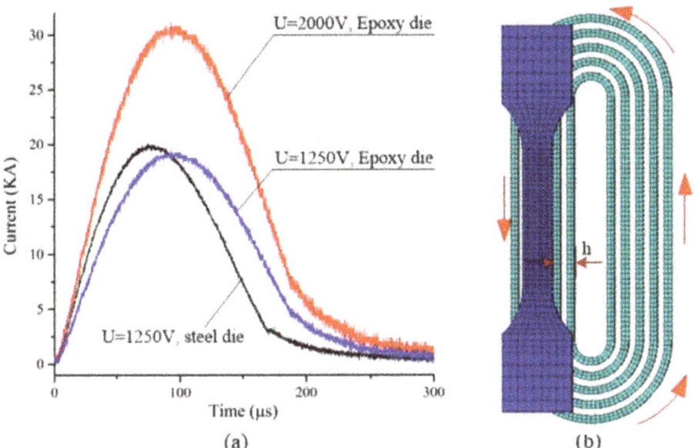

(a)

(b)

Figure 4. Current through the coil: (**a**) current curves measured by Rogowski coil and (**b**) counterclockwise current loaded on the coil.

Figure 5 shows the current density and current direction distributions on the sheet and die at 20 μs. In the case of the steel die, the sample and die constituted a clockwise current loop, the current in the middle of the sample was almost equal, and the current in the two side ends is slightly larger in the sample. In the case of the epoxy die, the induced current on the sample itself constituted a clockwise current loop within the sample because the epoxy die is not conductive. The current near the edge of the side arc was larger than that in the central area of the sheet, and the current directions on each side end of the sheet were opposite.

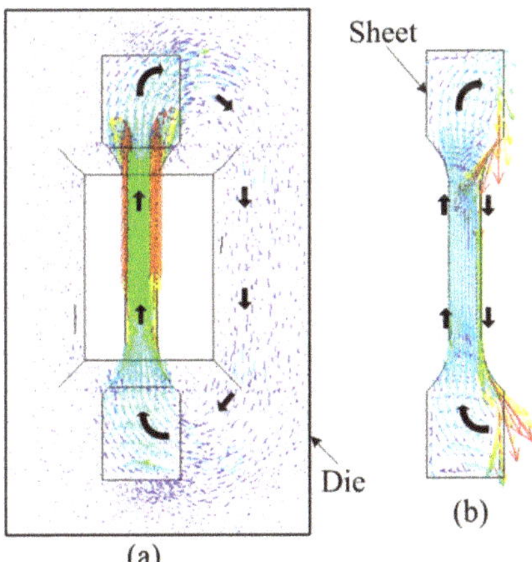

Figure 5. Current distribution after coil discharge at 20 μs: (**a**) using the conductive die and (**b**) using the nonconductive die.

Figure 6 shows the magnitude and direction of the magnetic force on the sheet at 20 μs. In the case of the steel die, the magnetic force acting on the sample was generally in the positive direction of the Z-axis, and the magnetic force on the central region of the sample was evenly distributed. In the case of the epoxy die, the magnetic force acting on the middle of the sample was too small. This is because the magnetic force is related to the current distribution on the sheet. According to the simulation results shown in Figure 5; Figure 6, a steel conductive die made of Q235 steel was used in the following simulation and experiment.

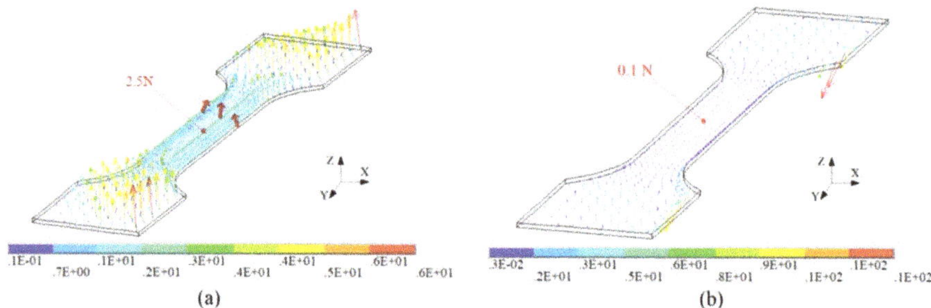

Figure 6. Magnetic force distribution on the sample at 20 μs: (**a**) using the conductive die and (**b**) using the nonconductive die.

3.2. Effect of the Relative Position of the Coil and Sample on the Deformation

Figure 7 shows the relative position of the sample to the coil. The width of the left half of the coil is 27 mm, and the width of the sample is 12.5 mm. If h = 14.5 mm, the left side of the straight edge of the sample overlaps with the leftmost side of the coil, as shown in Figure 7a. Figure 7e shows the current flow direction and current in the middle of the sample. The current density on the left side of sample was far less than that on the right side of sample. This is because there were more wires on the right side of the sample. If h = 0 mm, the right side of the straight edge of the sample coincides with the right side of the left part of the coil, as shown in Figure 7b. Figure 7f shows the current flow direction and intensity in the middle of the sample shown in Figure 7b. The current density on the left side of the sample was much higher than that on the right side of the sample. If h = 7.25 mm, the middle axis of the sample overlaps with the middle axis of the left half of the coil. Figure 7g shows the current flow direction and intensity in the middle of the sample, using the condition shown in Figure 7c. The current density on the left side of the sample was higher than that on the right side of the sample. This is because the right side of the coil was close to the left side of the coil, and the current flowing through the right side of the coil was opposite to the current flowing through the left side of the coil; therefore, the current density on the left side of the sample was higher than that on the right side of the sample. When h = 9.45 mm, the current densities on both sides of the straight edge of the sample were the same and higher than the current density in the middle of the sample. Therefore, the entire current density was symmetrically distributed on the sheet, as shown in Figure 7d,h.

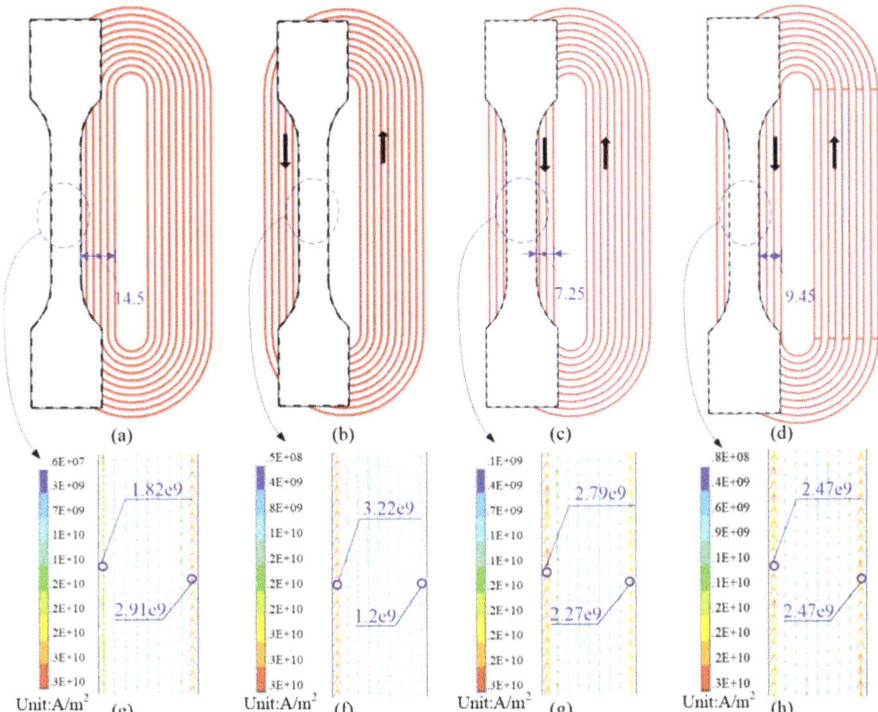

Figure 7. Effect of the relative position of the coil and sample on the current density with (**a**) $h = 14.5$ mm, (**b**) $h = 0$ mm, (**c**) $h = 7.25$ mm, and (**d**) $h = 9.45$ mm (**e**) current value with $h = 14.5$ mm (**f**) current value with $h = 0$ mm, (**g**) current value with $h = 7.25$ mm, (**h**) current value with $h = 9.45$ mm.

Figure 8 shows the relative position of the sample and coil and the effect of the magnetic force on path 1. The distribution of magnetic force was similar to the current density shown in Figure 7. Therefore, the higher the current density, the higher the magnetic force acting on the sample. Only when $h = 9.45$ mm, the magnetic force on Path 1 became symmetrically distributed.

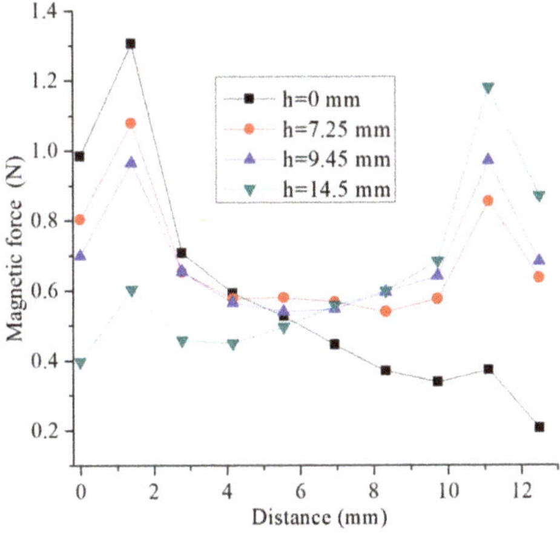

Figure 8. The relative position of the sample affected by the magnetic force on path 1.

Figure 9 shows the effect of the relative positions of the sheet and coil on the final sheet shape. When h = 0, 7.25, and 14.5 mm, because of the asymmetric distribution of magnetic force on the sheet, the sheet shape was distorted. In particular, when h = 0 mm, the magnetic force acting on the left side of the sheet was significantly higher than that on the right side of the sheet, and the deformation on the left side of the strip was significantly higher than that on the right side of the strip. When h = 14.5 mm, the deformation on the right side was significantly higher than that on the left side. When h = 9.45 mm, the displacements of the left, middle, and right ends of the highest point of the sheet were 19.3, 19, and 19.3 mm, respectively. The central deformation of the sheet was ~1.5% smaller than that of the left and right ends of sheet.

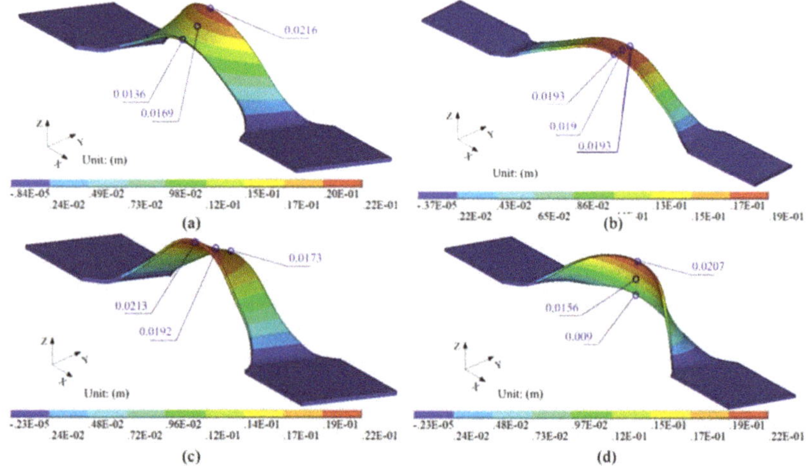

Figure 9. Effect of relative position on final shape: (**a**) h = 14.5 mm, (**b**) h = 9.45 mm, (**c**) h = 7.25 mm, (**d**) h = 0 mm.

Figure 10 shows the effect of the relative position of the sheet and coil on path 1. When h = 0, 7.25, and 14.5 mm, the deformation on Path 1 was linear. When h = 9.45 mm, the sheet was symmetrically distended on both sides, and the deformation was symmetrical.

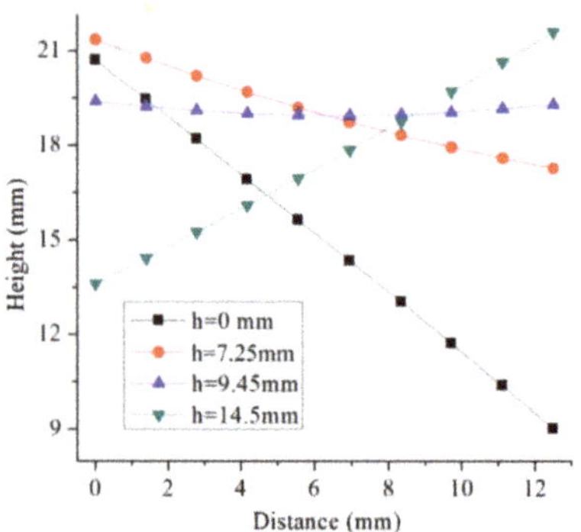

Figure 10. Deformation profiles on Path 1.

3.3. Comparison of Experimental and Simulation Results

Based on the above simulation results, the steel conductive die made of Q235 steel and the relative distance h = 9.45 mm were chosen. Figure 11 shows the profiles of path 2 and the maximum heights obtained from the experiment and simulation at different discharge voltages. The profiles between the experiment and simulation showed only a slight deviation and a small error at the sheet peak under different discharge voltages. Therefore, the simulation method had a high accuracy to predict the dynamic deformation from electromagnetic uniaxial tension.

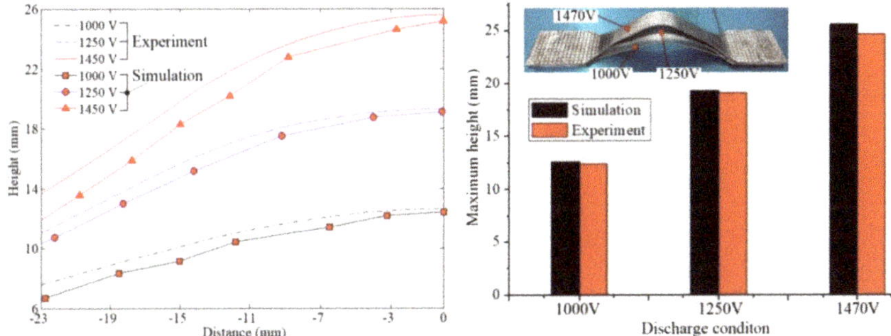

Figure 11. Experimental and simulation results: (**a**) the final profiles on path 2 and (**b**) the maximum height of the sample peak.

4. Discussion

At a discharge voltage of 1250 V, Figure 12a shows the equivalent plastic strain distribution on the sample when the deformation was terminated. The maximum equivalent plastic strain was 0.193. To facilitate subsequent analysis, the special nodes and elements shown in Figure 12a were extracted. Figure 12b–d shows the changes in the equivalent plastic strain, deformation height, and velocity with time at the special nodes. At time t = 240 µs, the equivalent plastic strain at special nodes reached a maximum and no longer changed. It can be considered that t = 240 µs represents the end of deformation. However, after t = 240 µs, the velocity and displacement of nodes on the sample still varied slightly.

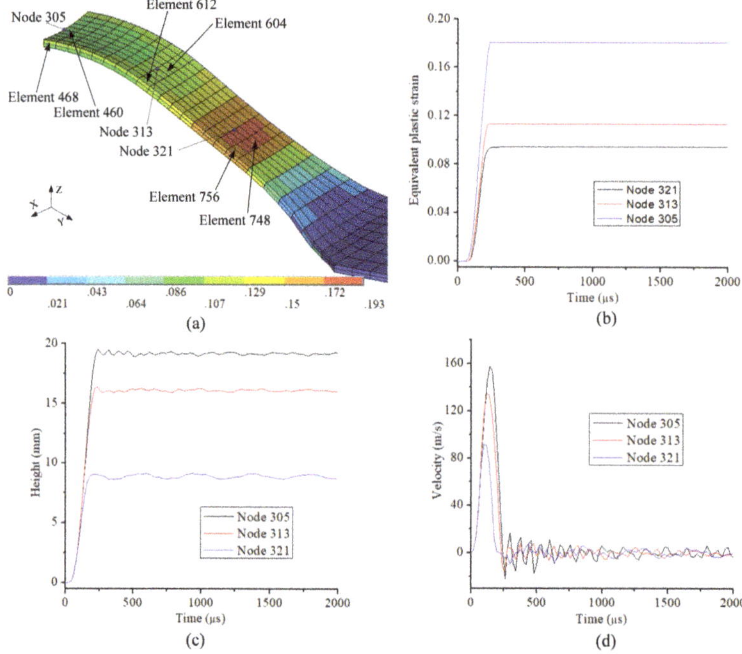

Figure 12. Dynamic deformation: (**a**) 3D shape and equivalent plastic strain, (**b**) plastic strain with time, (**c**) height with time, and (**d**) velocity with time.

Figure 13 shows the changes in the three principal stresses at special nodes with time. Some phenomena can be found at 0–240 µs: (1) the first principal stress at the three special nodes gradually increased, and the direction of the first principal stress is the tangential direction on the sample; (2) the second principal stress at the three special nodes was very small, and the direction of second principal stress was the width direction on the sample; (3) the third principal stress at the three special nodes was less than zero due to the magnetic pressure acting on these nodes. The magnitude of the third principal stress was much less than the first principal stress and the direction of the third principal stress was normal to the sample. Due to the second and third principal stresses both being much smaller than the first principal stress, the deformation process can be approximately considered as a uniaxial tensile stress state. After 240 µs, the three main stresses at the special nodes showed significant oscillations.

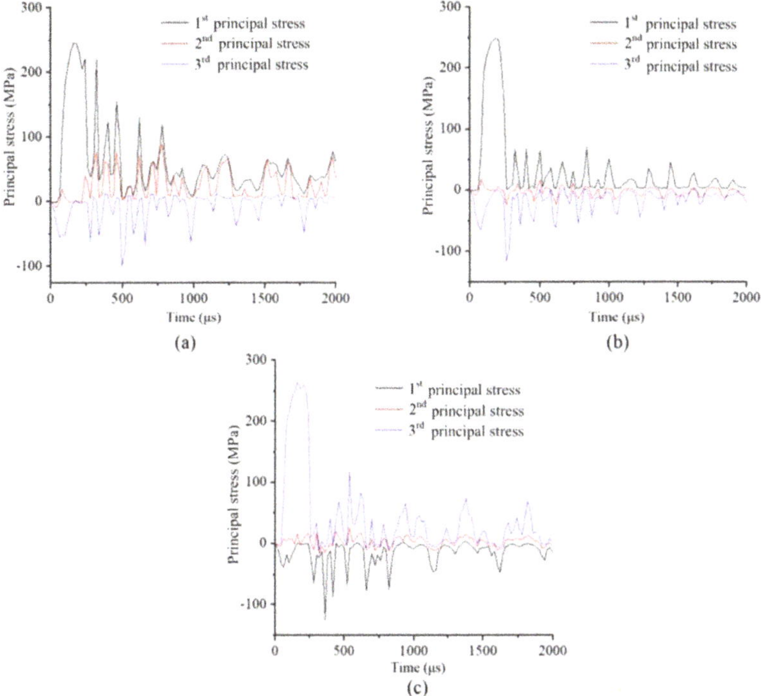

Figure 13. Relationship between time and principal stress at special nodes: (**a**) node 305, (**b**) node 313, and (**c**) node 321.

When the sheet was bent, the maximum stress of the sheet was usually at a tangent to the sheet. However, when the high-frequency shock occurred, the stress on the plate was disordered. To analyze the magnitude and direction of the principal stress on the sheet at different times more clearly, six elements in Figure 12a were extracted for the following analysis. Figure 14 shows the changes in stresses with time at six special elements. At a time of 180 µs, all the six elements were subjected to tangential tensile stress above 240 MPa, and the stresses in the width and thickness directions were very small. Therefore, the sheet deformation can be considered as a unidirectional tensile state. At 260 µs, the sheet top element (ELEM 468, 460) was still subjected to tangential tensile stress; however, the tangential tensile stress decreased to 33.5 MPa and 30 MPa. However, the elements (ELEM 612, 604, 756, and 748) were subjected to a higher tangential compressive stress. At 320 µs, the sheet top elements (ELEM 468 and 460) were subjected to the tangential tensile stress; however, the tangential tensile stress increased to 205 MPa and 219 MPa, respectively. At 360 µs, the stress direction of top element (ELEM 468 and 460) changed from tensile stress to compressive stress; however, the tangential compressive stresses were 84 and 27 MPa, respectively. At 360–500 µs, the stress on the top elements (ELEM 468 and 460) was still a lengthwise compressive stress; however, the value of the compressive stress continuously changed. Comparing the other elements (ELEM 612, 604, 756, and 748), the stress direction on each unit still alternately changed from tensile to compressive stress, and the magnitude also changed constantly. The dynamic changes in the stress generally significantly decreased the lengthwise stress on the sheet compared with that at 180 µs. At 2000 µs, the residual lengthwise stress on the sheet was not only small, but also both lengthwise tensile and compressive stresses existed on the outer surface of the sheet.

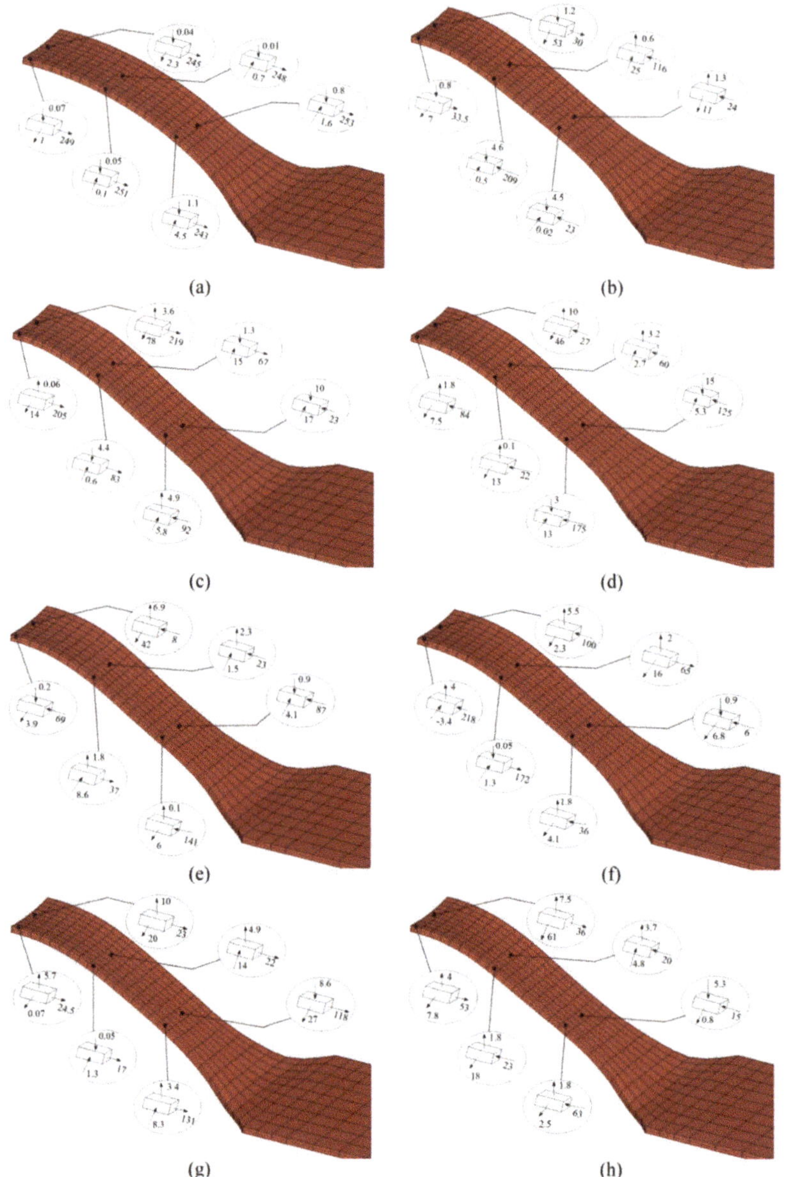

Figure 14. Stress and direction at different times: (**a**) 180 μs, (**b**) 260 μs, (**c**) 320 μs, (**d**) 360 μs, (**e**) 420 μs, (**f**) 500 μs, (**g**) 1000 μs, and (**h**) 2000 μs.

5. Conclusions

In this study, a 3D sequential coupling method was developed to analyze the factors affecting on-sheet deformation inhomogeneity using runway coils. Compared with the experimental results under different voltages, the simulation results had high accuracy and precision. Some conclusions can be obtained as follows:

(1) The die has a significant effect on the current distribution in the sample. Compared with a nonconductive die, the same current direction with similar current intensity in the sample can be obtained with a steel die.

(2) The symmetrical distribution of magnetic force on path 1 can be obtained by adjusting the relative position of the sample and coil. Subsequently, the sheet deforms with a uniform 3D shape.

(3) The deformation process can be approximately considered as a uniaxial tensile stress state before 240 μs. After 240 μs, the three main stresses at the special nodes showed significant oscillations.

Author Contributions: The research was conceived by X.C. and Z.Z.; X.C. and Z.Z. planned and performed all the experiments; Z.Z. collected all data; Theoretical and experimental analysis were performed by H.Y., Y.C. and X.X.; The manuscript was reviewed by X.C. and H.Y., The manuscript was written by X.C.; with support from all co-authors.

Funding: This work was supported by the National Natural Science Foundation of China (Grant Nos. 51775563 and 51405173), Innovation Driven Program of Central South University (Grant number: 2019CX006), State Key Laboratory of Materials Processing and Die & Mould Technology, Huazhong University of Science and Technology (No. P2017-013) and the Project of State Key Laboratory of High Performance Complex Manufacturing, Central South University (ZZYJKT2017-03), Guangzhou Science and Technology Plan Project (201707010472) and The project was supported by Open Research Fund of State Key Laboratory of High Performance Complex Manufacturing, Central South University (No. Kfkt2018-02).

Conflicts of Interest: The authors declare no conflict of interest.

References

1. Psyk, V.; Risch, D.; Kinsey, B.L; Tekkaya, A.E.; Kleiner, M. Electromagnetic forming—A review. *J. Mater. Process. Technol.* **2011**, *211*, 787–829. [CrossRef]

2. Imbert, J.M.; Winkler, S.L.; Worswick, M.J.; Oliveira, D.A.; Golovashchenko, S. The effect of tool-sheet interaction on damage evolution in electromagnetic forming of aluminum alloy sheet. *J. Eng. Mater. Technol.* **2005**, *127*, 145–153. [CrossRef]

3. Seth, M.; Vohnout, V.J.; Daehn, G.S. Formability of steel sheet in high velocity impact. *J. Mater. Process. Technol.* **2005**, *168*, 390–400. [CrossRef]

4. Meng, Z.H.; Huang, S.Y.; Hu, J.H.; Huang, W.; Xia, Z.L. Effects of process parameters on warm and electromagnetic hybrid forming of magnesium alloy sheets. *J. Mater. Process. Technol.* **2011**, *211*, 863–867. [CrossRef]

5. Jiang, H.W.; Ni, L.; Xiong, Y.Y.; Li, Z.G.; Liu, L. Deformation behavior and microstructure evolution of pure Cu subjected to electromagnetic bulging. *Mater. Sci. Eng. A* **2014**, *593*, 127–135. [CrossRef]

6. Li, F.Q.; Mo, J.H.; Li, J.J.; Zhou, H.Y. Formability of Ti-6Al-4V titanium alloy sheet in magnetic pulse bulging. *Mater. Des.* **2013**, *52*, 337–344. [CrossRef]

7. Fang, J.X.; Mo, J.H.; Cui, X.H.; Li, J.J.; Zhou, B. Electromagnetic pulse-assisted incremental drawing of aluminum cylindrical cup. *J. Mater. Process. Technol.* **2016**, *238*, 395–408. [CrossRef]

8. Li, G.Y.; Deng, H.K.; Mao, Y.F.; Zhang, X.; Cui, J.J. Study on AA5182 aluminum sheet formability using combined quasi-staticdynamic tensile processes. *J. Mater. Process. Technol.* **2018**, *255*, 373–386. [CrossRef]

9. Li, C.; Liu, D.; Yu, H.; Ji, Z. Research on formability of 5052 aluminum alloy sheet in a quasi-static–dynamic tensile process. *Int. J. Mach. Tools Manuf.* **2009**, *49*, 117–124. [CrossRef]

10. Xu, J.R.; Yu, H.P.; Li, C.F. An Experiment on Magnetic Pulse Uniaxial Tension of AZ31 Magnesium Alloy Sheet at Room Temperature. *J. Mater. Eng. Perform.* **2013**, *22*, 1179–1185. [CrossRef]

11. Xu, J.R.; Lin, Q.Q.; Cui, J.J.; Li, C.F. Formability of magnetic pulse uniaxial tension of AZ31magnesium alloy sheet. *Int. J. Adv. Manuf. Technol.* **2014**, *72*, 665–676. [CrossRef]

12. Cui, X.H.; Mo, J.H.; Li, J.J.; Huang, L.; Zhu, Y.; Li, Z.W.; Zhong, K. Effect of second current pulse and different algorithms on simulation accuracy for electromagnetic sheet forming. *Int. J. Adv. Manuf. Technol.* **2013**, *69*, 1137–1146. [CrossRef]
13. Cui, X.H.; Mo, J.H.; Han, F. 3D Multi-physics field simulation of electromagnetic tube forming. *Int. J. Adv. Manuf. Technol.* **2012**, *59*, 521–529. [CrossRef]

 © 2019 by the authors. Licensee MDPI, Basel, Switzerland. This article is an open access article distributed under the terms and conditions of the Creative Commons Attribution (CC BY) license (http://creativecommons.org/licenses/by/4.0/).

Article

Numerical Prediction of Forming Car Body Parts with Emphasis on Springback

Peter Mulidrán [1], Marek Šiser [2], Ján Slota [2,*], Emil Spišák [1] and Tomáš Sleziak [1]

[1] Department of Mechanical Technology and Materials, Faculty of Mechanical Engineering,
Technical University of Košice, Mäsiarska 74, 040 01 Košice, Slovakia; peter.mulidran@tuke.sk (P.M.);
emil.spisak@tuke.sk (E.S.); tomas.sleziak@tuke.sk (T.S.)

[2] Department of Computer Support of Technology, Faculty of Mechanical Engineering, Technical University
of Košice, Mäsiarska 74, 040 01 Košice, Slovakia; marek.siser@tuke.sk

* Correspondence: jan.slota@tuke.sk; Tel.: +421-55-602-3545

Received: 27 April 2018; Accepted: 6 June 2018; Published: 8 June 2018

Abstract: Numerical simulation is an important tool which can be used for designing parts and production processes. Springback prediction, with the use of numerical simulation, is essential for the reduction of tool try-outs through the design of the forming tools with die compensation, therefore, increasing the dimensional accuracy of stamped parts and reducing manufacturing costs. In this work, numerical simulation was used for performing the springback analysis of car body stamping made of aluminium alloy AA6451-T4. The finite element analysis (FEM) based software PAM-STAMP 2G was used for performing the forming and springback simulations. These predictions were conducted with various combinations of material models to achieve accurate springback prediction results. Six types of yield functions (Barlat89, Barlat2000, Vegter-Lite, Hill90, Hill48 isotropic, and Hill48 orthotropic) were used in combination with the Voce hardening model. Springback analysis was conducted in three sections of the formed part; the numerical results were compared with the experimental values. It was found that the combinations of Barlat's yield functions and the Voce hardening law were most accurate in terms of springback prediction. Additionally, it was found that the phenomena that were investigated, which are required for the determination of the kinematic hardening model, such as the change of Young's modulus E, the transient behaviour, work-hardening stagnation, and permanent softening, were not observed in the aluminium alloy studied.

Keywords: springback; numerical simulation; yield function; aluminium alloy formability

---

1. Introduction

Automobile manufacturers have started to use new types of high strength steels (HSS, AHSS, and UHSS) at the end of the last century, with the aim of increasing the passive safety of vehicles and to reduce vehicle weight to decrease fuel consumption [1–3]. However, these types of steels have a lower formability in comparison with steels used for deep drawing. The main reason for this is the higher values of the yield strength and lower ductility of high strength steels [2]. In addition, aluminium alloys are now widely used in the automotive industry due to advantages, including the low density, high specific strength, good corrosion resistance, exceptional specific stiffness, and so forth [1]. The implementation of aluminium alloys in car body production can reduce fuel consumption and emissions [4]. Both high strength steels and aluminium alloys are more prone to wrinkling and springback than mild steels [1,5].

Springback in the present refers to a change of shape which is elastically driven. Springback occurs following a sheet-forming operation when the forming loads are removed from the workpiece—sheet metal blank. It is usually unwanted, causing problems in the next forming operations, in assembly, and in the final product. These problems usually degrade the accuracy, appearance, and quality of the products being manufactured [2,3,6]. The most common counter measurement against the springback of car body parts is to design a forming tool with anticipation of springback, thus compensating springback by die design. However, the amount of compensation is a difficult question even for skilled tool designers. In practice, this compensation of die is still sometimes done by the "trial and error" method. This method can be replaced by FEA (finite element analysis)—numerical simulation. With the use of FEA, it is possible to achieve a more accurate prediction of springback [6–8]. There are other counter measurements against springback, for example, the stiffening of pressings (use of beads or embossing), crash forming with pressure pads, the use of variable blank holder force, and so forth [7].

In general, two types of methods are used for springback prediction—finite element analysis and the analytical model. For example, the analytical model for springback prediction of aluminium alloys can be found in the work by Gau and Kinzel [9]. Analytical methods usually use simplified models of real processes. Thus, analytical models are usually not as accurate in predicting springback as numerical simulations and their use is limited, especially for stampings with complex geometry [10]. The finite element method (FEM) is a well-known tool for the prediction and analysis of sheet metal deformation. Springback prediction with the use of numerical simulation is not limited by the geometrical complexity of the stamped part like in the case of the analytical model. However, the numerical simulation of springback is more sensitive to the accuracy of the input data than the analytical method. Thus, it is very important to choose the correct input and numerical parameters in the FEA analysis of springback [11].

Numerical parameters involve the through-thickness integration scheme (which can be implicit, explicit, or a combination of both), the number of integration points, the used elements (type, size, and count), and so forth. Trzepiecinski and Lemu [12] studied the effect of a number of integration points and integration rules on the springback amount. Their results indicate that at least 5 integration points must be used to achieve accurate springback prediction. The input parameters involve geometry (sheet thickness, tool and sheet dimensions, and so forth), process conditions (type of forming method, tribology, forming forces, forming temperature and speed, and so forth), and material characteristics (Young's modulus, yield strength, hardening behaviour, yield function, and so forth) [10,11]. Slota, Siser, and Dvorak [13] studied the effects of yield functions (isotropic and orthotropic) on the springback prediction accuracy of aluminium alloys. Their results showed that the orthotropic yield function is more accurate in predicting springback than the isotropic function. In addition, the effect of the die radius on springback was studied. They found out that the increase of the bending radius caused a higher springback of the bend materials. Seo et al. [14] conducted a study to evaluate the effect of constitutive equations on the springback prediction accuracy. They used two yield functions, Hill48 and Yld2000, in combination with the Yoshida-Uemori hardening model in the finite element (FE) simulation to predict the springback of the U-bend part and drawn T-shape part. Both parts were made of TRIP steel. They found out that it is essential to choose the right yield function to get an accurate prediction of springback. Mulidran et al. [15] conducted a numerical simulation of the drawing hat-shaped part made of the DP600 and DC04 steels with the use of two forming methods: drawing with a blank holder and crash forming with a pressure pad. They studied the effect of forming methods and the various process parameters on springback amount. Their results indicated that the higher blank holder and pad pressure have positive effects on reducing springback. Additionally, crash forming with a pressure pad showed lower springback in comparison with drawing with a blank holder. The work by Jung et al. [16] aimed at studying anisotropic hardening behaviour and the springback of AHSS steels. They proposed the modified isotropic-kinematic hardening model, which they used in the simulation of U-bending. Their model showed better results in the predicting springback in comparison with the isotropic hardening model.

The novelty of this work lies in findings which indicate that the isotropic hardening model is sufficient for predicting the springback of formed parts made of aluminium alloy. This model is simpler, and does not need cyclic shear tests. In addition, isotropic hardening models do not use as many parameters as kinematic hardening models. The accuracy of springback prediction with use of the isotropic hardening model was high. In addition, we found out that the phenomenon of the degradation of Young´s modulus is not present for aluminium alloys which are precipitation hardened, and that the degradation is not as significant as in AHSS steels (max. 2% degradation of Young´s modulus for aluminium alloys). Villuendas et al. [17] and Roca et al. [18] studied the effect of plastic deformation on the changes of Young´s modulus of metallic alloys. They reported that, in aluminium alloys, there were no appreciable changes in the E value. This conclusion is consistent with our findings. These changes are related to the dislocation density changes. However, even though the dislocation density is high, the values of parameter l (length of dislocations) are very low, due to the interaction between nanometric precipitates and dislocations in the aluminium alloys. The Mott model then shows that the change of Young´s modulus E is very small. Kinematic hardening models also take into account other phenomena, such as the transient behaviour, work-hardening stagnation, and the permanent softening. These phenomena were not observed in the material studied in this work.

These findings have a significant financial impact. For example, it is not necessary to conduct time-consuming tests on special (expensive) equipment, which are used to determine the parameters for kinematic hardening models.

In addition, the detailed analysis of a complex shaped part made of aluminium alloy with a significant thickness of 3 mm, mainly used in car production, was conducted. In most of the studied literature, the research was done on simply shaped parts.

In this research work, a FEM was used to predict the springback of a car body part made of aluminium alloy AA6451-T4. The finite element analysis (FEA) was conducted to investigate the influence of the used yield functions in the numerical simulation on springback prediction accuracy. Three sections were used for springback evaluation; in these sections, the thickness and part profile were measured and compared with the experimental results. The experimental results were given by the automobile manufacturer. Additionally, the computation times for the various yield functions were compared.

2. Materials and Methods

In the presented work, aluminium alloy AA6451-T4 with a thickness of 3.00 mm was used as the blank. Mechanical properties were measured by uniaxial and biaxial tensile tests. To obtain the required data for the FEM model, the specimens for the uniaxial tensile test were cut in three different orientations (0°, 45°, and 90° to the rolling direction). Specimens for the uniaxial tensile test were produced according to the EN 10002-1:2002-11 standards. Several specimens were tested for each orientation, and the average values of the basic mechanical properties (displayed in Table 1) were obtained by the formula

$$X_{av} = \frac{X_0 + 2X_{45} + X_{90}}{4} \tag{1}$$

where X is the mechanical parameter, and the subscripts denote the orientation of the specimen with respect to the rolling direction of the sheet. The elastic mechanical properties of the aluminium alloy are shown in Table 2.

Table 1. The uniaxial tensile test data of the AA6451-T4 sheet [19].

Orientation	Yield Strength σ_y (MPa)	Normal Anisotropy r (–)
0°	151.28	0.62
45°	171.20	0.33
90°	163.60	0.80
Average value	164.32	0.52

Table 2. The elastic mechanical properties of AA66451-T4 [19].

Sample	Density, ρ (g·cm^{-3})	Young´s Modulus, E (GPa)	Poisson´s Ratio, v
AA6451-T4	2.7	70.0	0.3

The true stress–strain curves obtained in three different orientations (0°, 45°, and 90° with respect to the rolling direction) are shown in Figure 1. The tension-compression test (Figure 2) was started by the tension load as the first part of the full cycle. After a specific crosshead stroke corresponding to a defined pre-strain level, the load was reversed to compression until it reached the crosshead displacement according to a given compression strain. Next, reloading in tension direction was introduced until the crosshead stroke was equal to that in the first tension. The investigated phenomena, such as the change of Young's modulus E, transient behaviour, work-hardening stagnation, and permanent softening, were not observed in the material studied in this work, as we can see from Figure 2.

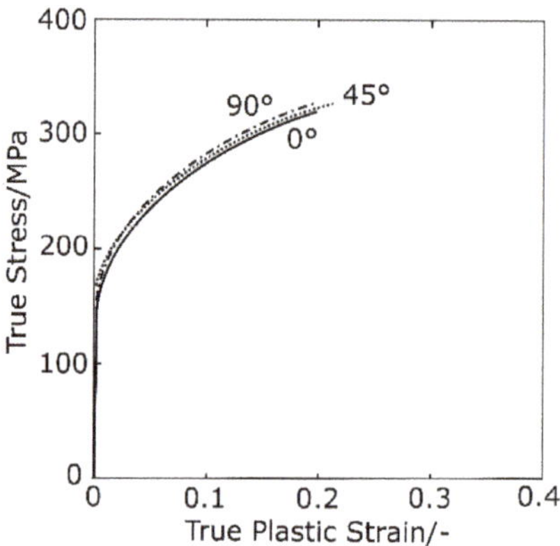

Figure 1. The experimental stress-strain curves from the tensile test.

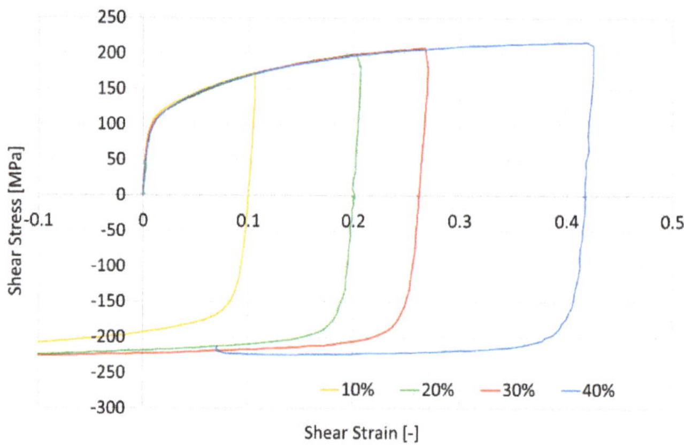

Figure 2. The cyclic tension-compression experimental curve of AA6451-T4 [19].

The equal biaxial tensile yield stress and the biaxial anisotropy r_b are given in Table 3. The parameters obtained from this test are necessary to determine the advanced yield functions.

Table 3. The equal biaxial tension test data [19].

Biaxial Yield Stress σ_b (MPa)	Biaxial Anisotropy r_b (-)
153.60	0.55

3. Numerical Model

The springback computation was performed using the dynamic explicit code in the PAM-Stamp 2G software. The tool setup imported in the simulation software is shown in Figure 3. The tool consists of punch, blankholder, and die. The tool is aligned with the global z- axis without a plane of symmetry. The blank was positioned between the die and blankholder.

Figure 3. The tool geometry in the PAM-Stamp 2G software.

The blank was meshed by the quadrilateral shell elements which were 23 mm in size. The refinement level of the elements was set to 4 so that the smallest elements after refinement had a size of 2.875 mm. The number of integration points was set to 11, which is recommended for springback computation. The friction coefficient was set to 0.08, which responded to the grease lubrication. The initial meshed blank with a rolling direction is illustrated in Figure 4.

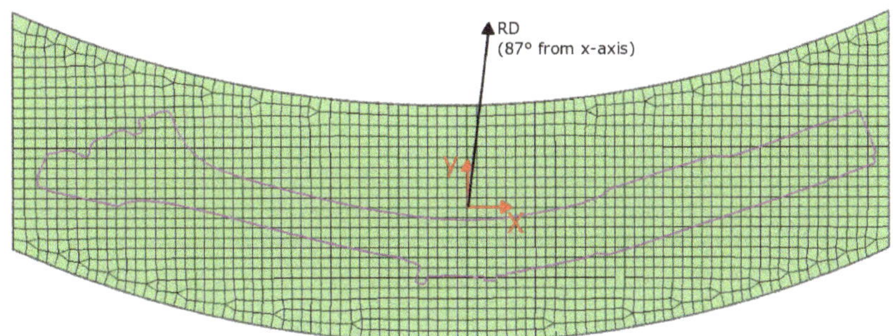

Figure 4. The rolling direction on the initial meshed blank.

The obtained values of the mechanical properties were used as the basic input for the material model in the FEM simulation. The accuracy of the springback prediction for several yield functions was investigated by the FEM simulation. The yield function describes the material transition from the elastic state to the plastic state. It can be described as a function of the area that limits the elastic area of the multi-axis stress plane. In 1948, Hill introduced the concept of material anisotropy in yield functions. According to Hill's plasticity conditions [20], in case of uniaxial load, a local thickness reduction occurs in a direction sensitive to the sample load. Hill assumed that the direction of the compression is in line with the direction of zero extension and, therefore, the deformation of the narrowed areas is only reflected as a reduction in thickness. This is assumed for the plane strain (σ_1—major stress, σ_2—minor stress, and $\sigma_3 = 0$). If we assumed that the anisotropy axes are identical with the main guideline strain tensor ($\sigma_x = \sigma_1$, $\sigma_y = \sigma_2$, $\tau_{xy} = 0$), it is possible to describe the Hill48 yield function by the formula

$$2f(\sigma) = (G+H)\sigma_{xx}^2 + (F+H)\sigma_{yy}^2 - 2H\sigma_{xx}\sigma_{yy} + 2N\sigma_{xy}^2 = 1 \tag{2}$$

where σ_{xx}, σ_{yy}, and σ_{zz} are stresses in the RD (x), TD (y), and thickness (z) directions, respectively; σ_{xy}, σ_{yz}, and σ_{zx} are the shear stresses in xy, yz, and zx directions. Parameters F, G, H, and N are material parameters that describe the anisotropy of the material. If $F = G = H = 1$, and $N = 3$, the Hill48 function is reduced to the von Mises criterion, or as it is called in FEM code, the Hill48 isotropic criterion. A more common description is based on normal anisotropy in the $0°$, $45°$, and $90°$ directions to the rolling direction. Then, the material parameters F, G, H, and N can be described by

$$F = \frac{r_0}{r_{90}(r_0+1)}, \quad G = \frac{1}{r_0+1}, \quad H = \frac{r_0}{r_0+1}, \quad N = \frac{(r_0+r_{90})(1+2r_{45})}{2r_{90}(1+r_0)}. \tag{3}$$

For orthotropic hardening law and the values of anisotropy under 1.0, the Hill90 yield function is more suitable. This function is considered to be more suitable for aluminium alloys, and it is based on a non-quadratic transition function. In order to construct this function, the values from the uniaxial tensile test are deficient. For a complete description of this function, the biaxial test data are also required. The function can be described as

$$\left(\frac{\sigma_1}{\sigma_2}\right)^2 + \left(\frac{\sigma_2}{\sigma_{90}}\right)^2 + \left[(p+q+c) - \frac{p\sigma_1 + p\sigma_2}{\sigma_b}\right]\left(\frac{\sigma_1\sigma_2}{\sigma_0\sigma_{90}}\right) = 1, \tag{4}$$

where σ_0 is uniaxial tensile stress in the rolling direction, σ_{90} is uniaxial tensile stress in the direction normal to the rolling direction, σ_b is the stress under the balanced biaxial stress, and the c, p, and q parameters are defined as follows [21]:

$$c = \frac{\sigma_0}{\sigma_{90}} + \frac{\sigma_{90}}{\sigma_0} - \frac{\sigma_0\sigma_{90}}{\sigma_b^2}, \tag{5a}$$

$$\left(\frac{1}{\sigma_0} + \frac{1}{\sigma_{90}} - \frac{1}{\sigma_b}\right)p = \frac{2R_0(\sigma_b - \sigma_{90})}{(1+R_0)\sigma_0^2} - \frac{2R_{90}\sigma_b}{(1+R_{90})\sigma_{90}^2} + \frac{c}{\sigma_0}, \tag{5b}$$

$$\left(\frac{1}{\sigma_0} + \frac{1}{\sigma_{90}} - \frac{1}{\sigma_b}\right)q = \frac{2R_{90}(\sigma_b - \sigma_0)}{(1+R_{90})\sigma_{90}^2} - \frac{2R_0\sigma_b}{(1+R_0)\sigma_0^2} + \frac{c}{\sigma_{90}}, \tag{5c}$$

where R_0 is the anisotropy value for the uniaxial tension in the rolling direction and R_{90} is the anisotropy value for the uniaxial tension in the in-plane direction, perpendicular to the rolling direction.

The Barlat's material models describe the plastic behaviour of a material in a more detailed way than Hill's functions, but the higher number of parameters increases the calculation time. The Barlat89 model needs three parameters for its complete formulation, by which it is possible to describe the plane strain behaviours. Those parameters are defined in Table 4. The formulation is the following:

$$f = a|k_1 + k_2|^M + a|k_1 - k_2|^M + (2-a)|2k_2|^M = 2\sigma_e^M, \tag{6}$$

where M is the exponent related to the crystallographic structure of the material, and k_1 and k_2 can be described as

$$k_1 = \frac{\sigma_x + h\sigma_y}{2}, \quad k_2 = \left[\left(\frac{\sigma_x - h\sigma_y}{2}\right) + p^2\tau_{xy}^2\right]^{1/2}, \tag{7}$$

where a, h, and p are the material model parameters.

Table 4. The material constants for the Barlat89 yield function (m = 8.0).

a	c	h	p
1.3033	0.9556	0.9247	0.8465

A more precise function was presented by Barlat in 2003, called Barlat2000, where the linear transformation method was used. In the FEM software, this function is described by the eight parameters shown in Table 5. The formulation for this model is as follows:

$$\phi = \phi'\left(X'\right) + \phi''\left(X''\right) = 2\overline{\sigma}^a,$$ (8)

where a is an exponent related to the crystallographic structure of the material and ϕ' and ϕ'' are two isotropic functions described as follows [22]:

$$\phi' = \left|X_1' - X_2'\right|^\alpha; \quad \phi'' = \left|2X_2'' + X_1''\right|^\alpha + \left|2X_1'' + X_2''\right|^\alpha.$$ (9)

Table 5. The material constants for the Barlat2000 yield function (a = 8.0).

a_1	a_2	a_3	a_4	a_5	a_6	a_7	a_8
1.065173	0.841891	0.960059	0.958652	1.034037	1.027112	0.838988	0.877033

According to several works [23–25], the Vegter yield function should be more suitable for special steels and aluminium alloys due to its more convenient results. The Vegter criterion describes the yield locus more accurately from a series of physically tested points. According to Vegter, it is possible to establish the first quadrant of the yield function on the basis of the basic experimental measurement. To construct the ellipses, the Bezier interpolations between each point need to be performed. Every point requires three parameters to be defined, the main stresses σ_1 and σ_2, and the strain vector $\rho = d\varepsilon_2/d\varepsilon_1$. For a complete description of planar anisotropy, it is necessary to obtain 17 parameters from 9 mechanical tests. The mathematical expression of this function is

$$\begin{pmatrix} \sigma_1 \\ \sigma_2 \end{pmatrix} = (1-\lambda)^2 \begin{pmatrix} \sigma_1 \\ \sigma_2 \end{pmatrix}_i^r + 2\lambda(1-\lambda) \begin{pmatrix} \sigma_1 \\ \sigma_2 \end{pmatrix}_i^h + \lambda^2 \begin{pmatrix} \sigma_1 \\ \sigma_2 \end{pmatrix}_{i+1}^r,$$ (10)

where λ is the parameter for the Bezier interpolation subscript, i refers to the first reference point, r and h refer to a reference point and hinge point, respectively [26].

It is possible to use a simplified formula—Vegter-Lite. For this optional model, only 7 parameters from three mechanical tests (uniaxial tensile test, hydraulic bulge test, and the measurement of anisotropy) need to be defined. In this model, the second order Bezier interpolation is replaced by the second order NURBS interpolation, and the weight factor w—that controls the position of the curve between the points—is introduced. The formula for this model is

$$\begin{pmatrix} \sigma_1 \\ \sigma_2 \end{pmatrix} = \frac{(1-\lambda)^2 \begin{pmatrix} \sigma_1 \\ \sigma_2 \end{pmatrix}_i^r + 2\lambda(1-\lambda) \begin{pmatrix} \sigma_1 \\ \sigma_2 \end{pmatrix}_i^h + \lambda^2 \begin{pmatrix} \sigma_1 \\ \sigma_2 \end{pmatrix}_{i+1}^r}{(1-\lambda)^2 + 2w\lambda(1-\lambda) + \lambda^2}.$$ (11)

To fully define the material behaviour, the hardening curve of the material is also required. The Voce hardening curve gives the best correlation with the experimental results at an orientation of

0° from the rolling direction. This law provides a sufficient description of the elastic behaviour for aluminium alloys. The Voce hardening law is given by the equation

$$\sigma_y(\varepsilon_p) = A - Be^{-C\varepsilon_p}, \tag{12}$$

where A, B, and C are parameters defined in Table 6.

Table 6. The parameters for the Voce hardening curve.

A (MPa)	B (MPa)	C (–)
359.093260	169.310139	9.374256

To determine the failure criteria, Keller´s and Goodwin´s forming limit curve (FLC) model was used [27]. This empirical formula was obtained from experimental trials, and requires only two parameters: the thickness of the material and the strain hardening coefficient. The formula can be written as follows:

$$\varepsilon_{10} = \frac{(23.3 + 14.13t_0)n}{0.21}, \tag{13}$$

where t_0 is the initial thickness of the sheet and n is the strain hardening coefficient.

The simulation process consisted of three operations (stamping, trimming, and springback). Stamping was carried out as one continuous process in which the die moved at a speed of 100 mm/s. The blank was positioned between the die and the blankholder during holding. The die movement was set in the $-z$-direction at 300 mm until the blank was clamped. Subsequently, a blankholding force of 1900 kN was applied. The die and blankholder moved in the $-z$-direction until the tool was closed. After the part was fully formed, the trimming operation was performed. The trimming curve is shown in Figure 5.

Figure 5. The trimming curve on the punch.

4. Results

The results obtained from the numerical simulation were compared with the experimental ones. The springback was measured using locked nodes of the model. The stamped part profile and thickness were evaluated in three sections displayed in Figure 6. Section 1 is located on the right side of the part. This section passes through the hole, and due to the complex shape of the stamped part, the effect of the springback is quite significant in this area. The hole also runs through Section 2; this section is located on the left side of the part. The third section is located between section one and two. Figure 6 also shows the centres of the coordination systems used for profile evaluation. These centres were key for assembly purposes.

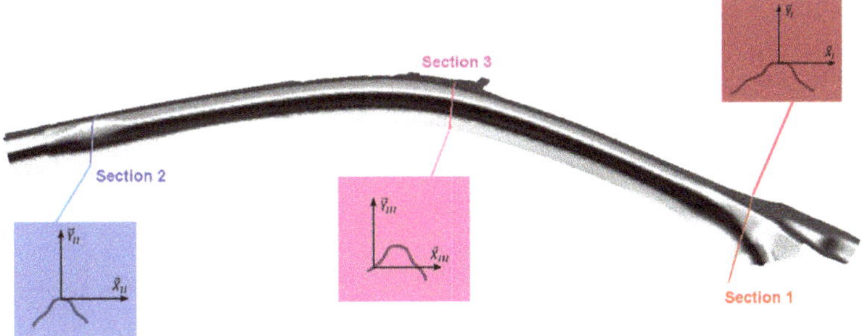

Figure 6. The sections used for profile and thickness evaluation.

4.1. Profile Analysis

The profile of the part was measured before and after springback. Since the stamped part copies the shape of the die, the profiles after stamping were almost identical for every material model. Good correlation of the experimental and numerical results can be observed after stamping. The comparison of the experimental results and the FEM simulation after springback for each section is shown in Figure 7. Subsequently, the FEM results for each yield function after springback were also compared.

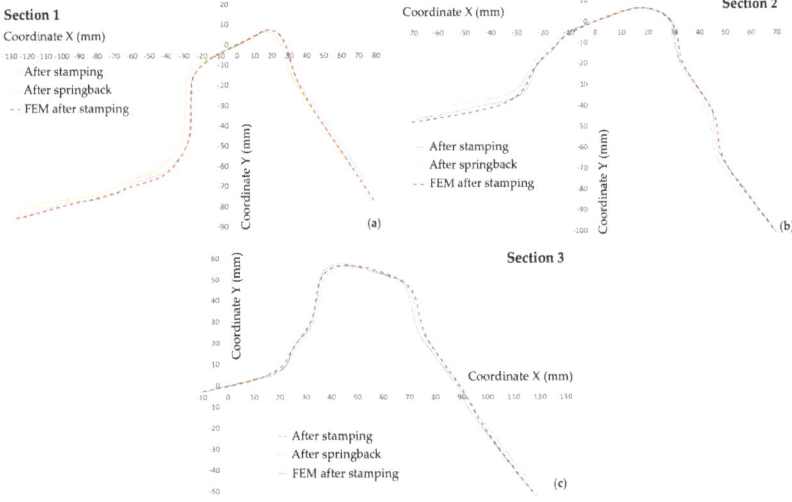

Figure 7. The comparison of the part profile from FEM simulation after springback with the experimental results after stamping and springback in (**a**) Section 1; (**b**) Section 2; (**c**) Section 3.

The deviations from the experimental results were measured on the left and on the right side of the sections. For assembly purposes, the deviation was measured in mm. Due to the difficulty of using conventional methods for springback measurement, the MATLAB system was implemented into the evaluation process. A simple program in the MATLAB environment was developed by the authors, which made it possible to measure the springback more accurately and easily. At first, a section was imported into the program, and the coordinate system was defined. Then, by selecting one point on the arm, a straight line parallel to the X-axis was created. Further points could only be created on this line, and the distance between each point was evaluated.

Figure 8 shows the experimental results and results from the numerical simulation in Section 1. Deviation from the experimental results for every yield function is shown in Table 7. A positive value means that the numerical simulation predicted a higher springback value than the experimental ones, and a negative value means that the simulation value was lower than the experimental ones. From these values, the average value of springback was calculated by the following equation:

$$X_{av} = \frac{|X_L| + |X_R|}{2},\tag{14}$$

where X_L is the offset for the left side of the part and X_R is the offset for the right side of the part.

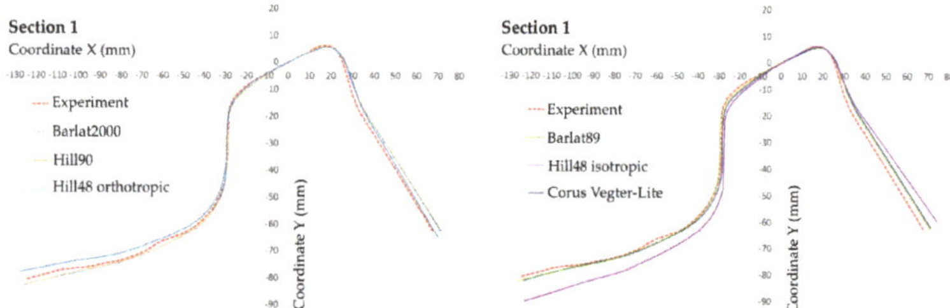

Figure 8. The comparison of the yield functions in Section 1.

Table 7. The results of springback in Section 1.

Yield Function	Left Side Offset (mm)	Right Side Offset (mm)	Yield Function	Left Side Offset (mm)	Right Side Offset (mm)
Barlat2000	+3.36	+0.88	Barlat89	−1.28	+1.15
Hill90	−2.14	+3.77	Hill48 isotropic	−9.29	+3.53
Hill48 orthotropic	−1.85	+4.51	Vegter-Lite	−1.67	+1.59

The lowest average deviation of 1.21 mm was measured when the Barlat89 yield function was used. Yield functions Barlat2000 and Vegter-Lite had average deviations of 2.12 and 1.63, also showing very good correlation with the experimental results. The highest average deviation of 5.44 mm was measured for the isotropic Hill48 yield function. In this function, the anisotropy of the material was not considered.

In Section 2 (Figure 9), greater deviation of the numerical springback values from the experimental ones can be seen on the right side of the profile. The reason is the shape of the part. Springback did not appear so significantly on the left side of the part where the material is compressed. The results are displayed in Table 8. In this section, the material model with the Vegter-Lite yield function showed the lowest average value of deviation: 2.65 mm. The Barlat89 yield function with 2.68 mm average deviation also shows good correlation. The other material models show very similar results, but the isotropic Hill48 method shows the highest deviation (3.51 mm) from the experimental results.

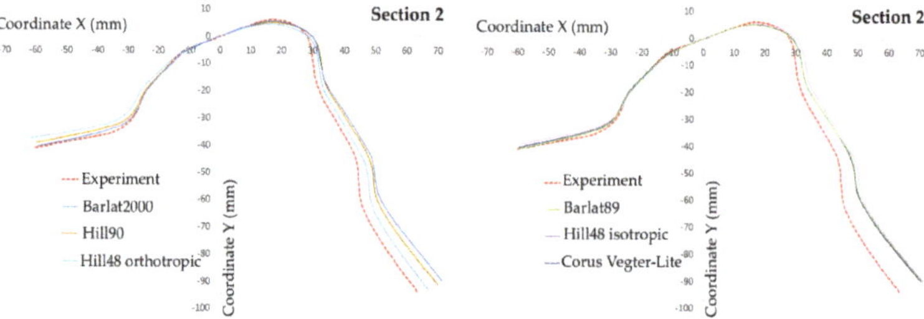

Figure 9. The comparison of yield functions in Section 2.

Table 8. The results of springback in Section 2.

Yield Function	Left Side Offset (mm)	Right Side Offset (mm)	Yield Function	Left Side Offset (mm)	Right Side Offset (mm)
Barlat2000	+0.45	+5.79	Barlat89	−0.20	+5.15
Hill90	+1.62	+4.48	Hill48 isotropic	+1.56	+5.46
Hill48 orthotropic	+3.78	+2.21	Vegter-Lite	+0.04	+4.90

In the third section, the deviation was measured only on the right side because, on the left side, the deviation was too low, as shown in Figure 10. In this section, the best correlation was achieved with the Barlat yield functions, where the Barlat2000 deviation was 0.21 mm and the Barlat89 deviation was 0.32 mm. The highest deviation of 5.26 mm was measured for the isotropic Hill48 yield function. The results for all the material models are shown in Table 9.

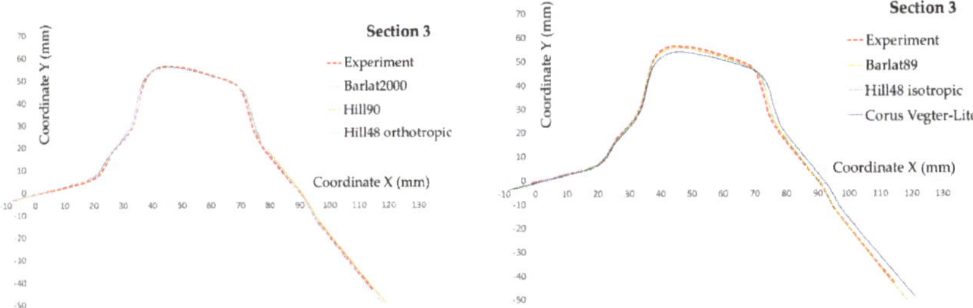

Figure 10. The comparison of yield functions in Section 3.

Table 9. The results of springback in Section 3.

Yield Function	Right Side Offset (mm)	Yield Function	Right Side Offset (mm)
Barlat2000	−0.21	Barlat89	−0.32
Hill90	−0.75	Hill48 isotropic	+5.26
Hill48 orthotropic	+0.37	Vegter-Lite	+1.97

4.2. Thickness Analysis

The next investigated parameter in this work was thickness, which was also measured in mentioned sections. The thickness of the material can significantly influence the accuracy of the FEM prediction. Since shell elements, which were used in this work, are suitable for thicknesses up to 1 mm, the volume element should give better results. However, in the FEM software, it is possible to define the volume elements for only the Hill48 and Barlat2000 yield functions. For comparison purposes of all previously mentioned yield functions, the models must have shell elements. Figures 11–13 show comparisons of the experimental and FEM results in each section.

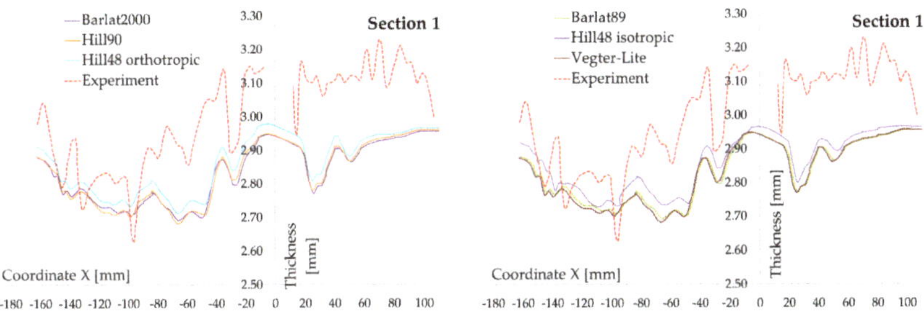

Figure 11. The thickness distribution in Section 1.

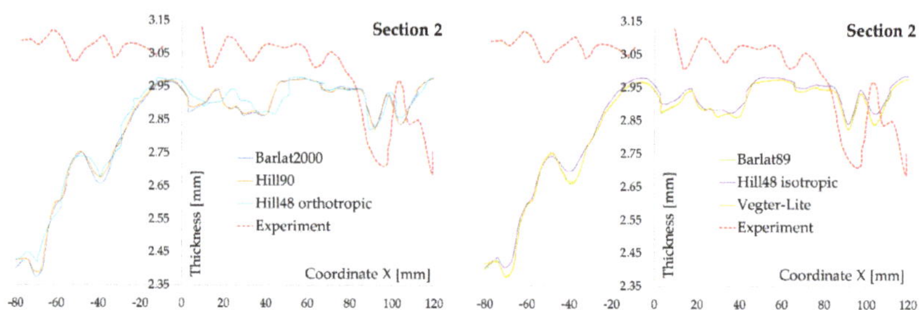

Figure 12. The thickness distribution in Section 2.

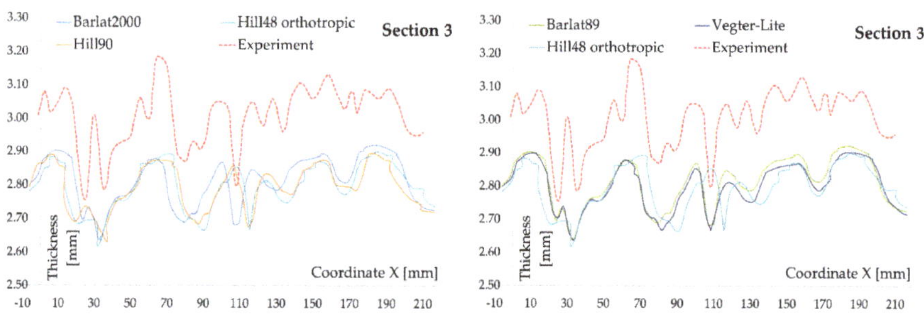

Figure 13. The thickness distribution in Section 3.

From the results of the thickness distribution in the individual sections, it is clear that the material models with the Barlat (Barlat89 and Barlat2000) yield functions, and the Vegter-Lite yield function shows very similar results. Although the description of the yield function and the amount of data needed for their definition is different between those models, the results of the thickness distribution were practically the same. The Hill48 model, either isotropic or orthotropic, exhibited significant deviations. The average values of thickness are shown in Table 10. The experimental results show a higher average thickness than the thickness data obtained by the FEM simulation.

Table 10. The comparison of the average thickness of each section.

	Barlat2000	Barlat89	Vegter-Lite	Hill90	Hill48 Orthotropic	Hill48 Isotropic	Experiment
Section 1	2.825	2.828	2.826	2.828	2.834	2.858	2.980
Section 2	2.856	2.847	2.853	2.852	2.847	2.859	2.991
Section 3	2.809	2.810	2.792	2.798	2.797	2.809	2.990

4.3. Computation Time

With the increased complexity of the yield function formulation and, thus, with the increased number of necessary variables, the calculation time was increased. For the Hill48 isotropic model, where the anisotropy of material was not considered, the computation took around 13 h and 47 min. For the Hill48 orthotropic and Hill90 models, where the anisotropy of the material was considered, the computation time took 14 h. In the Barlat yield functions, where the material's crystallographic structure was considered, the computation time increased significantly to around 15 h and 20 min. The Vegter-Lite yield function achieved a similar accuracy and computation time as the Barlat material models. Figure 14 shows the comparison of the computation times for each yield function.

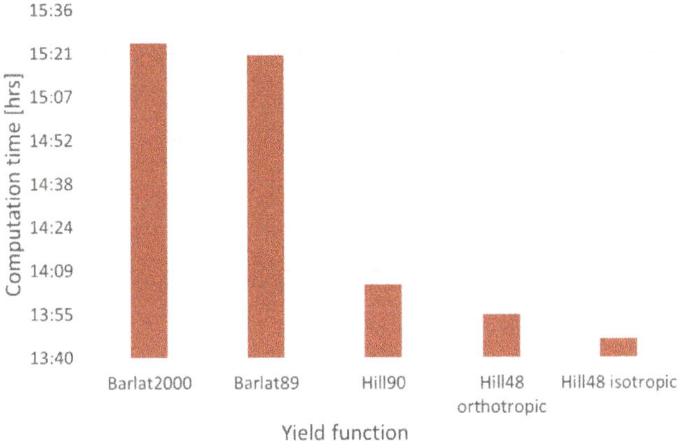

Figure 14. The computation time comparison.

5. Conclusions

The accuracy of the springback prediction is one of the most challenging problems in the numerical simulation of forming processes. In the present article, the influence of the yield function on the accuracy of springback prediction with the use of numerical simulation was investigated. Three sections were defined on the formed part. In these sections, the thicknesses of the profile of the stamped part after springback calculation were evaluated. After the stamping operation, for all the examined yield functions, the sheet metal copied the shape of the die. Visible differences can be seen after cutting and springback calculation. The results of the Barlat2000, Barlat89, and Vegter-Lite yield functions were in good correlation with the experimental results. Hill's yield functions (Hill90, isotropic Hill48, and anisotropic Hill48) were not as accurate as the yield functions mentioned above. Barlat's yield functions takes into account the material's crystallographic structures. The Barlat89 yield function, with an average deviation of 1.40 mm from experimental results of springback, is not suitable for materials with strong anisotropy [26]. Additionally, this model cannot capture the change of yield stress and the Lankford coefficient values. However, the advantage of this function lies in its simple mathematical description, and in the ability of the accurate plastic behaviour prediction (yield locus) of aluminium alloys, thus, the results of the Barlat89 model are more accurate than the results obtained with the use of Hill's yield functions. The Barlat2000 yield function is an improved version of the Barlat89 model, but the description of this improved model in the numerical simulation is more difficult. This is the reason why this function is not used as much in industrial practice. The experimental thickness values were higher than the predicted ones in all cases. From the results of the thickness distribution in the individual sections, it is clear that the yield functions of the Barlat's family and Vegter-Lite yield function show very similar results. Although the description of the yield function and the amount of data needed for their definition is different between those models, the results of the thickness distribution were practically the same. The Hill48 model, either isotropic or orthotropic, exhibited significant deviations. This can be attributed to the shell elements used in the numerical simulation or to the Voce isotropic hardening law. However, the Voce hardening law exhibited good correlation with the experimentally measured FLC (forming limit curve). One of the results of the work is that the combination of the isotropic hardening law with the isotropic yield function did not achieve accurate springback prediction results. The combination of more advanced yield functions (Barlat2000, Barlat89 and Vegter-Lite) with the isotropic Voce hardening law improved the accuracy of the springback prediction, but the computation time was increased by approximately an hour. Additionally, it was found out that the investigated phenomena which are required for the determination of the kinematic hardening model, such as the change of Young's modulus E, transient behaviour, work-hardening stagnation, and permanent softening, were not observed in the aluminium alloy studied in this work. Our research confirmed that in aluminium alloys, there are no appreciable changes in the Young's modulus value.

Author Contributions: M.Š. and J.S. performed the numerical simulations; E.S., J.S., M.Š., P.M., and T.S. performed the results analysis; E.S., J.S, M.Š., and P.M. wrote the article; E.S. and J.S. designed the experiments.

Funding: This research was funded by the Slovak Research and Development Agency under grant number APVV-14-0834 and the Grant Agency under grant number VEGA 1/0441/17.

Acknowledgments: The authors are grateful for the support of experimental works to the Slovak Research and Development Agency under project APVV-14-0834 "Increasing the quality of cut-outs and effectiveness of cutting electric sheets" and the grant agency for the support of the project VEGA 1/0441/17 "Application of high-strength materials for exterior car body parts".

Conflicts of Interest: The authors declare no conflict of interest.

References

1. Jeswiet, J.; Geiger, M.; Engel, U.; Kleiner, M.; Schikora, M.; Duflou, J. Metal forming progress since 2000. *CIRP J. Manuf. Sci. Technol.* **2008**, *1*, 2–17. [CrossRef]
2. Banu, M.; Takamura, M.; Hama, T.; Naidim, O.; Teodosiu, C.; Makinouchi, A. Simulation of springback and wrinkling in stamping of a dual phase steel rail-shaped part. *J. Mater. Process. Technol.* **2006**, *173*, 178–184. [CrossRef]
3. Chongthairungruang, B.; Uthaisangsuk, V. Springback prediction in sheet metal forming of high strength steels. *Mater. Des.* **2013**, *50*, 253–266. [CrossRef]
4. Lei, D.; Xinyun, W.; Junsong, J.; Liangjun, X. Springback and hardness of aluminum alloy sheet part manufactured by warm forming processs using non-isothermal dies. *Procedia Eng.* **2017**, *207*, 2388–2393. [CrossRef]
5. Mahabunphachai, S.; Koc, M. Investigations on forming of aluminum 5052 and 6061 sheet alloys at warm temperatures. *Mater. Des.* **2010**, *31*, 2422–2434. [CrossRef]
6. Wagoner, R.H.; Lim, H.; Lee, M.G. Advanced Issues in springback. *Int. J. Plast.* **2013**, *45*, 3–20. [CrossRef]
7. Yoshida, T.; Sato, K.; Hashimoto, K. *Springback Problems in Forming of High-Strength Steel Sheets and Countermeasures*; Nippon Steel Technical Report No. 103; Nippon Steel & Sumitomo Metal Corporation: Tokyo, Japan, May 2013.
8. Toros, S.; Polat, A.; Ozturk, F. Formability and springback characterization of TRIP800 advanced high strength steel. *Mater. Des.* **2012**, *41*, 298–305. [CrossRef]
9. Gau, J.T.; Kinzel, G.L. A New Model for Springback Prediction for Aluminum Sheet Forming. *J. Eng. Mater. Technol.* **2005**, *127*, 279–288. [CrossRef]
10. Li, M.; Gan, W.; Wagoner, R.H. Sheet springback: Prediction and design. In Proceedings of the HSIMP 2007, High Speed Manufacturing Process, Senlis, France, 13–15 November 2007; pp. 1–7.
11. Wagoner, R.H. *Report: Advanced High Strength Workshop 2006*; The Ohio State University: Columbus, OH, USA, 2006.
12. Trzepiecinski, T.; Lemu, H.G. Effect of Computational Parameters on Springback Prediction by Numerical Simulation. *Metals* **2017**, *7*, 380. [CrossRef]
13. Slota, J.; Siser, M.; Dvorak, M. Experimental and Numerical Analysis of Springback Behavior of Aluminum Alloys. *Strength Mater.* **2017**, *49*, 565–574. [CrossRef]
14. Seo, K.Y.; Kim, J.H.; Lee, H.S.; Kim, J.H.; Kim, B.M. Effect of Constitutive Equations on Springback Prediction Accuracy in the TRIP1180 Cold Stamping. *Metals* **2017**, *8*, 18. [CrossRef]
15. Mulidran, P.; Spisak, E.; Majernikova, J.; Sleziak, T.; Gres, M. Influence of forming method and process conditions on springback effect in the sheet metal forming simulation. *Int. J. Eng. Sci.* **2017**, *6*, 62–67. [CrossRef]
16. Jung, J.B.; Jun, S.; Lee, H.S.; Kim, B.M.; Lee, M.G.; Kim, J.H. Anisotropic Hardening Behaviour and Springback of Advanced High-Strength Steels. *Metals* **2017**, *7*, 480. [CrossRef]
17. Villuendas, A.; Jorba, J.; Roca, A. The Role of Precipitates in the Behavior of Young's Modulus in Aluminum Alloys. *Metall. Mater. Trans. A* **2014**, *45*, 3857–3865. [CrossRef]

18. Roca, A.; Villuendas, A.; Meíja, I.; Benito, J.B.; Llorca-Isern, N.; Lluma, J.; Jorba, J. Can Young's modulus of metallic alloys change with plastic deformation? *Mater. Sci. Forum* **2014**, *783–786*, 2382–2387. [CrossRef]

19. Allen, M.; Oliveira, M.; Hazra, S.; Adetoro, O.; Das, A.; Cardoso, R. Benchmark 2—Springback of a Jaguar Land Rover Aluminium. *J. Phys. Conf. Ser.* **2016**, *734*, 022002. [CrossRef]

20. Dasappa, P.; Inal, K.; Mishra, R. The effects of anisotropic yield functions and their material parameters on prediction of forming limit diagrams. *Int. J. Solids Struct.* **2012**, *49*, 3528–3550. [CrossRef]

21. Bruschi, S.; Altan, T.; Banabic, D.; Bariani, P.F.; Brosius, A.; Cao, J.; Ghiotti, A.; Khrasheh, M.; Merklein, M.; Tekkaya, A.E. Testing and modelling of material behaviour and formability in sheet metal forming. *CIRP Ann.-Manuf. Technol.* **2014**, *63*, 727–749. [CrossRef]

22. Barlat, F.; Brem, J.C.; Yoon, J.W.; Chung, K.; Dick, R.E.; Lege, D.J.; Pourboghrat, F.; Choi, S.H.; Chu, E. Plane stress yield function for aluminum alloy sheets—Part 1: Theory. *Int. J. Plast.* **2003**, *19*, 1297–1319. [CrossRef]

23. Slota, J.; Siser, M. Advanced Material Models for Stamping of AW 5754 Aluminum Alloy. *Strength Mater.* **2016**, *48*, 487–494. [CrossRef]

24. Novy, J.; Vache, V.; Sobotka, J. Influence of used yield function in deep drawing simulation of highly anisotropic aluminum alloy. In Proceedings of the IDDRG, Zurich, Switzerland, 2–5 June 2013; pp. 273–277.

25. Vegter, H.; Horn, C.; Abspoel, M. The Corus-Vegter Lite Material model: Simplifying advanced material modeling. *Int. J. Mater. Form* **2009**, *2*, 511–514. [CrossRef]

26. Slota, J.; Spisak, E. Comparison of the forming-limit diagram (FLD) models for drawing quality (DQ) steel sheets. *Metabk* **2005**, *4*, 249–253.

27. Banabic, D. Advanced anisotropic yield criteria. In *Sheet Metal Forming Processes*, 1st ed.; Springer: Heidelberg/Berlin, Germany; Dordrecht, The Netherlands, 2010; pp. 76–87. ISBN 978-3-540-88112-4.

© 2018 by the authors. Licensee MDPI, Basel, Switzerland. This article is an open access article distributed under the terms and conditions of the Creative Commons Attribution (CC BY) license (http://creativecommons.org/licenses/by/4.0/).

Article

Predictive Simulation of Plastic Processing of Welded Stainless Steel Pipes

Riccardo Rufini, Orlando Di Pietro and Andrea Di Schino *

Dipartimento di Ingegneria, Università di Perugia, Via G. Duranti 93, 06125 Perugia, Italy;
ruf.riccardo@gmail.com (R.R.); orlando.dipietro@gmail.com (O.D.P.)
* Correspondence: andrea.dischino@unipg.it; Tel.: +39-328-331-1875

Received: 11 June 2018; Accepted: 3 July 2018; Published: 5 July 2018

Abstract: Metal forming is the most used technique to manufacture complex geometry pieces in the most efficient way, and the technological progress related to the various application fields requires increasingly higher quality standards. In order to achieve such a requirement, people are forced to perform quality and compliance tests finalized to guarantee that these standards are met; this often implies a waste of material and economic resources. In the case of welded stainless steel pipes, several critical points affecting the general trend of subsequent machining need to be taken into account. In this framework, the aim of the paper is to study the effects of different process parameters and geometrical characteristics on various members of the stainless steel family during finite elements method (FEM) simulations. The analysis of the simulation outputs, such as stress, strain, and thickness, is reported through mappings, in order to evaluate their variation, caused by the variation of the simulation input parameters. The feasibility of the simulated process is evaluated through the use of forming limit diagrams (FLD). An experimental validation of the model is performed by comparison with real cases. Major parameters that mainly guide the outcome of the simulations are highlighted.

Keywords: numerical simulation; modeling; hardening; anisotropy; parameters identification; damage; mechanical properties

---

1. Introduction

Stainless steels represent a quite interesting material family, both from the scientific and commercial point of view, following to their excellent combination in terms of strength and ductility, together with corrosion resistance [1–5]. Thanks to such properties, stainless steels have been indispensable for the technological progress during the last century, and their annual consumption has increased with a rate of 5% during the last 20 years, faster than other materials [6]. They find application in all these fields, requiring good corrosion resistance together with ability to be worked into complex geometries [7,8].

Metal forming is the most used technique to manufacture complex geometry pieces in the most efficient way, and the technological progress related to the various application fields requires increasingly higher quality standards. In order to face such a requirement, people are forced to perform quality and compliance tests, finalized to guarantee that these standards are met; this often implies a waste of personnel, time, and resources, both material and economic. From the prospective of plastic forming, the plastic processing of welded pipes is characterized by a poor homogeneity of the behavior, especially in the case of ferritic steels [9]. This involves a certain percentage of unreliability in the tests, carried out on random samples, because of the nature of the steel itself, whose behavior can be completely modified by defects and inhomogeneity. Therefore, these checks, generally carried out by means of tensile tests according to specifications, are not sufficient to guarantee the required standards.

Many researchers are focused on solving such problems through the implementation of predictive simulations using a finite element method (FEM) numerical analysis [10–12], to predict the behavior of various pipes' geometries in different processing areas, such as hydroforming and bending, or the cold metal forming of steel sheets. Many relevant industrial applications, where a proper procedure of pipe bending and a correct simulation of pipe yielding after bending turns out to be critical, are found in the literature (e.g., [13,14]). All of these applications require the analysis of the steel mechanical properties, both at the macroscopic level, and at the crystalline structure and grain level, such as stress–strain curves and hardening, and with particular attention to the anisotropic characteristics [15] caused by the plastic processing that led to the pipe manufacturing.

As far as the pipes manufactured are concerned, starting from rolled and welded steel plates, several critical points affecting the general trend of subsequent machining need to be taken into account, especially regarding high strength materials for application in the structural field. For example, the geometry of the pipe itself or the operating parameters that are used during the plastic processing, such as used speed and bending angle, have a strong impact on the influence of the aforementioned defects and on the final outcome of the process carried out on the same type of stainless steel.

In this framework, the aim of the paper is to study the effects of different process parameters and geometrical characteristics on various types of stainless steel (both ferritic and austenitic).

2. Material Properties and Modelling

2.1. Materials

The following steel grades and pipe geometries are considered:

- AISI 304 and AISI 316 (austenitic stainless steel)

  - Diameter: 35, 40, 50 and 60 mm
  - Thickness: 1.0, 1.2, 1.5 mm

- AISI 409 and AISI 441 (ferritic stainless steel)

  - Diameter: 35, 40, 50 and 60 mm
  - Thickness: 0.8, 1.0, 1.2, 1.5 and 1.8 mm

The mean experimental stress–strain curves at room temperature for the considered materials are reported in Figure 1.

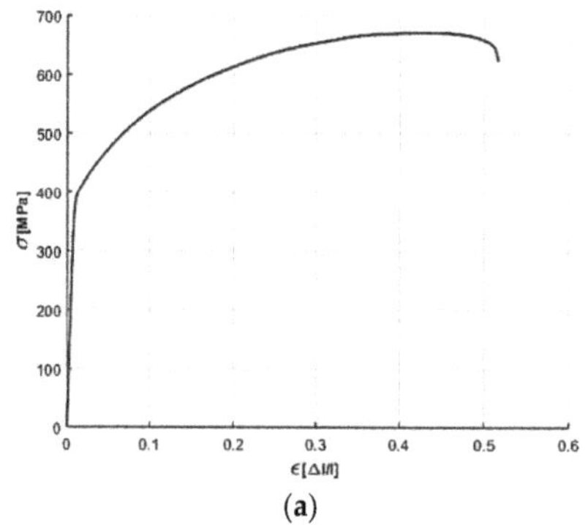

(a)

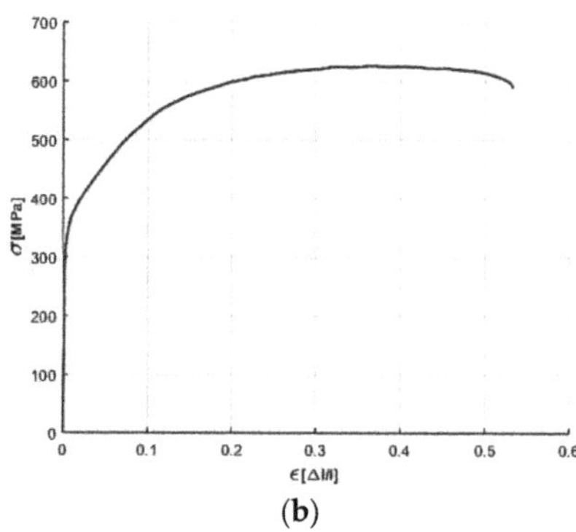

(b)

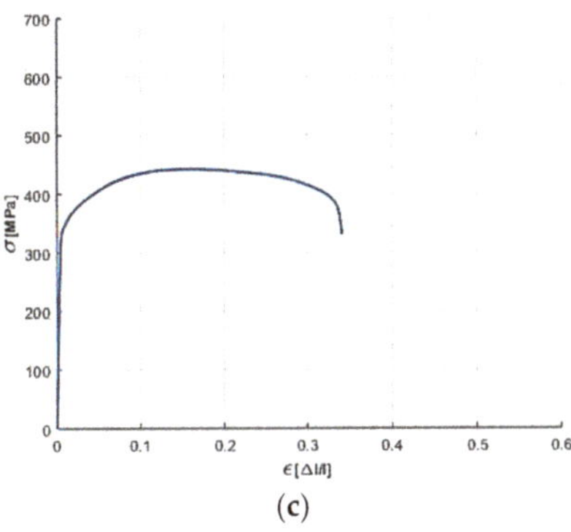

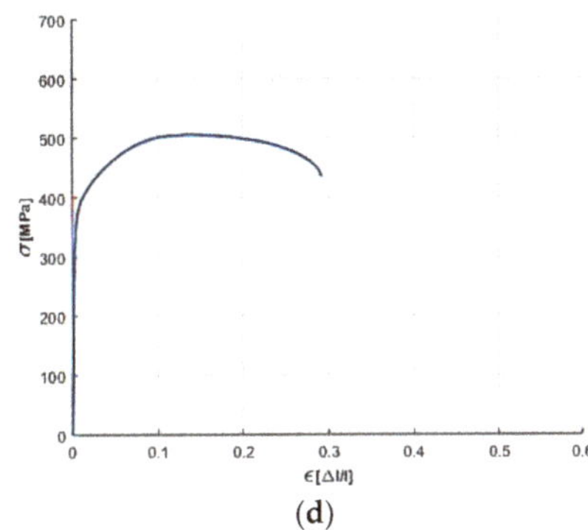

(c)   (d)

Figure 1. Examples of experimental stress–strain curves for the considered materials, obtained from tensile tests at room temperature on the pipes for the following grades of stainless steel: (**a**) AISI 304; (**b**) AISI 316; (**c**) AISI 409; and (**d**) AISI 441.

All of the considered steels were characterized by a sigma–epsilon curve calculated according to the specification for the tensile tests for pipes (UNI EN ISO 6892), for each combination of diameter and thickness.

2.2. Model

A commercial software package integrated with its own solver, commonly used by automotive engineers, was adopted for the numerical calculations. Inside the software framework [16,17] the Hill 48' yield function was adopted, ideal for small-sized tubular geometries [11], as a constitutive equation for stainless steels' behavior.

The following parameters are taken into account in order to simulate the bending process:

- Bending radius
- Bending angle
- Rotational speed

Based on the above assumptions/inputs, it is possible to simulate the pipe bending behavior. A typical model for such an approach is as reported in Figure 2.

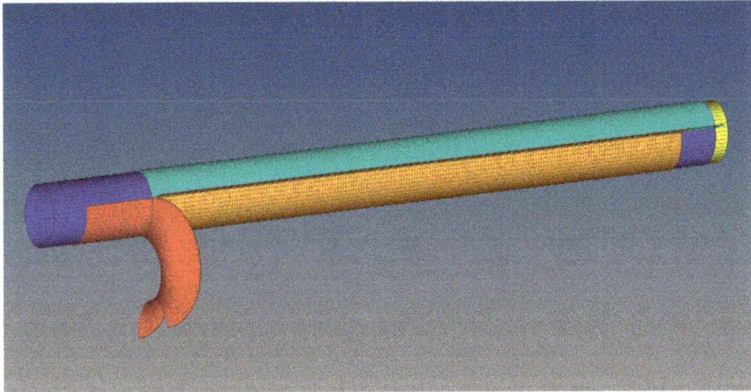

Figure 2. Geometry and working parts of the machinery simulated inside the software.

2.3. Methodology of Analysis

The analysis of the simulation outputs is carried out through mappings of the values calculated by the solver (such as internal stress, thinning, and deformation). Specifically, the stress analyzed is an equivalent stress calculated by the solver on the basis of the Hill criterion, which is a mathematical form optimized for the Finite Element Analysis (FEA) and developed starting from the Von Mises criterion. In order to analyze the influence that the parameters have on the final process and on the feasibility, the maximum values obtained on the mappings will be considered in order to consider the critical points on the geometry.

Usually, however, above all in the industrial field, a diagram defined as formability limit diagram (FLD) is used to describe the deformation paths along the whole sample. This diagram, as shown in the Figure 3, contains the formability limit curve (FLC), showing the maximum capacity of a material to be deformed, calculated by carrying out repeated Nakazima tests. The strains obtained from this test are measured with a conventional grid method, creating a pattern of circles on the surface of the specimen, which are deformed during the process obtaining ellipses. On these ellipses, the strains on the minor and major dimensions are measured, identifying the deformation state points of the material on the FLD diagram.

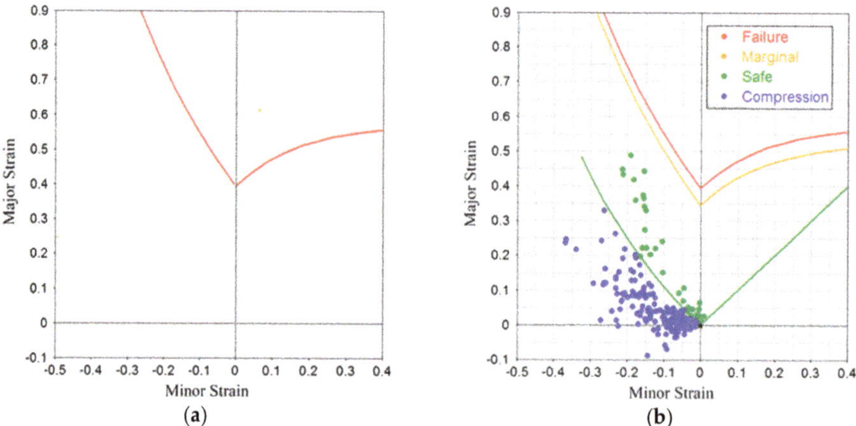

Figure 3. Typical shape of the formability limit curve (**a**); example of Nakazima's experimental test results on the formability limit diagram (FLD) with deformation state points for an AISI 304 stainless steel sheet (**b**).

3. Results and Discussion

The influence of the different input parameters are reported below.

3.1. Diameter Influence

The typical ratio between the curvature radius and the pipe diameter (R/D) for industrial application is in the range between 1.0 and 1.5. Therefore, considering the various diameter cases, it was decided that R/D = 1 as a constant value, which aimed to consider cases that are actually representative of the real processes, and to have results, in terms of stress, that can be compared with each other.

Figure 4 shows the stress mapping for smaller and larger diameters, while Figure 5 shows the stress reached for each diameter.

The results show that by varying the diameter and always keeping the ratio R/D = 1, a variation of the maximum stresses in a range between −2% and 3% is found; such an effect can be considered negligible. Furthermore, it is also evident that the distribution of the internal stresses is not particularly modified with the increase in diameter.

Additionally, in order to evaluate the deformation capacity of the various samples, the FLD diagrams were compared, and Figure 6 shows the extreme cases of the analyzed range of stainless steel AISI 304 with a 35 and 60 mm diameter. Furthermore, from the FLD diagrams, it can be deduced that the deformation path of the various elements of the geometry are not particularly modified by this parameter.

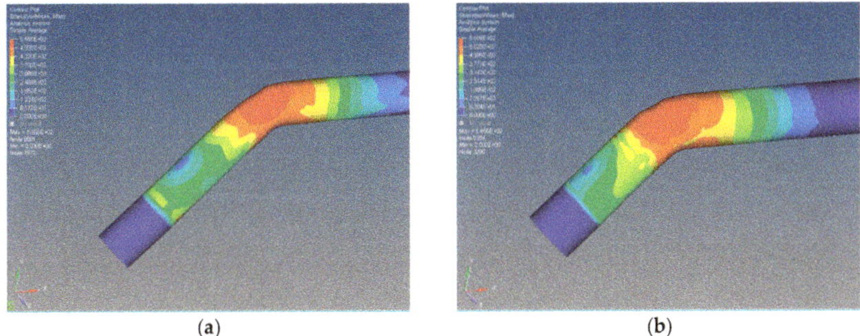

Figure 4. Equivalent stress mapping for diameters of 35 mm (**a**) and 60 mm (**b**) for AISI 304, 1.5 mm thickness.

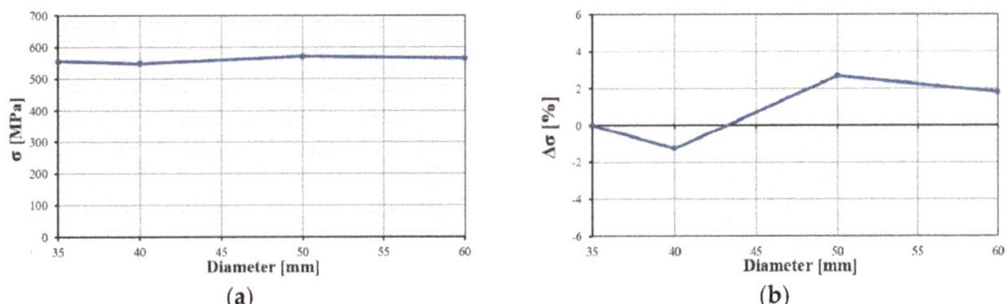

Figure 5. Trend of maximum equivalent stress according to the diameter for AISI 304, 1.5 mm thickness (**a**); percent variation of maximum equivalent stress according to the diameter for AISI 304, 1.5 mm thickness (**b**).

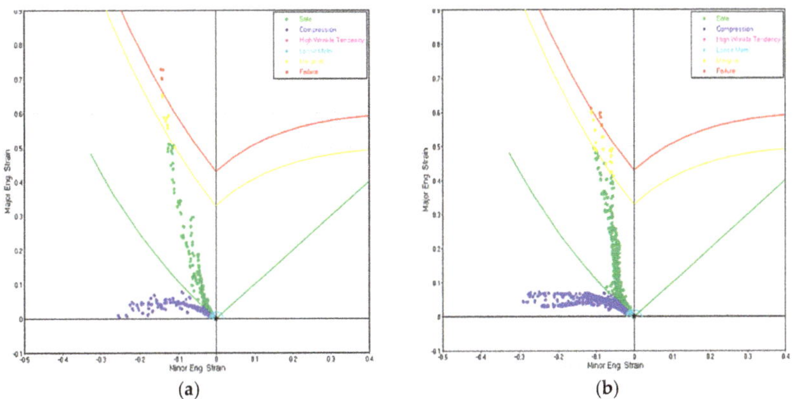

Figure 6. FLD diagrams of 35 mm (**a**) and 60 mm (**b**) for AISI 304, 1.5 mm thickness.

3.2. Influence of Thickness

As for the study carried out for the diameter, the same cases have been analyzed, however varying the thickness of the tube and keeping all the other parameters constant and equal to the case studies of the diameter, such as R/D = 1. Figure 7 shows stress mapping for smaller and larger thickness and in Figure 8 their distribution as before.

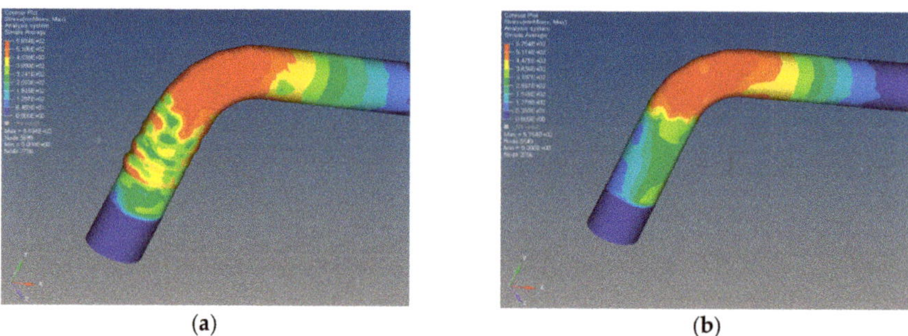

(a)   **(b)**

Figure 7. Equivalent stress mapping for a thickness of 1 mm (**a**) and 1.8 mm (**b**) for AISI 304, 50 mm diameter.

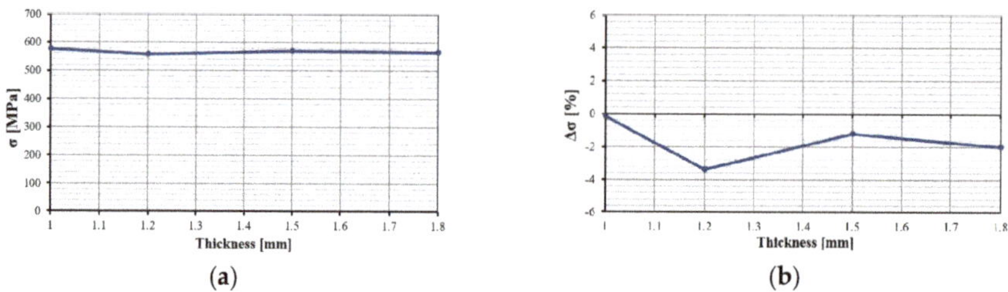

(a)   **(b)**

Figure 8. The trend of maximum equivalent stress according to thickness for AISI 304, 50 mm diameter (**a**); percent variation of maximum equivalent stress according to the thickness for AISI 304, 50 mm diameter (**b**).

While from the graphs of the maximum stresses there is not a substantial difference, except for an oscillating decreasing trend, depending on the thickness, it is noticed in the images of the distributions that there is a processing failure for the thickness of 1 mm. Therefore, to deepen the analysis, the thinning caused by the working on the tube geometry was considered, as shown in Figure 9. From these graphs, it is more evident how the initial thickness of the geometry has a strong impact on the success of the bending process. In fact, we can see, first of all, the decreasing trend, even if it does not appear to be monotonic, confirming what was supposed before, but above all, the variation that this parameter involves, having a percentage difference of about 30% in the final thickness of the most stressed area.

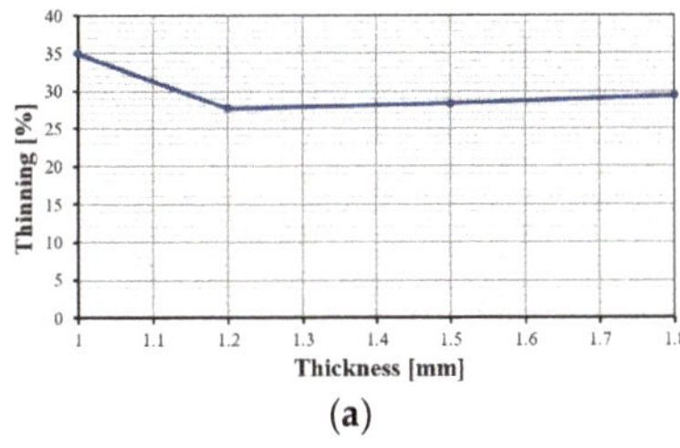
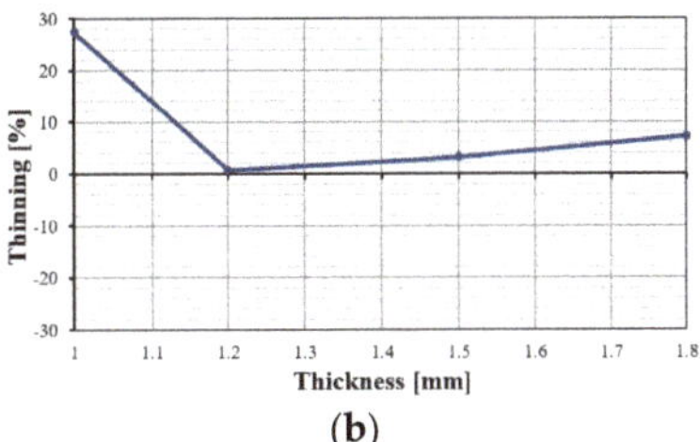

(a) **(b)**

Figure 9. Trend of maximum thinning according to thickness for AISI 304, 50 mm diameter (**a**); percent variation of maximum thinning according to thickness for AISI 304, 50 mm diameter (**b**).

3.3. Influence of the R/D Ratio

The R/D ratio, defined as the radius of curvature/diameter, as previously mentioned, is a parameter not present within the simulation model, but is widely used in industrial applications. It is primary used as a feasibility index for processing and is generally between 1.0 and 1.5 in standard conditions. Already, in the case R/D = 1 (radius of curvature equal to diameter), the tube undergoes a considerable stress, and for lower values, there is a breakage of the piece in almost all of the cases with the standard processing conditions; therefore, in this case, particular precautions are necessary. For this reason, no cases of R/D < 1 have been studied and values of R/D greater than 1.5 are instead used for larger pipes, ducts, or special cases that are not present in the application fields of the pipes produced by the company.

Then, by performing the simulations (with a constant diameter and then varying the radius of curvature) and analyzing the results, we see how, in Figure 10, the stresses achieved do not vary significantly by increasing the R/D ratio, but, as also seen for the other case studies, the maximum internal stress results are not sufficient to correctly describe the influence of the parameter. In fact, as can be seen in Figure 11, as the R/D ratio increases, the maximum thinning obtained decreases, therefore, although one would expect a decrease in stress, an increase in the ratio (obtained by increasing the radius of curvature or decreasing the diameter) causes minor forces in the tube and, at the same time, a minor thinning (as seen in Figure 11), thus significantly attenuating their decrease. It is also useful to observe the FLD diagrams in order to be able to make further considerations. Therefore, in Figure 12, it can be observed that both for the austenitic and ferritic steels, the increase in the R/D ratio leads to an increase in the feasibility of bending, which is very important, especially for AISI 409, because of its low plastic deformability compared with other families of the stainless steels considered.

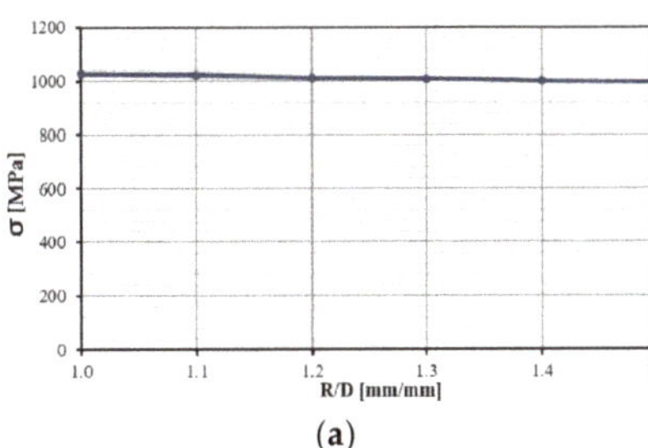
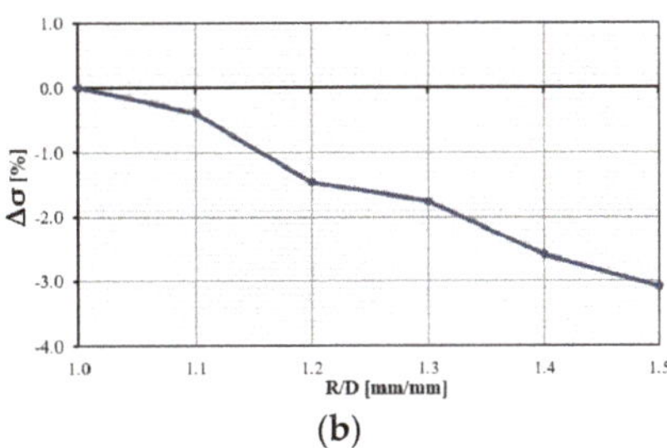

(a) **(b)**

Figure 10. Trend of maximum equivalent stress according to the curvature radius and the pipe diameter (R/D) ratio for AISI 304 (**a**); percent variation of maximum equivalent stress according to the R/D ratio for AISI 304 (**b**).

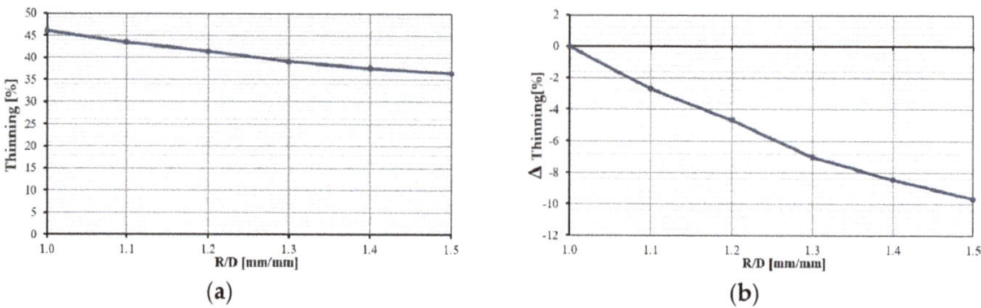

Figure 11. Trend of maximum thinning according to the R/D ratio for AISI 304 (**a**); percent variation of maximum thinning according to the R/D ratio for AISI 304 (**b**).

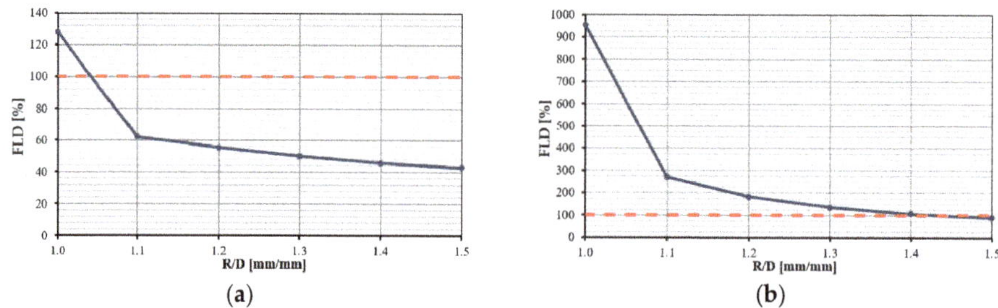

Figure 12. Percentage reached of the formability limit (red dashed line) according to the R/D ratio for AISI 304 (**a**) and AISI 409 (**b**).

3.4. Influence of Velocity Based on the Bend Angle

In this case, for the study of velocity, the effect of its variation for each bend angle, which is generally set during the common forming processes (specifically, angles between 30° and 90°), has been analyzed. Furthermore, angles of more than 90° have been considered for completeness, and in order to verify the trend that involves the variation of speed, up to a maximum of 180°.

So, it can be initially observed, in Figure 13, how the percentage reached of the formability limit is mapped for each combination of the speed and angle.

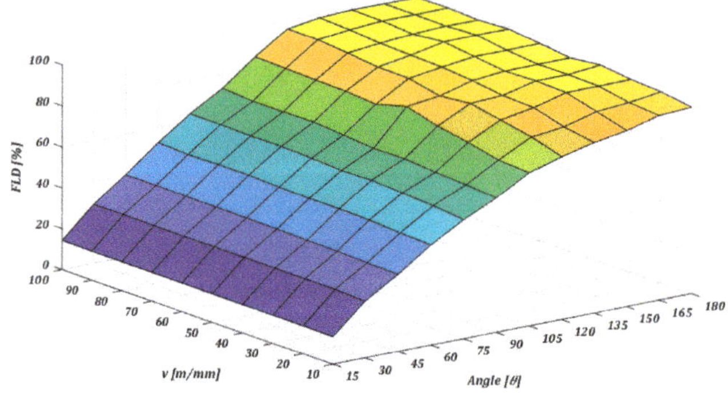

Figure 13. Percentage reached of the formability limit for every combination of speed and angle for AISI 304.

Moreover, in order to more advantageously evaluate the influence of the speed variation, it was decided to report the variation between percentage obtained for minimum and maximum speed of each considered angle, according to the Equation 1, in a graph (Figure 14), thus obtaining a graph purely from the variation itself.

$$\Delta FLD = FLD_{v\text{-}max} - FLD_{v\text{-}min} \tag{1}$$

Therefore, as can be observed from Figure 14, it is noted that for the region of angles between 30° and 90°, which is the range of interest for common industrial processes, the influence of speed variation assumes an approximate linear trend against the bend angle. Instead, considering the whole range of study, it can be observed that the variation tends to reach a maximum at about 120° of bending, then returning to decrease with the increasing angle. The motivations that lead to this particular behavior can be studied in more detail, but currently, we can hypothesize, also observing Figure 15, that this is due to the concentration of stress that is more localized in the first 90° of bending, where the machine actually deforms the tube continuously and applies the forces involved in forming.

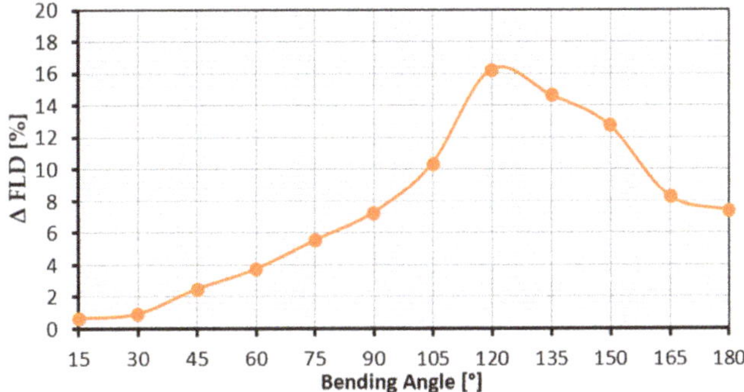

Figure 14. Percentage of formability limit variation in the minimum–maximum speed range as a function of the bending angle for AISI 304 steel.

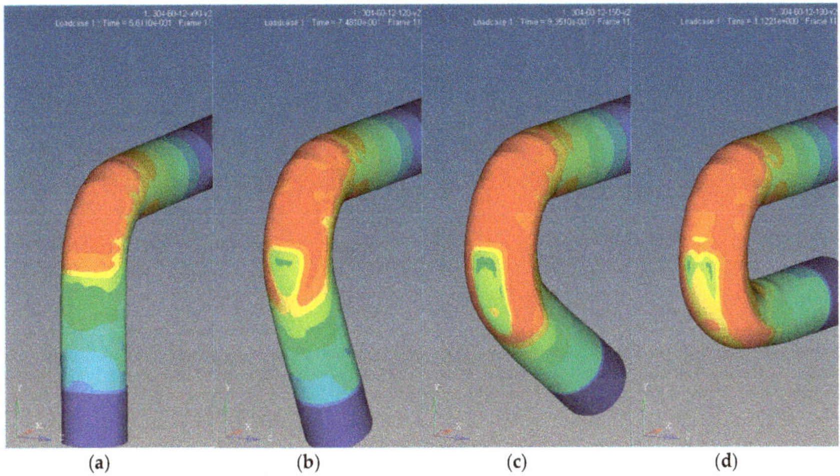

(a)   (b)   (c)   (d)

Figure 15. Equivalent stress mapping for bending angle of 90° (**a**); 120° (**b**); 150° (**c**); and 180° (**d**), at 2.7 rad/s for AISI 304.

3.5. Experimental Validation

3.5.1. Validation Methods

In order to be able to consider these results correctly, it is necessary to validate the simulation model by comparing it with the corresponding real case in the same conditions. Six samples of tubes, for each of the following stainless steels families, were then examined:

- AISI 304
- AISI 316
- AISI 409
- AISI 441

All of the samples have the same dimensions corresponding to the simulation performed, a diameter 60 mm and thickness 1.2 mm. For each steel family, one of the six samples, for each group, were used to obtain, through tensile tests as before, the specific stress–strain curves, in order to eliminate the uncertainty due to the use of a mean curve. Both the tests and the simulations have been carried out with the settings, the rotational velocity and bending radius, that are actually used by the industries for the typical finishing process.

For the comparison, we decided to measure both of the thicknesses reached during the bend along the backbone at specific angles, as shown in Figure 16.

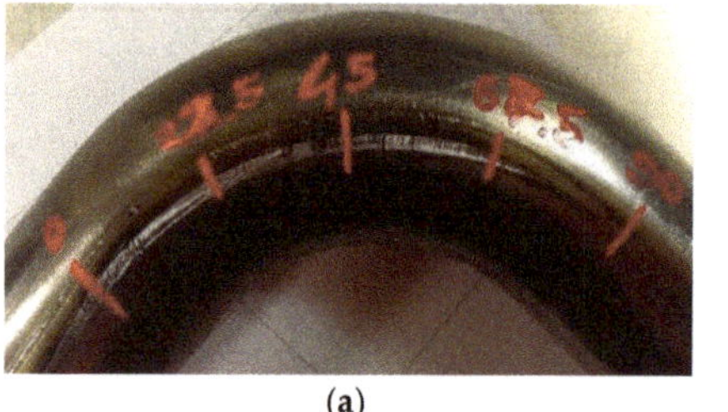
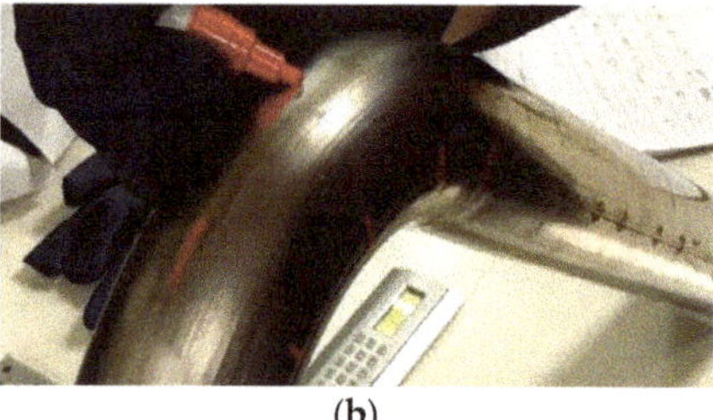

(a) (b)

Figure 16. Thicknesses measuring grid on the backbone (a,b).

3.5.2. Validation Results

The values of the thicknesses were measured for each marked angle, and the average has been calculated. The values obtained are shown in Tables 1–4, in millimeters.

The same measures were extracted from the simulation result in the same points. The thickness variation and the percentage variation of the simulation, with respect to the real case, were then calculated to have a first feedback on the goodness of the predictive model adopted. Tables 5–8 show the calculated data and Figure 17 shows the trends of the four families of stainless steels considered.

Table 1. Thickness for the AISI 304 samples.

Measurement Angle	Sample n° 1	Sample n° 2	Sample n° 3	Sample n° 4	Sample n° 5	Sample n° 6	Mean Value
0°	1.169	1.200	1.180	1.124	1.250	1.235	1.108
22.5°	1.003	1.019	1.026	1.023	1.023	1.101	1.037
45°	0.982	0.993	1.002	1.058	1.157	1.016	1.039
67.5°	1.086	1.016	1.050	1.029	1.166	1.052	1.042
90°	1.200	1.146	1.180	1.152	1.149	1.161	1.149

Table 2. Thickness for the AISI 316 samples.

Measurement Angle	Sample n° 1	Sample n° 2	Sample n° 3	Sample n° 4	Sample n° 5	Sample n° 6	Mean Value
0°	1.091	1.131	1.290	1.119	1.119	1.123	1.119
22.5°	1.023	1.010	1.028	1.014	1.048	1.032	1.026
45°	1.016	0.999	1.004	0.986	0.987	1.021	1.002
67.5°	1.028	1.015	1.021	1.030	1.071	1.017	1.030
90°	1.135	1.101	1.149	1.128	1.192	1.143	1.140

Table 3. Thickness for the AISI 409 samples.

Measurement Angle	Sample n° 1	Sample n° 2	Sample n° 3	Sample n° 4	Sample n° 5	Sample n° 6	Mean Value
0°	1.080	1.067	1.070	1.075	1.083	1.079	1.076
22.5°	0.959	0.961	0.950	0.963	0.942	0.962	0.956
45°	1.011	0.970	0.936	0.911	1.026	0.937	0.965
67.5°	0.987	0.971	0.968	0.990	1.010	0.989	0.986
90°	1.200	1.092	0.097	1.107	1.079	1.096	1.112

Table 4. Thickness for the AISI 441 samples.

Measurement Angle	Sample n° 1	Sample n° 2	Sample n° 3	Sample n° 4	Sample n° 5	Sample n° 6	Mean Value
0°	1.169	1.200	1.180	1.124	1.250	1.235	1.193
22.5°	1.003	1.019	1.026	1.023	1.123	1.101	1.049
45°	0.982	0.993	1.002	1.058	1.157	1.016	1.035
67.5°	1.086	1.016	1.050	1.029	1.166	1.052	1.067
90°	1.200	1.180	1.180	1.152	1.149	1.161	1.166

Table 5. Thickness for the AISI 304 samples.

Measurement Angle	Simulation Thickness [mm]	Sample mean Thickness [mm]	Δ Thickness [mm]	Percentage Variation [%]
0°	1.138	1.108	0.031	2.80
22.5°	0.958	1.037	−0.079	−7.62
45°	0.850	1.039	−0.188	−18.10
67.5°	0.880	1.042	−0.162	−15.55
90°	1.160	1.149	0.010	0.87

Table 6. Thickness for the AISI 316 samples.

Measurement Angle	Simulation Thickness [mm]	Sample mean Thickness [mm]	Δ Thickness [mm]	Percentage Variation [%]
0°	1.135	1.118	0.007	0.62
22.5°	0.952	1.025	−0.073	−7.12
45°	0.840	1.002	−0.160	−16
67.5°	0.800	1.030	−0.230	−22.3
90°	1.135	1.140	−0.005	−0.43

Table 7. Thickness for the AISI 409 samples.

Measurement Angle	Simulation Thickness [mm]	Sample Mean Thickness [mm]	Δ Thickness [mm]	Percentage Variation [%]
0°	1.170	1.076	0.094	8.7
22.5°	1.000	0.956	0.044	4.6
45°	0.880	0.965	−0.085	−8.8
67.5°	0.200	0.986	−0.786	−79.7
90°	1.170	1.112	0.058	5.2

Table 8. Thickness for the AISI 441 samples.

Measurement Angle	Simulation Thickness [mm]	Sample Mean Thickness [mm]	Δ Thickness [mm]	Percentage Variation [%]
0°	1.166	1.193	−0.027	−2.26
22.5°	1.013	1.049	−0.036	−3.43
45°	0.874	1.035	−0.161	−15.56
67.5°	0.814	1.067	−0.253	−23.71
90°	1.171	1.166	0.005	0.42

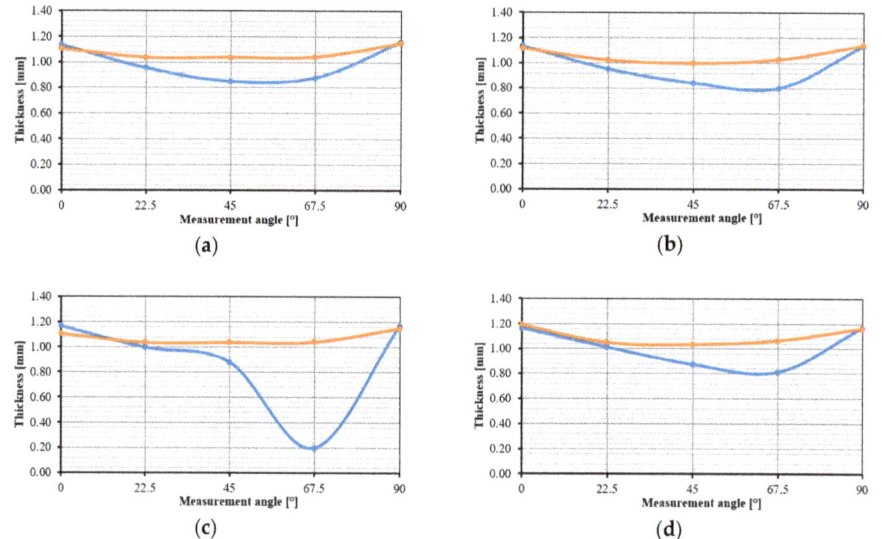

Figure 17. Thickness trend for the simulation case (blue) and real case (orange) for AISI 304 (**a**); AISI 316 (**b**); AISI 409 (**c**); and AISI 441 (**d**).

The above figures show that the simulation results are in agreement with the real case. It is noted that the major deviation occurs in all four of the cases for 67.5° of measurement, and a maximum deviation of −79.7% for AISI 409 and 18–24% for the other stainless steels is reached. As a matter of fact, it is important to note that there is a huge difference between the two cases, specifically for the 67.5° of AISI 409 ferritic stainless steel, but this is due to the fact that for this particular case there is a localized break near the considered angle on the simulation results, and furthermore, the maximum variation for AISI 304, 316 and 441 corresponds to a value in the range of 0.16–0.25 mm. The difference between realty and simulation, for both the failure of AISI 409 and for the general variation of the other three stainless steels families, is due to the presence of an additional support element present in most of the bending machines, called a 'booster', which was not present in the simulation model. Its function is precisely that of pushing the tube during bending, in order to avoid the deformations or failure caused by the friction between the element and the machine or by the concentration of stresses. Its action also affects the distribution of the thinning, caused by the deformation, in fact, of the tube being pushed by the booster, which will have more evenly distributed the stresses on itself, and consequently, the deformations and the thinning will take place on a wider area and will not lead to failure of the piece.

4. Conclusions

In this paper, the bending process of stainless steel pipes has been studied. The experimental investigations coupled with simulations highlighted the importance of each parameter, both operational and geometric, on the final results.

In particular, it has been observed that the pipe diameter does not prove to be a decisive parameter for the success of the working process, while the pipe thickness appears to be a determinant factor for failure and/or unwanted deformation of the formed piece. The R ratio is extremely important; in fact, its variation within the standard range (between 1.0 and 1.5) identified the transition between the failure and success of the operation, both for the AISI 304 austenitic stainless steel and for AISI 409 ferritic stainless steel, of which for the latter led to a 90% increase in feasibility.

The combined study of the rotational speed and bending angle allowed us to define a trend of influence for these operating parameters, showing how there is a linear increase in the influence of the speed in the range between $30°$ and $90°$ of bending, while for the angles higher than around $120°$, this tendency is reversed.

The overall experimental validation showed a deviation of the model from the reality, between an overestimation of 8% to an underestimation of 20%, with the maximum displacement generally located on the back of the bending, probably due to the presence in the experimental tests of an additional element support of the machinery, called a booster, which was not contemplated in the simulation model. Furthermore, the maximum deviation recorded corresponds to a deviation in thickness between the two cases in the order of 10^{-2} mm, thus resulting in a good starting point for the refinement and optimization of the final model.

Moreover, thanks to this analysis and the preliminary experimental tests, the FEM simulation has proved to be a useful tool in order to predict the industrial deformation processes, where there are currently no means to characterize the processes generally carried out on these components, but only of the empirical methods to define its overall feasibility.

Author Contributions: R.R. and O.D.P conceived, designed and performed the experiments and analyzed the data; R.R. wrote the paper; A.D.S. guided the research.

Funding: This research received no external funding.

Conflicts of Interest: The authors declare no conflict of interest.

References

1. Marshall, P. *Austenitic Stainless Steels*; Springer: Heidelberg, Germany, 1984; ISBN 978-0-85334-277-9.
2. Di Schino, A.; Barteri, M.; Kenny, J.M. Fatigue behavior of a high nitrogen austenitic stainless steel as a fuction of its grain size. *J. Mater. Sci. Lett.* **2003**, *22*, 1511–1513. [CrossRef]
3. Di Schino, A.; Kenny, J.M.; Salvatori, I.; Abbruzzese, G. Modelling recrystallization and grain growth in low nickel austenitic stainless steels. *J. Mater. Sci.* **2001**, *36*, 593–601. [CrossRef]
4. Di Schino, A.; Kenny, J.M.; Barteri, M. High temperature resistance of high nitrogen low nickel austenitic stainless steels. *J. Mater. Sci. Lett.* **2003**, *22*, 691–693. [CrossRef]
5. Corradi, M.; Di Schino, A.; Borri, A.; Rufini, R. A review of the use of stainless steel for masonry repair and reinforcement. *Constr. Build. Mater.* **2018**, *181*, 335–346. [CrossRef]
6. Badoo, N.R. Stainless steel in construction: A review of research, applications, challenges and opportunities. *J. Constr. Steel Res.* **2008**, *64*, 1199–1206. [CrossRef]

7. Lo, K.H.; Shek, C.H.; Lai, J.K.L. Recent development in stainless steels. *Mater. Sci. Eng. R Rep.* **2009**, *65*, 39–104. [CrossRef]

8. Gardner, L. The use of stainless steel in structures. *Prog. Struct. Eng. Mater.* **2005**, *7*, 45–52. [CrossRef]

9. Bong, H.J.; Barlat, F.; Lee, M.G.; Ahn, D.C. The forming limit diagram of ferritic stainless steel sheets: Experiments and modeling. *Int. J. Mech. Sci.* **2012**, *64*, 1–10. [CrossRef]

10. Yang, T.B.; Yu, Z.Q.; Xu, C.B.; Li, S.H. Numerical analysis for forming limit of welded tube in hydroforming. *J. Shanghai Jiaotong Univ.* **2011**, *45*, 6–10.

11. Zhang, H.; Liu, Y. The inverse parameter identification of Hill'48 yield function for small-sized tube combining response surface methodology and three-point bending. *J. Mater. Res.* **2017**, *32*, 2343–2351. [CrossRef]

12. Zhan, M.; Guo, K.; Yang, H. Advances and trends in plastic forming technologies for welded tubes. *Chin. J. Aeronaut.* **2016**, *29*, 305–315. [CrossRef]

13. Giannella, V.; Fellinger, J.; Perrella, M.; Citarella, R. Fatigue life assessment in lateral support element of a magnet for nuclear fusion experiment "Wendelstein 7-X". *Eng. Fract. Mech.* **2017**, *178*, 243–257. [CrossRef]

14. Citarella, R.; Giannella, V.; Lepore, M.A.; Fellinger, J. FEM-DBEM approach to analyse crack scenarios in a baffle cooling pipe undergoing heat flux from the plasma. *AIMS Mater. Sci.* **2017**, *4*, 391–412. [CrossRef]

15. Khalfallah, A.; Oliveira, M.C.; Alves, J.L.; Zribi, T.; Belhadjsalah, H.; Menezes, L.F. Mechanical characterization and constitutive parameter identification of anisotropic tubular materials for hydroforming applications. *Int. J. Mech. Sci.* **2015**, *104*, 91–103. [CrossRef]

16. Dama, K.K.; Snirivasulu, S.; Swaroop, D. Design for crashworthiness of an automotive sub-system using cae techniques. *Int. J. Mech. Eng. Technol.* **2018**, *9*, 21–27.

17. Cao, F.; Li, J.; Cui, M. Analysis of frame structure of medium and small truck crane. *AIP Conf. Proc.* **2018**, *1944*. [CrossRef]

© 2018 by the authors. Licensee MDPI, Basel, Switzerland. This article is an open access article distributed under the terms and conditions of the Creative Commons Attribution (CC BY) license (http://creativecommons.org/licenses/by/4.0/).

Article

Numerical Simulation of the Depth-Sensing Indentation Test with Knoop Indenter

Maria I. Simões [1], Jorge M. Antunes [2,*], José V. Fernandes [1] and Nataliya A. Sakharova [1]

[1] CEMMPRE—Department of Mechanical Engineering, University of Coimbra, Rua Luís Reis Santos, Pinhal de Marrocos, 3030-788 Coimbra, Portugal; isabel.simoes@dem.uc.pt (M.I.S.); Valdemar.fernandes@de.uc.pt (J.V.F.); nataliya.sakharova@dem.uc.pt (N.A.S.)

[2] Escola Superior de Tecnologia de Abrantes, Instituto Politécnico de Tomar, Rua 17 de Agosto de 1808—2200 Abrantes, Portugal

* Correspondence: jorge.antunes@dem.uc.pt; Tel.: +351-239-790-700

Received: 11 October 2018; Accepted: 26 October 2018; Published: 31 October 2018

Abstract: Depth-sensing indentation (DSI) technique allows easy and reliable determination of two mechanical properties of materials: hardness and Young's modulus. Most of the studies are focusing on the Vickers, Berkovich, and conical indenter geometries. In case of Knoop indenter, the existing experimental and numerical studies are scarce. The goal of the current study is to contribute for the understanding of the mechanical phenomena that occur in the material under Knoop indention, enhancing and facilitating the analysis of its results obtained in DSI tests. For this purpose, a finite element code, DD3IMP, was used to numerically simulate the Knoop indentation test. A finite element mesh was developed and optimized in order to attain accurate values of the mechanical properties. Also, a careful modeling of the Knoop indenter was performed to take into account the geometry and size of the imperfection (offset) of the indenter tip, as in real cases.

Keywords: depth-sensing indentation; Knoop indenter; hardness; Young's modulus; numerical simulation

---

1. Introduction

Depth-sensing indentation (DSI) tests are typically used to evaluate the hardness and Young's modulus of materials. They can also be used to extract the uniaxial mechanical properties of bulk and composite materials, such as the yield stress and the strain hardening parameter (see, e.g., [1–4]). The most common hardness testing methods were developed in the early twentieth century. They are typically performed using spherical, conical and pyramidal indenters with Vickers and Berkovich geometries. In addition, the Knoop pyramid geometry is also used [5].

Hardness tests with the Knoop indenter have been valuable in the mechanical characterization of some materials, such as thin coatings [6,7] and biological materials (e.g., dental tissue [8]). Another important application of this hardness tests is in the field of the gradient materials obtained by severe plastic deformation (see, e.g., [9–11]), for which it is required to determine the mechanical properties in thin samples and/or thin surface layers of the samples.

In fact, the Knoop indenter geometry leads to wider and shallower indentations for a given applied load than the Vickers and Berkovich geometries. This makes the Knoop hardness test particularly attractive in some cases as for the determination of the near-surface properties and the characterization of brittle materials. Moreover, the results obtained in the Knoop hardness test are sensitive to the indenter orientation, making it a useful tool to analyze the materials anisotropy (e.g., [12]).

Although Knoop indenter is relatively common to use, it has not received suitable attention with respect to the study of some of the peculiarities of the test itself. One of the first studies using numerical simulation of the Knoop indentation test was performed by Rabinovich and Sarin [13], in their study, the Knoop indentation was analyzed in the context of linear elasticity. Giannakopoulos [14] presented analytical results on the response of frictionless and adhesionless contact of flat, linear elastic and visco-elastic isotropic surfaces penetrated by pyramidal indenters including the Knoop geometry. Giannakopoulos and Zisis [15] studied the Knoop indentation of elastic and elastoplastic materials with and without strain hardening. More recently, Giannakopoulos and Zisis [16] presented a finite element study on the adhesionless contact of flat surfaces by Knoop indenter. Only a few experimental studies, such as those by Riester et al. [17,18] were performed in order to clarify the analysis procedure of DSI tests with the Knoop geometry. Recently, Ghorbal et al. [19] performed an experimental work on ceramic materials in order to compare the conventional Knoop and Vickers hardness.

In this context, the present study intends to contribute to clarify some aspects of the Knoop indentation, particularly those related to the analysis of depth sensing indentation (DSI) results. For this purpose, three-dimensional numerical simulations of the indentation tests were performed, using various pyramidal indenter geometries, from Vickers to Knoop. A systematic study is accomplished using materials with different mechanical properties whose ratio between the residual indentation depth after unloading (h_f) and the indentation depth at maximum load (h_{max}) is in the range $0.022 < h_f/h_{max} < 0.984$. The correction factor, β, required to determine the Young's modulus, is evaluated as a function of the indenter geometry, from Vickers to Knoop. In addition, numerical simulations using flat indenters with equivalent lozenge geometries were performed, in order to better understand the role of the pyramidal geometry.

2. Theoretical Aspects

The ability of the ultramicrohardness equipment to register the load versus the depth indentation, during the test, enables us to evaluate not only the hardness, but also other properties, such as the Young's modulus. Based on the Sneddon relationship [20] between the indentation parameters and Young's modulus, Doerner and Nix [21] have proposed an equation that relates the Young's modulus with the compliance, C, of the unloading curve at the point of the maximum load and the contact area, A_c, such as:

$$E_r = \frac{\sqrt{\pi}}{2\beta} \frac{1}{\sqrt{A_c}} \frac{1}{C}, \tag{1}$$

where β is the geometrical correction factor for the indenter geometry. The specimen's Young's modulus, E_s, is obtained using the equation:

$$\frac{1}{E_r} = \frac{\left(1 - \vartheta_s^2\right)}{E_s} + \frac{\left(1 - \vartheta_i^2\right)}{E_i}, \tag{2}$$

where E and ϑ are the Young's modulus and the Poisson's ratio, respectively, of the specimen (s) and of the indenter (i). When performing numerical simulations, the indenter can in a first approximation considered as rigid for simplicity (e.g., [22]), and so $\left(1 - \vartheta_i^2\right)/E_i = 0$. The accuracy of the Young's modulus results, obtained with the above equations, depends on the correct evaluation of the contact area and compliance. The contact area, A_c, can be evaluated by two different procedures. One procedure uses the contour of the indentation in the finite element mesh, in order to make the results independent of the formation of pile-up or sink-in. The other is the usual experimental procedure, which makes use of the compliance, C, evaluated by fitting the unloading part of the load–indentation depth curve, $(P - h)$, using the power law [22]:

$$P = P_0 + T(h - h_0)^m, \tag{3}$$

where T and m are constants obtained by fit and h_0 is the lower value of the indentation depth, h, used in the fitted region, corresponding to a load value P_0, during unloading. The upper part of the unloading curve taken into account in the fits is 70% [22]. Once the value of compliance, C, is known, the contact indentation depth, h_c, that allows the calculation of the contact area according to the geometry of the indenter $(A_c = f(h_c))$, is determined by the equation $h_c = h_{max} - \varepsilon P_{max}C$, where h_{max}

is the indentation depth at the maximum load, P_{max}, and ε is a parameter equal to 0.72 for conical or pyramidal (Berkovich [23], Vickers [22] and Knoop [18] indenters).

An approach that can be used for evaluating the correction factor, β, was proposed by Joslin and Oliver [24], combining the hardness and Young's modulus equations:

$$\frac{P}{S^2} = \frac{\pi}{4\beta^2}\frac{H_{IT}}{E_r^2},$$

(4)

where $H = P/A_c$ is the hardness, P the maximum applied load and S the stiffness ($S = 1/C$). The ratio between the maximum applied load and the square of the stiffness, P/S^2, is an experimentally measurable parameter that is independent of the contact area and so of the penetration depth [24].

3. Numerical Simulation and Materials

Three-dimensional numerical simulations of the hardness tests were carried out, using a finite element (FE) in-house code DD3IMP. This FE code, which has already been tested in the case of Vickers, Berkovich and conical indentation of bulk materials and thin films (see, e.g., [22,25,26]), allows simulating the hardness tests with any type of indenter shape, taking into account contact with friction between the indenter and the sample [22,27,28]. The mechanical model, which is the basis of the DD3IMP code, considers the hardness test as a quasi-static process that occurs in the large plastic deformations domain. In DD3IMP, the contact with friction problem is modeled using a classical Coulomb's law. To relate the static equilibrium problem to the contact with friction, an augmented Lagrangean method is applied to the mechanical formulation. This leads to a system of non-linear equations, where the kinematic (material displacements) and static variables (contact forces) are the final unknowns [27,28]. To solve this problem, the code uses a fully implicit Newton–Raphson-type algorithm. All non-linearities, induced by the elastoplastic behavior of the material and by the contact with friction, are treated in a single iterative loop [27,28].

In the current study, the friction between the indenter and the deformable body was assumed to have a friction coefficient of 0.16. This is a commonly used value and leads to a better description of the indentation process than if frictionless contact is assumed [22,26].

3.1. Indenters

The Knoop indenter has a pyramidal geometry, with a lozenge-shaped base having one diagonal (L) 7.11 times longer than the other (m). The angles between the edges (apical angles) are 172.5° for the long edges and 130° for the short edges, as shown in Figure 1.

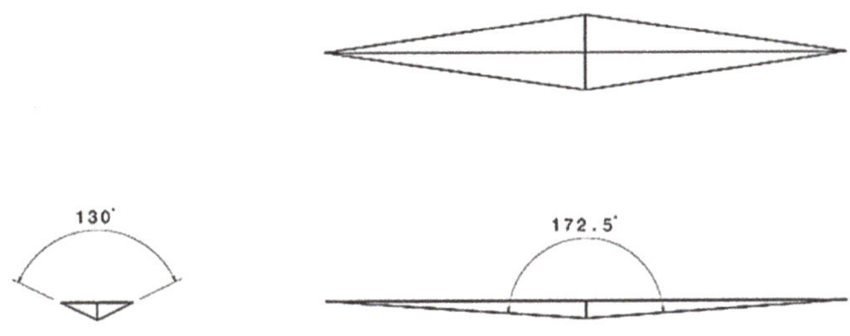

Figure 1. Schematic representation of the Knoop indenter geometry.

The ideal (in the absence of pile-up or sink-in) indentation contact area, A_c, of the Knoop indenter as function of the indentation contact depth is given by:

$$A_c = 2h_c^2 \tan\theta_1 \tan\theta_2 = 65.4h_c^2,$$

(5)

where h_c is the ideal indentation contact depth, $\theta_1 = 65°$ and $\theta_2 = 86.25°$ are the semi-apical angles of the Knoop indenter.

The Knoop indenter geometry was modeled using parametric Bezier surfaces, which allows a satisfactory description of the indenter tip, namely an imperfection such as occurs in the real geometry, similar to the case of offset in the Vickers indenter [25,29]. The modeled Knoop indenter, shown in Figure 2, has a tip imperfection consisting of a plane normal to the indenter axis with an area equal to 0.0032 µm^2 (this value is the same as the experimentally found, for the Vickers indenter, by Antunes et al. [29]).

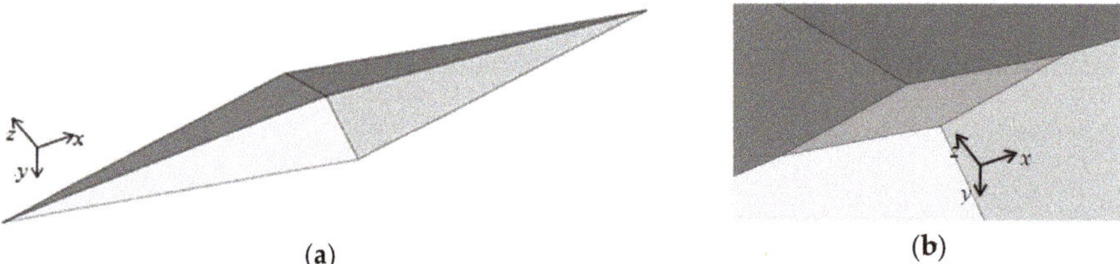

(a)   (b)

Figure 2. Knoop indenter modeled with Bezier surfaces: (**a**) general view; (**b**) detail of the indenter tip imperfection.

Due to the tip imperfection, the indenter area function does not match the ideal area function above mentioned (Equation (5)). Table 1 shows the area function of the Knoop indenter (ratio, R = L/m = 7.11) used in the numerical simulations. This table also shows four others area functions of pyramidal indenters, used in this study, with different values of the ratio, R, between the diagonals of the indenter (R = 1, 2.5, 4 and 5.5), where the ratio R = 1 corresponds to the Vickers indenter.

Table 1. Area function of the indenters used in the numerical simulations.

R = L/m	θ_1	θ_2	Area Function
1 (Vickers)	74.0546	74.0546	$A = 24.5000h^2 + 0.5600h + 0.0032$
2.5	69.1723	81.3478	$A = 34.5500h^2 + 0.6650h + 0.0032$
4	67.0462	83.9559	$A = 44.6000h^2 + 0.7556h + 0.0032$
5.5	65.8369	85.3366	$A = 54.6500h^2 + 0.8364h + 0.0032$
7.11 (Knoop)	64.8379	86.2199	$A = 65.4377h^2 + 0.9152h + 0.0032$

For pyramidal indenters, other than Vickers and Knoop, the angles θ_1 and θ_2 were chosen such that the tangents of θ_1 and θ_2 follows a quasi-linear evolution with R, as shown in Figure 3. In this way it is possible to study the extent to which the deviation of Vickers geometry towards the Knoop geometry influences the indentation results obtained with DSI experiments.

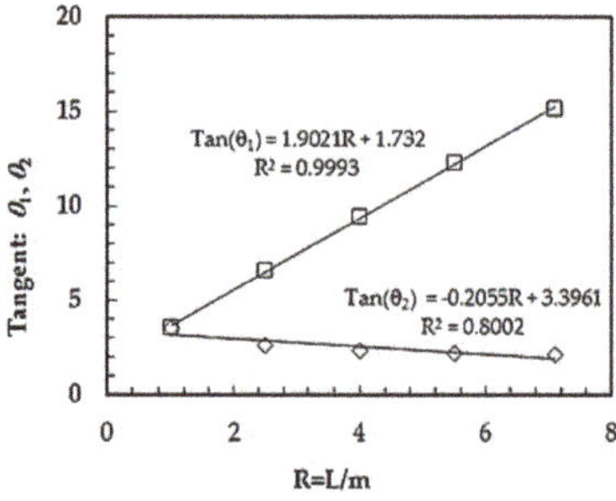

Figure 3. Evolution of the tangents of θ_1 and θ_2 as function of the R-ratio values.

In order to better understand some aspects related with the influence of the indenter geometry on the mechanical properties evaluation, numerical simulations using lozenge-shaped flat indenters were also performed. The flat indenters were modulated with five values of the ratio, R = L/m, between the diagonals of the lozenge, as for the pyramidal indenters (R = 1, 2.5, 4, 5.5, and 7.11).

3.2. Finite Element Mesh

The test sample used in the numerical simulations has both radius and thickness equal to 40 μm. Its discretization was performed using three-linear eight-node isoparametric hexahedrons. Due to geometrical and material symmetries in the X = 0 and Z = 0 planes, only a quarter of the sample was used in the numerical simulation, as shown in Figure 4.

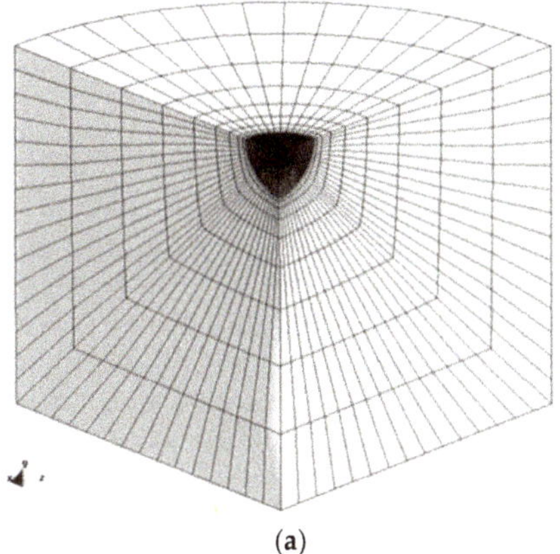

(a)

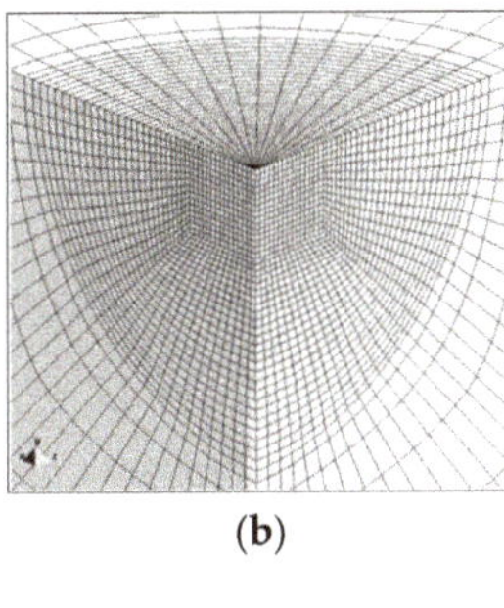

(b)

Figure 4. Finite element mesh used in the numerical simulations: (**a**) global view; (**b**) detail of the central region where indentation occurs.

The FE mesh is composed by 17,850 elements. The size of the elements in the indentation region is 0.055 μm. This refinement has proven to provide accurate values of the indentation contact area, when measured using the contour of the indentation, in case of Vickers and Berkovich geometries, with equal value of offset area (see, e.g., [22,26]). In fact, the Young's modulus values obtained from the indentation contact area, evaluated using the contour of nodes in the FE mesh in contact with the indenter at maximum load presents an error less than 1%, when compared with the values used as input in the numerical simulation (e.g., [26]). In the present study, the size of the central region of the finite element mesh, especially refined, is larger than in these cases, and thus the total number of elements is approximately three times greater, in order to take into account the elongated geometry of the Knoop indenter.

3.3. Materials

Three-dimensional numerical simulations of depth-sensing indentation with pyramidal indenters were carried out using 45 fictitious materials, whose mechanical properties are shown in Table 2. In order to cover a wide range of materials used in engineering applications, five values of yield stress (0.2, 2, 6, 10, and 20 GPa), three values of Young's modulus (70, 200 and 400 GPa) and of strain hardening parameter of the Swift law (0.01, 0.15 and 0.3), were taken into account.

The plastic behavior of the materials is described by the von Mises yield criterion and the Swift hardening law: $\sigma = k(\varepsilon + \varepsilon_0)^n$ where σ and ε are the equivalent stress and plastic strain, respectively, and k, ε_0 and n (strain hardening coefficient) are the material parameters (the yield stress is: $\sigma_y = k\varepsilon_0^n$); the parameter ε_0 was considered to be equal to 0.005. The elastic behaviour is isotropic and described by the generalised Hooke's law; the Poisson's ratio, ϑ is 0.3, for all simulations.

Table 2. Mechanical properties of the materials used in the numerical simulations.

Materials	Studied Cases	n	σ_y (GPa)	E (GPa)
Without strain hardening	5			70
	5	≈ 0		200
	5			400
With strain hardening	5	0.15	0.2, 2, 6, 10 and 20	70
	5			200
	5			400
	5			70
	5	0.30		200
	5			400

4. Results

4.1. Indentation Geometry and Equivalent Plastic Strain Distributions

A study of the indentation geometry with the Knoop indenter (R = 7.11) was performed using only three of the materials in Table 2, covering different values in the possible range of h_f/h_{max}. Table 3 shows the mechanical properties of these materials, which have two Young's modulus values, 200 and 400 GPa, yield stress of 0.2, 6, and 20 GPa, and two values of the strain hardening parameter, n $\approx$ 0 and n = 0.3. Table 3 also includes the value of the ratio between the indentation depth after unloading and the indentation depth at maximum load, h_f/h_{max}. This ratio, which can be easily obtained from the experimental load-unloading curve, is independent of the maximum indentation depth, for a given material in cases of conical [30] and Vickers [22] indentations. Its range of values is from 0 to 1, which corresponds to materials with purely elastic and rigid-plastic behaviors, respectively. All numerical simulations of the hardness test were performed up to the same maximum indentation depth, $h_{max} = 0.2$ μm.

Table 3. Mechanical Properties of The Materials Used in the Study of The Knoop Indentation Geometry.

Material	σ_y (GPa)	n	E (GPa)	ν	h_f/h_{max}
M1	0.2	0.01			0.97
M2	6		200	0.3	0.40
M3	20	0.3	400		0.25

Figure 5 shows the indentation profiles, obtained for the three materials at maximum load (Figure 5a,c) and after unloading (Figure 5b,d), along the two diagonals, the shorter, m, and the longer, L, respectively. Figure 5a shows that for the shorter diagonal, the indentation profile obtained at the maximum load depends on the material mechanical properties. In case of materials M2 and M3 the indentation surface sink-in. This behavior can be associated with the value of the ratio h_f/h_{max} that is equal to 0.40 and 0.25 (for the materials M2 and M3, respectively), and with the high value of the strain hardening parameter, n. In case of M1 material, with a ratio h_f/h_{max} equal to 0.97, i.e., close to 1, the indentation surface does not present sink-in or pile-up. On the other hand, the indentation profile obtained at maximum load for the longer diagonal, almost does not depends on the materials mechanical properties (Figure 5c).

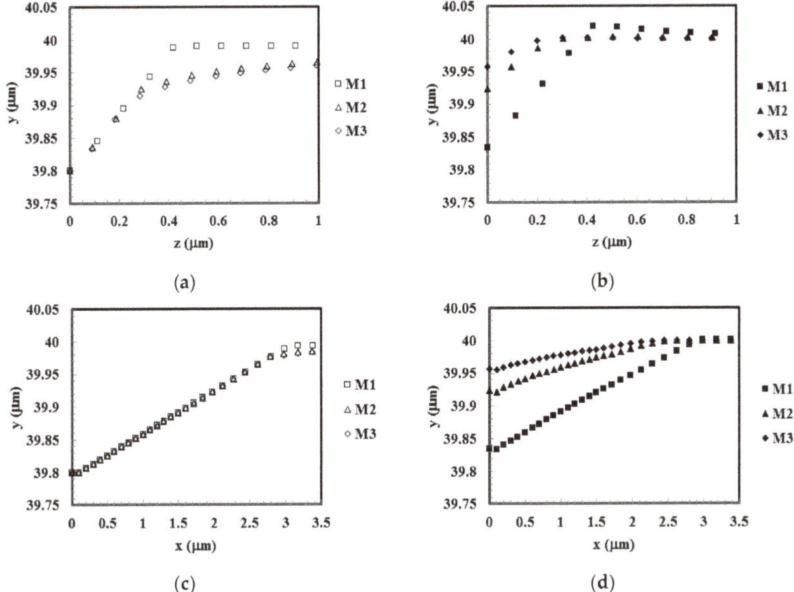

Figure 5. Surface indentation profiles obtained along the two diagonals, the shorter, m (*z*-axis) and the longer, L (*x*-axis), respectively: (**a**,**c**) obtained at maximum load; (**b**,**d**) after unloading.

After unloading, the indentation profiles corresponding to the materials M2 and M3 shown elastic recovery in both diagonals (Figure 5b,d). Moreover, the amount of elastic recover increases with the decrease of the ratio h_f/h_{max}. In case of M1 material, with a ratio $h_f/h_{max} = 0.97$ and without strain hardening, the indentation profile along the short diagonal shows pile-up (Figure 5b). In fact, for other indenter geometries (conical and Vickers), the pile-up appears for values of the ratio h_f/h_{max} higher than 0.8 in materials without strain hardening (see, e.g., [22,30]). It should be noted that, for a given material, the h_f/h_{max} ratio correlates with the value of the H/E ratio between the hardness and the Young's modulus, and slightly depends on the strain hardening parameter.

Figure 6 shows the comparison between the Knoop indentation diagonals at maximum load. To make possible this comparison, the indentation profiles were normalized by considering the value of the R-ratio between the indenter diagonals (R equal to 7.11). Figure 6a shows that in the material M1 the two diagonal exhibit a similar behavior. In the case of the materials M2 and M3 sink-in is observed along the short diagonal. Moreover, the sink-in slightly increases with the value of the ratio h_f/h_{max}.

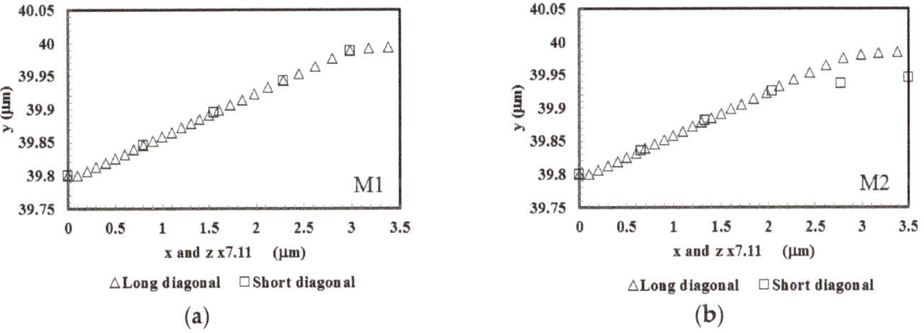

Figure 6. *Cont.*

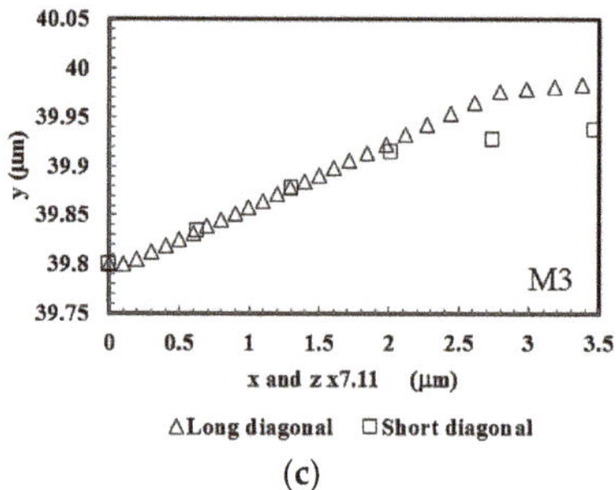

Figure 6. Surface indentation profiles at maximum load, obtained from the results of Figure 5a,c, where z is multiplied by R = 7.11, in order to easily compare the profiles along the two diagonals, the longer, L, and the shorter, m: (**a**) Material M1; (**b**) Material 2; (**c**) Material M3.

The effect of the indenter geometry on the equivalent plastic strain distribution of the indentations was also studied. Figure 7a,c shows the equivalent plastic strain distributions obtained at the maximum load in the numerical simulations of the materials M1 ($h_f/h_{max} = 0.97$) and M3 ($h_f/h_{max} = 0.25$) using the Knoop indenter. For comparison, the same figure also shows the same distributions obtained with the Vickers indenter (Figure 7b,d).

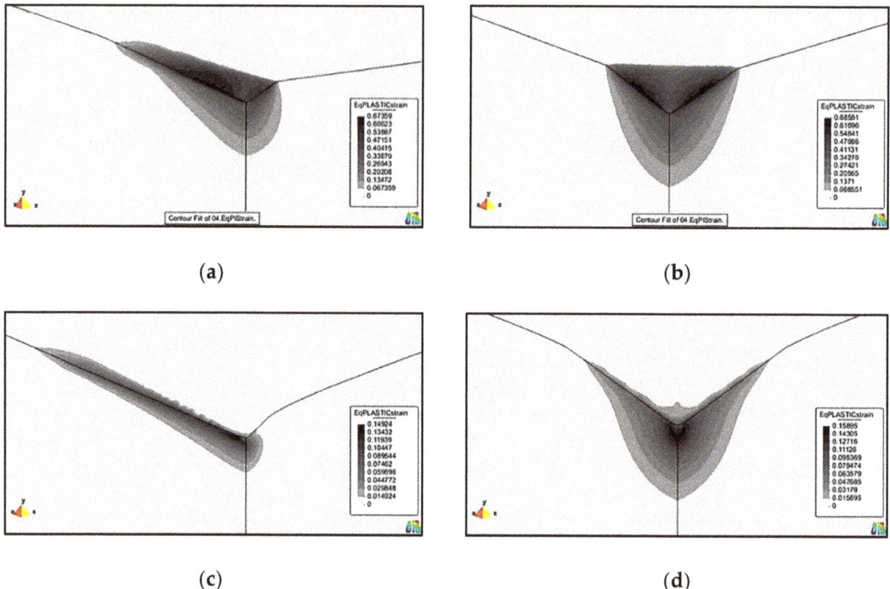

(**a**)   (**b**)

(**c**)   (**d**)

Figure 7. Equivalent plastic strain distributions obtained at maximum load in the numerical simulations using the Knoop and Vickers indenters: (**a**,**b**) Material M1; (**c**,**d**) Material M3.

For each material, the maximum values of the equivalent plastic strain are quite similar for the Knoop and Vickers indenters. However, in the case of Knoop, the equivalent plastic strain distributions are asymmetric with respect to the indenter axes. The maximum values of the equivalent plastic strain are observed along the longest diagonal. For M1 material ($h_f/h_{max} = 0.97$), the maximum value of the equivalent plastic strain is slightly higher in the Vickers indentation (≈ 0.686) than for Knoop (≈ 0.674). The maximum plastic strain region is located just at the surface in the edge regions of the indentation (Figure 7a,b). In case of the M3 material ($h_f/h_{max} = 0.25$), as shown in Figure 7c,d, the maximum value of equivalent plastic strain is also somewhat higher for the Vickers indentation (≈ 0.159), than for the Knoop (≈ 0.149). However, for this material, the region with maximum equivalent plastic strain is located beneath the indentation surface, whatever the indentation geometry.

4.2. Indentation Contact Area and Young's Modulus

The results of the numerical simulation of the hardness tests with the Knoop indenter, for all materials in Table 2, were used to study the influence of the mechanical properties on the evaluation of the indentation contact area and, consequently, on the Young's modulus. As mentioned above, the indentation contact area was determinate using two different procedures: one of them uses the load-unloading curve as in the experimental DSI procedure for evaluating this area, A_{h_c}, and the other considers the contour of nodes of the FE mesh in contact with the indenter at maximum load, A_{FE} (numerical contact area).

Figure 8 shows the evolution of the indentation contact areas, A_{h_c} and A_{FE}, as a function of the ratio h_f/h_{max}. These contact areas are normalized with respect to the reference area, A_{REF}, corresponding to the obtained with the area function of the indenter (Equation (5), with h_c equal to the indentation depth, which does not take into account the pile-up or sink-in formation). Figure 8a shows that the contact area, A_{h_c}, is independent of the strain hardening parameter and Young's modulus, whatever the value of the ratio h_f/h_{max}. Moreover, the ratio A_{h_c}/A_{REF} is always less than 1. Figure 8b shows that the numerically calculated contact area, A_{FE}, is nearly independent from the strain hardening parameter and Young's modulus, for the ratio $h_f/h_{max} < 0.6$. However, for $h_f/h_{max} > 0.6$ the normalized contact area depends on the strain hardening parameter. The ratio A_{FE}/A_{REF} is even higher than 1 after a value of h_f/h_{max} that depends on the strain hardening parameter (h_f/h_{max} equal to 0.85, 0.90 and 0.95 for n equal to 0, 0.15 and 0.3, respectively), indicating pile-up formation. For both cases, A_{h_c}/A_{REF} and A_{FE}/A_{REF}, similar behaviors were previously observed for the case of the conical, Vickers and Berkovich indenters (see, e.g., [22,26,30]).

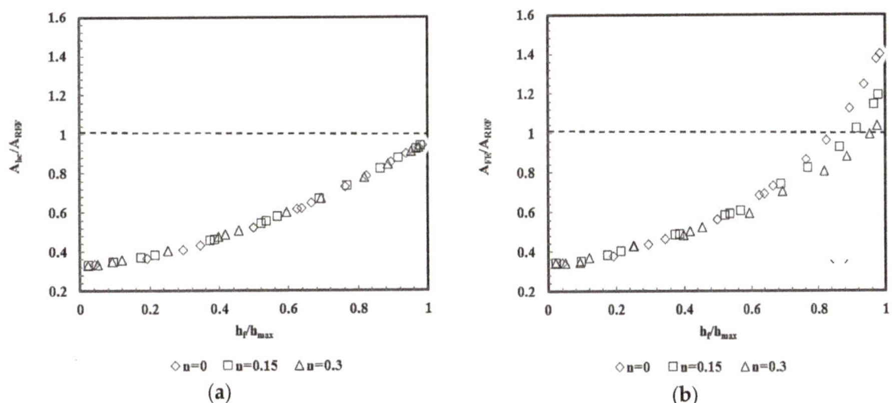

Figure 8. Normalized contact area results obtained in the numerical simulation of the materials with the values of the strain hardening parameter, yield stress and Young's modulus shown in Table 2, using the Knoop indenter. (**a**) Contact area A_{h_c}/A_{REF}; (**b**) Contact area A_{FE}/A_{REF}.

Figure 9 shows the evolution of the Young's modulus E_{h_c} and E_{FE} normalized by the input value used in the numerical simulation, E_{REF}, as a function of the ratio h_f/h_{max}. The average ratio of E_{h_c}/E_{REF} is close to 1.419, except for values of h_f/h_{max} approaching 1, for which the ratio A_{h_c}/A_{REF} increases for low values of the strain hardening parameter of the materials (n = 0 and 0.15). This is a consequence of the pile-up formation during the indentation of these materials, as was also observed for Vickers indenters [25]. The ratios of E_{FE}/E_{REF} are slightly higher than those of E_{h_c}/E_{REF}, being in average close to 1.469. Slight differences between both ratios, E_{h_c}/E_{REF} and E_{FE}/E_{REF}, were already observed for the Vickers indenter [25]. Under these conditions, in the experimental indentation tests, a value for the correction factor β of 1.419 is required to be used for the contact area evaluation (h_c) from the unloading curve, in case of the Knoop indentation. This value is essentially different from other indenter geometries such as conical, Vickers and Berkovich, which require β values around 1.034, 1.055, 1.081, respectively (see, e.g., [22,26]).

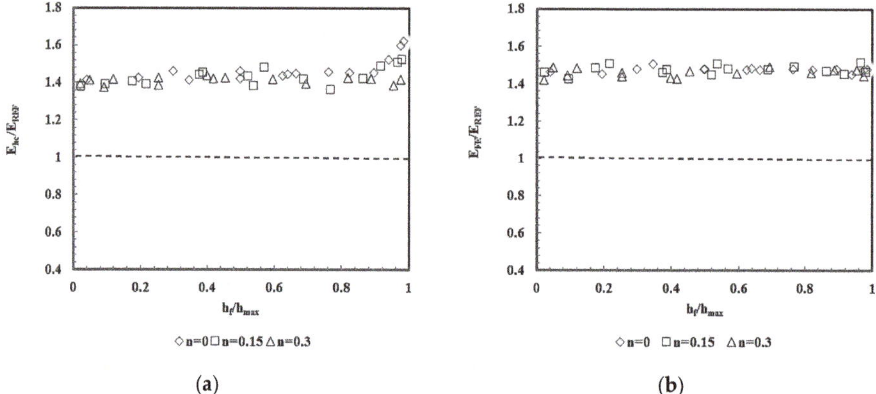

(a)   (b)

Figure 9. Normalized Young's modulus results obtained in the numerical simulation of the materials with the values of the strain hardening parameter, yield stress and Young's modulus shown in Table 2, using the Knoop indenter: (**a**) Young's modulus E_{h_c}/E_{REF}; (**b**) Young's modulus E_{FE}/E_{REF}.

Equation (4) was used in order to confirm the above correction factor β. Figure 10 shows the ratio P/S^2 versus H/E_r^2 obtained by numerical simulation of all materials in Table 2, for the case of the Knoop and Vickers indenters. The reduced Young's modulus, E_r, was determined considering the input Young's modulus and Poisson ratio, H, determined using the contact area evaluated from the contour of the indentations.

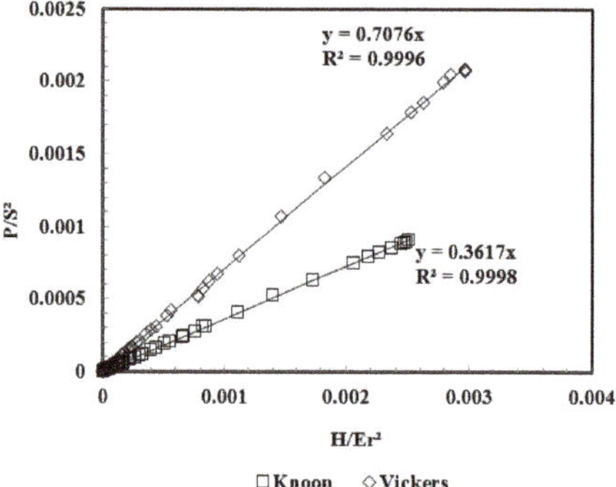

Figure 10. Evolution of the ratio P/S^2 as a function of H/E_r^2, obtained in the numerical simulations of all materials in Table 2, using the Knoop and Vickers indenters.

The straight-lines in Figure 10 pass through the origin of the axes as indicated by Equation (4) (independently of the indenter geometry, curves match for $H/E_r^2 = 0$, i.e., for materials with rigid-plastic behaviour, which corresponds to the ratio $h_f/h_{max} = 1$). The β factor is evaluated from the slope, ρ, of the straight line, related with β through $\rho = (\pi/4\beta^2)$. Using this procedure, for the values of the R-ratio of 1, 2.5, 4, 5.5, and 7.11, the β correction factor obtained were 1.054, 1.141,

1.232, 1.351, and 1.473, respectively. The value of 1.473 for the Knoop indenter is very close to that mentioned above (1.469), obtained from the ratios E_{FE} / E_{REF} (Figure 9b).

4.3. Flat Indenter

In order to understand if the value of the β factor is affected by the plastic deformation beneath the indenter, five flat-ended punches were considered having the same R-ratio values between the diagonals (L and m) as for pyramidal indenters (Table 1). Also, the flat-ended punches were modeled using parametric Bezier surfaces and the finite element mesh is shown in Figure 4. This study allows an understanding of to what extent the evolution of the β factor values with R is related to the pyramidal geometry of the indenter and/or to the loss of symmetry of the flat punches, when R evolves from 1 to 7.11.

Using flat indenters, numerical simulations were performed up to a 0.025 µm displacement imposed so that only elastic deformation occurs in the material beneath the indenter. Five values of Young's modulus were used: 30, 200, 400, 600 and 800 GPa. Figure 11 shows the evolution of the load as a function of the indentation depth, h_e, obtained in the numerical simulations with the flat indenters for values of the R-ratio equal to 1 (as for the Vickers indenter), 4 and 7.11 (as for the Knoop indenter). The figure shows that, for a given indentation depth, the load increases with the increase of the Young's modulus and the R-value.

As above mentioned, Equation (1) comes from the Sneddon's equation [20] that relates the applied load, P, with the elastic deflection of the surface of the material, h_e, and can be written as follows:

$$P = \frac{2\beta E}{1 - \vartheta^2} a h_e, \tag{6}$$

where E and ϑ are de Young's modulus and Poisson ratio of the material, respectively; a is the radius of the rigid Sneddon cylindrical flat indenter, or an equivalent value for other rigid indenter geometry with the same area; and β is a parameter that takes into account the geometry of the indenter ($\beta = 1$, for cylindrical flat indenter).

The Young's modulus results obtained by the data such as in Figure 11 indicate that, in case of the flat punches, the value of the β parameter in Equation (6) is different from 1 and depends on the R-ratio, whatever the Young's modulus, as shown in Table 4.

Table 4. Values obtained for the correction factor β in the numerical simulations with flat indenters.

R = L/m	E (GPa)					Average Values of β
	30	200	400	600	800	
			β			
1.00	1.055	1.054	1.053	1.054	1.054	1.054
2.50	1.125	1.123	1.124	1.125	1.124	1.124
4.00	1.215	1.214	1.214	1.214	1.215	1.214
5.50	1.269	1.266	1.267	1.267	1.266	1.267
7.11	1.374	1.372	1.371	1.371	1.372	1.372

The values of β for R = 1 (as for the Vickers indenter) is in agreement with those obtained in previous studies (see, e.g., [22,26]). In the case of R = 7.11 (as for the Knoop indenter), β is equal to 1.372. The increase of the value of β with R is certainly related with the loss of symmetry of the flat punches with the increase of the ratio R. Indeed, the same was observed in case of the conical, Vickers and Berkovich indenters, whose values of the factor β increase in this order (1.034, 1.055, 1.081, respectively), although on a smaller scale [26].

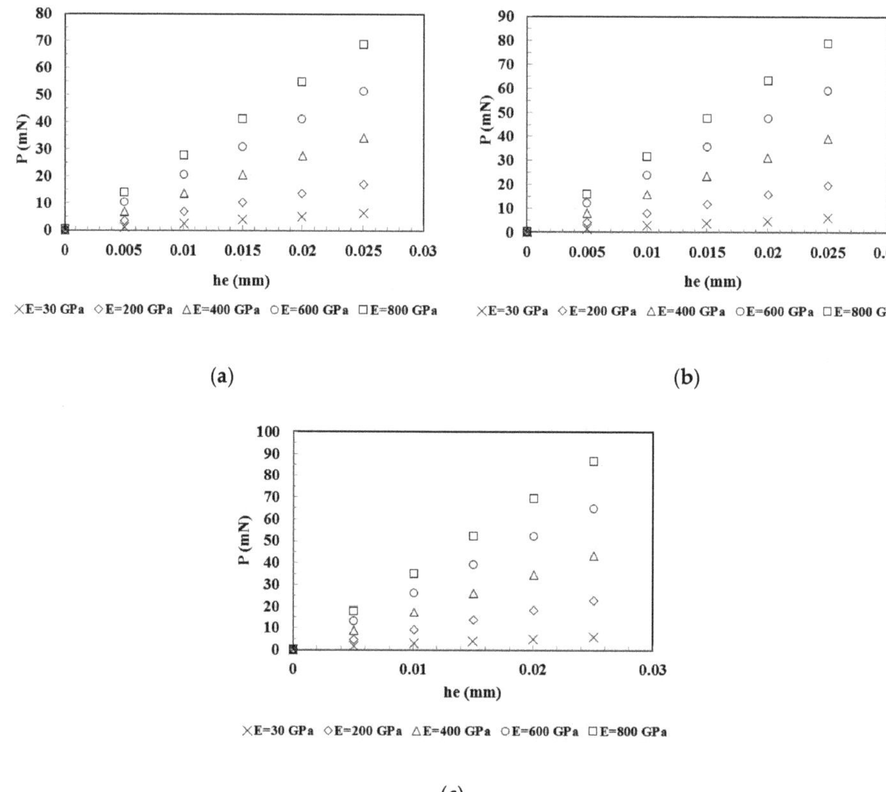

(a)

(b)

(c)

Figure 11. Evolution of load as a function of the elastic indentation depth obtained in the numerical simulations using flat indenters with different ratio R: (**a**) R = 1; (**b**) R = 4; (**c**) R = 7.11.

Figure 12 shows that the evolution of β with R for the flat punches is a quasi-linear function, quite close to that observed for the case of the pyramidal indenters (obtained with Equation (4): 1.054, 1.141, 1.232, 1.351 and 1.473). Moreover, the values of β for the pyramidal indenters are slightly higher than those determined for the flat indenters, except the R-ratio equal to 1.

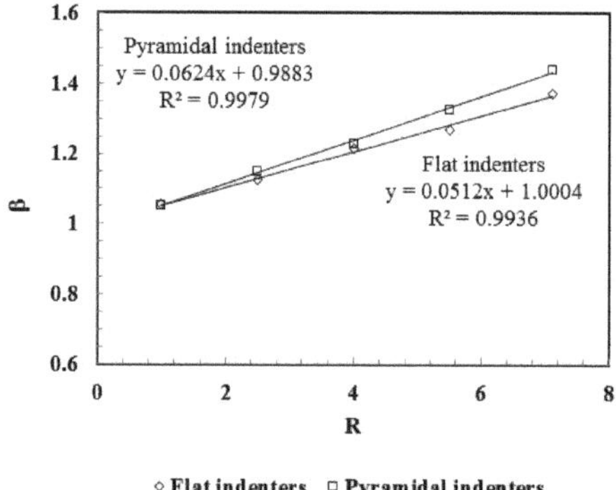

Figure 12. Evolution of β as a function of the R-ratio obtained in numerical simulations using the five flat and pyramidal indenters with different values of the R-ratio.

The dissimilarity between the β values obtained for the flat and the pyramidal indenters is certainly because, in the case of pyramidal indenters, not only elastic deformation occurs, but also plastic deformation appears, which significantly distorts the material surface. Moreover, the increase of the value of R-ratio leads to the lowering of the symmetry of the plastic strain distribution along two axes of the indenter (see Figure 7).

4.4. Correlation with Experimental Results

In order to check the performance of the correction factor β proposed in this study, experimental results of five materials were used to calculate the Young's modulus. They are fully dense brittle materials, covering a wide range of mechanical properties, whose experimental data were obtained from bibliography [19]. Table 5 presents the Young's modulus results of these materials determined by Grobal et al. [19], using two different methods proposed by: (i) Grobal et al. [19] (E_G); (ii) and Marshall et al. [31] and Riester et al. [18] (E_M). The results obtained by the method proposed in the current study (E), using β equal to 1.419 as defined above for the Knoop indenter, are also shown in this table. For comparison, nominal values of the Young's modulus, E_{nom}, evaluated by ultrasonic method [19] are presented and used to calculate the errors. To evaluate the Young's modulus, Grobal et al. [19] make use of the plot of the contact stiffness, C, as a function of $1/\sqrt{A_C}$, whose values of the slopes of the fitted straight lines ($(\sqrt{\pi}/2\beta E_r)$ in Equation (1)), are also used in current study to assess the value of E, for all materials.

Table 5. Experimental Young's modulus results.

Materials	E_{nom} (GPa) [19]	E_G (GPa) [19]	Error (%)	E_M (GPa) [19]	Error (%)	$\frac{\sqrt{\pi}}{2\beta}\frac{1}{E_r}$ (μm/N^2) [19]	E (GPa)	Error (%)
Si$_3$N$_4$	317 ± 4	316.5 ± 4.24	-0.16	300 ± 20.0	-5.36	2.548 ± 0.029	302.8 ± 4.40	-4.48
Ceramic-glass	82 ± 2	85.0 ± 0.36	3.66	85 ± 4.0	3.66	7.930 ± 0.031	82.1 ± 0.35	0.15
Alumina	385 ± 6	386.0 ± 7.75	0.26	380 ± 18.5	-1.30	2.233 ± 0.032	359.4 ± 6.90	-6.64
β-TCP	130 ± 2	129.0 ± 0.85	-0.77	142 ± 14.0	9.23	5.568 ± 0.032	120.7 ± 0.70	-7.12
Fused silica	68 ± 1	65.0 ± 0.30	-3.00	70 ± 4.0	2.94	9.221 ± 0.034	69.9 ± 0.25	2.80
Average of the absolute value of the error			1.57		4.50			4.27

The results in Table 5 show that the proposed β coefficient enables relatively good accuracy of the Young's modulus. The average of the absolute value of the error is equal to 4.27%. This is higher than that obtained by the method of Grobal et al. [19] and slightly lower than that achieved using the method by Marshall et al. [31].

5. Conclusions

A finite element study using the three-dimensional numerical simulation of hardness tests of elastic–plastic materials is performed. Pyramidal indenters with geometry from Vickers to Knoop were used. Also flat-ended punches were considered. This allowed us to obtain important information about the geometry of the indentation surface (sink-in and pile-up formation) and the distribution of plastic deformation beneath the pyramidal indenters. Both types of tests, with pyramidal and flat punch indenters, allow an assessment of the values of the geometrical parameter β to be used in the Sneddon's equation [20], for flat-ended punches, and the Doerner and Nix equation [21], for pyramidal indenters. This permits to determine the reduced Young's modulus of the indented material when using depth-sensing indentation (DSI) equipment. In the case of the Knoop indenter, the correction factor β of at about 1.419 is required, in order to accurately obtain the Young's modulus. This makes the procedure for analyzing the results of the Knoop indenter more expeditious than it currently is.

Author Contributions: M.I.S. carried out the numerical simulations, with the support of N.A.S.; M.I.S., J.M.A., J.V.F. and N.A.S. performed the formal analysis; J.M.A. wrote the original manuscript; all the authors participate in the writing-review and editing process.

Acknowledgments: This research work is sponsored by national funds from the Portuguese Foundation for Science and Technology (FCT) via Projects UID/EMS/00285/2013 and CENTRO-01-0145-FEDER-000014 (MATIS), by UE/FEDER funds through Program COMPETE2020. One of the authors, N. A. Sakharova, was supported by a grant for scientific research from the Portuguese Foundation for Science and Technology (SFRH/BPD/107888/2015). All supports are gratefully acknowledged. The authors also express their thanks to Professor Marta Oliveira for her support in numerical simulation with the finite element in-house code DD3IMP.

Conflicts of Interest: The authors declare no conflict of interest. The funders had no role in the design of the study; in the collection, analyses, or interpretation of data; in the writing of the manuscript, or in the decision to publish the results.

References

1. Fernandes, J.V.; Trindade, A.C.; Menezes, L.F.; Cavaleiro, A. Influence of substrate hardness on the response of W-C-Co-coated samples to depth-sensing indentation. *J. Mater. Res.* **2000**, *15*, 1766–1772. [CrossRef]

2. Dao, M.; Chollacoop, N.; Van Vliet, K.J.; Venkatesh, T.A.; Suresh, S. Computational modelling of the forward and reverse problems in instrumented sharp indentation. *Acta Mater.* **2001**, *49*, 3899–3918. [CrossRef]

3. Chollacoop, N.; Dao, M.; Suresh, S. Depth-sensing instrumented indentation with dual sharp indenters. *Acta Mater.* **2003**, *51*, 3713–3729. [CrossRef]

4. Antunes, J.M.; Menezes, L.F.; Fernandes, J.V.; Chaparro, B.M. A New Approach for Reverse Analyses in Depth-sensing Indentation Using Numerical Simulation. *Acta Mater.* **2007**, *55*, 69–81. [CrossRef]

5. Knoop, F.; Peters, C.G.; Emerson, W.B. A sensitive pyramidal-diamond toll for indentation measurements. *J. Res. Natl. Bur. Stand.* **1939**, *23*, 39–61. [CrossRef]

6. Hasan, M.F.; Wang, J.; Berndt, C.J. Determination of the mechanical properties of plasma-sprayed hydroxyapatite coatings using the Knoop indentation technique. *J. Therm. Spray Technol.* **2015**, *24*, 865–877. [CrossRef]

7. Zhou, B.; Liu, Z.; Tang, B.; Rogachev, A.V. Catalytic effect of Al and AlN interlayer on the growth and properties of containing carbon films. *Appl. Surf. Sci.* **2015**, *326*, 174–180. [CrossRef]

8. Sinavarat, P.; Anunmana, C.; Muanjit, T. Simplified method for determining fracture toughness of two dental ceramics. *Dent. Mater. J.* **2016**, *35*, 76–81. [CrossRef] [PubMed]

9. Lu, K. Making strong nanomaterials ductile with gradients. *Science* **2014**, *345*, 1455–1456. [CrossRef] [PubMed]

10. Vu, V.Q.; Beygelzimer, Y.; Kulagin, R.; Toth, L.S. Mechanical modelling of the plastic flow machining process. *Materials* **2018**, *11*, 1218. [CrossRef] [PubMed]

11. Kulagin, R.; Beygelzimer, Y.; Ivanisenko, Y.; Mazilkin, A.; Hahn, H. Instabilities of interfaces between dissimilar metals induced by pressure torsion. *Mater. Lett.* **2018**, *222*, 172–175. [CrossRef]

12. Borc, J. Study of Knoop microhardness anisotropy on the (001), (100) and (010) faces of gadolinium calcium oxyborate single crystals. *Mater. Chem. Phys.* **2014**, *148*, 680–685. [CrossRef]

13. Rabinovich, V.L.; Sarin, V.K. Three-dimensional modelling of indentation fracture in brittle materials. *Mater. Sci. Eng. A* **1996**, *206*, 208–214. [CrossRef]

14. Giannakopoulos, A.E. Elastic and viscoelastic indentation of flat surfaces by pyramid indentors. *J. Mech. Phys. Solids* **2006**, *54*, 1305–1332. [CrossRef]

15. Giannakopoulos, A.E.; Zisis, T. Analysis of Knoop indentation. *Int. J. Solids Struct.* **2011**, *48*, 175–190. [CrossRef]

16. Giannakopoulos, A.E.; Zisis, T. Analysis of Knoop indentation of cohesive frictional materials. *Mech. Mater.* **2013**, *57*, 53–74. [CrossRef]

17. Riester, L.; Blau, P.J.; Lara-Curzio, E.; Breder, K. Nanoindentation with a Knoop indenter. *Thin Solid Films* **2000**, *377–378*, 635–639. [CrossRef]

18. Riester, L.; Bell, T.J.; Fischer-Cripps, A.C. Analysis of depth-sensing indentation tests with a Knoop indenter. *J. Mater. Res.* **2001**, *16*, 1660–1667. [CrossRef]

19. Ghorbal, G.B.; Tricoteaux, A.; Thuault, A.; Louis, G. Comparison of conventional Knoop and Vickers hardness of ceramic materials. *J. Eur. Ceram. Soc.* **2017**, *37*, 2531–2535. [CrossRef]

20. Sneddon, I.N. The relation between load and penetration in the axisymmetric Boussinesq problem for a punch of arbitrary profile. *Int. J. Eng. Sci. Technol.* **1965**, *3*, 47–57. [CrossRef]

21. Doerner, M.F.; Nix, W.D. A method for interpreting the data from depth-sensing indentation instruments. *J. Mater. Res.* **1986**, *1*, 601–609. [CrossRef]

22. Antunes, J.M.; Menezes, L.F.; Fernandes, J.V. Three-Dimensional Numerical Simulation of Vickers indentation Tests. *Int. J. Solids Struct.* **2006**, *43*, 784–806. [CrossRef]

23. Oliver, W.C.; Pharr, G.M. An improved technique for determining hardness and elastic-modulus using load and displacement sensing indentation experiments. *J. Mater. Res.* **1992**, *7*, 1564–1583. [CrossRef]

24. Joslin, D.L.; Oliver, W.C. New method for analysing data from continuous depth-sensing microindentation tests. *J. Mater. Res.* **1990**, *5*, 123–126. [CrossRef]

25. Antunes, J.M.; Menezes, L.F.; Fernandes, J.V. Influence of the Vickers Tip Imperfection on the Sensing Indentation Tests. *Int. J. Solids Struct.* **2007**, *44*, 2732–2747. [CrossRef]

26. Sakharova, N.A.; Fernandes, J.V.; Antunes, J.M.; Oliveira, M.C. Comparison between Berkovich, Vickers and conical indentation tests: A three-dimensional numerical simulation study. *Int. J. Solids Struct.* **2009**, *46*, 1095–1104. [CrossRef]

27. Menezes, L.F.; Teodosiu, C. Three-dimensional numerical simulation of the deep-drawing process using solid finite elements. *J. Mater. Process. Technol.* **2000**, *97*, 100–106. [CrossRef]

28. Oliveira, M.C.; Alves, J.L.; Menezes, L.F. Algorithms and Strategies for Treatment of Large Deformation Frictional Contact in the Numerical Simulation of Deep Drawing Process. *Arch. Comput. Meth. Eng.* **2008**, *15*, 113–162. [CrossRef]

29. Antunes, J.M.; Cavaleiro, A.; Menezes, L.F.; Simões, M.I.; Fernandes, J.V. Ultra-microhardness testing procedure with Vickers indenter. *Surf. Coat. Technol.* **2002**, *149*, 27–35. [CrossRef]

30. Bolshakov, A.; Pharr, G.M. Influences of pileup on the measurement of mechanical properties by load and depth sensing indentation techniques. *Int. J. Solids Struct.* **1998**, *13*, 1049–1058. [CrossRef]

31. Marshall, D.B.; Noma, T.; Evans, A.G. A simple method for determining elastic-modulus-to-hardness ratios using Knoop indentation measurements. *J. Am. Ceram. Soc.* **1982**, *65*, 175–176. [CrossRef]

 © 2018 by the authors. Licensee MDPI, Basel, Switzerland. This article is an open access article distributed under the terms and conditions of the Creative Commons Attribution (CC BY) license (http://creativecommons.org/licenses/by/4.0/).

Article

Physical Modelling and Numerical Simulation of the Deep Drawing Process of a Box-Shaped Product Focused on Material Limits Determination

Miroslav Tomáš [1],*, Emil Evin [1], Ján Kepič [2] and Juraj Hudák [1]

[1] Faculty of Mechanical Engineering, Institute of Technology and Materials Engineering, Technical University of Košice, Mäsiarska 74, 040 01 Košice, Slovakia; emil.evin@tuke.sk (E.E.); hudak@pdruz.in (J.H.)

[2] Institute of Materials Research, Slovak Academy of Sciences, Watsonova 47, 040 01 Košice, Slovakia; jkepic@saske.sk

* Correspondence: miroslav.tomas@tuke.sk; Tel.: +421-55-602-3524

Received: 26 August 2019; Accepted: 26 September 2019; Published: 28 September 2019

Abstract: Similitude theory helps engineers and scientists to accurately predict the behaviors of real systems through the application of scaling laws to the experimental results of a scale model related to the real system by similarity conditions. The theory was applied when studying the deep drawing process of a bathtub made from cold rolled low carbon aluminum-killed steel from the point of view of material limits. The bathtub model was created on the basis of geometric, physical, and mechanical similarity on a scale of 1:5. Thus, simulations and physical models were created. The simulation model was used to verify the combination yield locus/hardening law on the basis of comparing the thickness change. As a result, Hill 48/Krupkowski showed the minimal deviation by comparing data evaluated from numerical simulations and that measured on the physical model. Additionally, material anisotropy was modelled when virtual materials were defined from experimentally measured values of the plastic strain ratio. As an outcome, extra deep drawing quality steel with an average plastic strain ratio of $r_m \geq 1.47$ and an average strain hardening exponent of $n_m \geq 0.23$ must be used for the deep drawing of the bathtub.

Keywords: similitude; the bathtub model; numerical simulation; physical experiment; yield locus; hardening law; anisotropy

---

1. Introduction

The deep-drawing process is widely used in automotive, transport, household, and other industries when metal sheets are processed. The optimization of process parameters, such as the blank shape, material, lubrication, gaps, drawbead dimensions, etc. requires good knowledge of the process and parameters influencing the quality of the drawn part [1]. Thus, the die design process for the drawing of stampings with complicated shapes is time- and cost-consuming due to testing of the concepts designed [2].

Manufacturing processes are tested either on physical models or by numerical simulations [3]. The real production process is difficult to test during its exploitation, so physical experiments are usually based on the scale model [4]. Similitude theory helps engineers or scientists to predict the behavior of a researched system through a scaled model. Langhaar [5] presented the general definition of similarity in mathematical terms for two functions. Szucs [6] widened the theory of similarity from functions to systems. Coutinho summarized the state-of-the-art knowledge on similitude theory and methodologies used to create reduced scale models, including those based on the use of dimensional analysis, differential equations, and energetic methods [7]. The theory of similarity and dimensional analysis in mechanics was also elaborated by Sedov [8].

Focusing on the metal forming processes, analysis of the similarity concept based on scaled model testing and dimensional analysis has taken place. Gronostajski proposed the plastic similarity condition for physical modelling of the axisymmetric backward extrusion of lead [9]. Davey introduced a novel scaling methodology using transport equations for the scaling of mass, momentum, energy, and entropy, as well as any associated material-constitutive relationships [10]. Al-Tamimi applied and validated the approach by means of scaled experimental and numerical and analytical solutions of scaled cold upsetting tests for cylindrical and ring samples for three trial materials [11]. Krishnamurthy applied the theory for the hot forging of the disk using numerical simulations of the real process and the scaled one. All the characteristics of the metal flow remained identical at all stages of the process [12]. Keran presented a study of the correlation between the workpiece size and forming force for a case of cylindrical body upsetting by using a numerical simulation. As a result, a difference of less than 3% compared with the results of the calculation using similarity theory was reached [13]. Ajiboye used the dimensional analysis based on the Buckingham π theorem for a sensitivity study of the frictional behavior in cold forging [14].

Accordingly, the deep drawing processes of the model need to be researched and designed under strictly defined conditions. The similarity criteria that must be considered when physically modelling the deep drawing process based on scaled models are

(a) the geometry similarity—the corresponding dimensions of the model and object have to be proportional; thus, the length scale factor is constant;

(b) the mechanical similarity—equal pressures, strain rates, press ram weights, and deformation works; and

(c) the physical similarity—the same material chemical composition, structure, temperature, friction, distribution of stress, strain, etc. [3,10,15].

Numerical simulations were used when designing the manufacturing processes in the 1990s to reduce time and costs. Silva utilized Pam-Stamp software to re-evaluate the stamping process for a rear seat and a structural reinforcement when changing the blank thickness and conventional steel to high strength steel [16]. Choudhury optimized the die geometry and determined the safe limit of the blankholder forces for plain carbon steel, reinforced steel, and aluminum alloy [17]. Padmanabhan numerically simulated the deep drawing process of bottles for LPG (bottled gas) by using DD3IMP FE code. The optimization of the variable blank holder force and friction condition at specific locations during deep drawing resulted in an increased minimum thickness in the deep drawn part [18]. Vafaeesefat proposed an algorithm to predict the initial blank shape from the desired part in the sheet metal forming process using LS-DYNA software [19]. Fracz used eta/Dynaform software when optimizing sheet metal forming of a cylindrical part made from AMS 5512 steel [20]. Čada used Dynaform software to evaluate the influences of the shape, size, and location of rectangular and semicircular draw beads on the sheet-metal forming process [21]. Labergere employed Abaqus software to propose and validate a global methodology to simulate the stamping of the embossed sheet and the capacity of the model to predict severe folds and the final shape of the part [22]. Using LS-Dyna software, Schrek researched the deep drawing process of tailor welded DP 600 and BH220 materials in tools with an elastic blankholder. They determined the values and points of application of the blankholder forces to achieve minimal movement of the weld interface [23].

As Roll presented in [15], it is important to describe material behavior and tribological factors in numerical simulations using proper mathematic models. This includes both yield loci definition and the hardening law. The use of the standard von Mises model to describe material behavior is not enough, and it needs to be widened to describe effects such as anisotropy and kinematic hardening. The problem is more emphasized when the simulation of springback is performed and new advanced high-strength steels or aluminum alloys are applied. Consequently, correct models for the material and friction coefficient must still be verified by experiments. Neto et al. numerically simulated the anisotropic behavior of the mild steel sheet used in the reverse deep drawing process of a cylindrical cup. The effect of the yield criterion on the numerical results was analyzed using three yield functions—von

Mises, Hill'48, and Barlat Yld'91—combined with the Swift hardening law. The cup wall thickness distribution was strongly influenced by the yield criteria [24]. Other work by Neto et al. focused to the experimental and numerical analysis of a rail component made of mild steel and dual phase steel. They used the Swift hardening law to describe isotropic hardening and the Frederick–Armstrong law to describe the kinematic part of the work hardening combined with the Hill'48 yield criterion to describe the orthotropic plastic behavior of both metallic sheets. The results showed that the wrinkling behavior was strongly affected by the blank's material as well as by the symmetry conditions ($\frac{1}{4}$ of blank, full blank) defined in the numerical model [25]. Mulidran et al. focused on the springback prediction of a car body stamping made from aluminum alloy. The springback simulations were conducted with six yield functions (Barlat89, Barlat2000, Vegter-Lite, Hill90, Hill48 isotropic, and Hill48 orthotropic) combined with the Voce hardening model. Springback analysis was done in three sections, and the results were compared with the experimental values [26].

Chen researched the bathtub deep drawing process using numerical simulation [27]. It is supposed that he used the Hill isotropic definition of the yield locus and the point definition of the hardening curve. As a result, he determined an optimum drawbead distribution on the die face to avoid the formation of both fractures and wrinkles. Hojny [28] numerically simulated the stamping of the W1200 bathtub when the effects of the blank holder pressure and friction on the occurrence of fracture and wrinkling were investigated. He used the anisotropic Barlat model and point defined flow stress to define the material behavior.

In the article, numerical simulations based on the finite element method were performed to evaluate the influences of both the yield locus and the hardening law when deep drawing the model of box-shaped pressing—the bathtub model. The model pressing designed on the principle of similitude theory was numerically simulated for verification. The combination of the yield locus/hardening law was validated experimentally when compared with the thickness change in the selected sections. Simulations and experiments on the drawing quality of mild steel for enameling were performed, which is used in the production of real box-shaped pressing. Additionally, material anisotropy was modelled and verified using numerical simulation to determine the limit value from the point of view of fractures during deep drawing.

2. Materials and Methods

2.1. The Press-Die-Pressing System

Bathtub model pressing was considered here as a box-shaped pressing. The model was designed according to similitude theory in order to verify the material parameters when deep drawing the real bathtub produced by Festap Ltd., Filakovo, Slovakia. The real bathtub pressing was scaled to the model (Figure 1) to meet the similarity criteria shown in Table 1. During real production, the steel sheet thickness decreases over the years from 2.5 to 1.63 mm; thus, the formability criteria of the material should be verified.

Based on the model dimensions and drawing die used for the real bathtub, the drawing die for the model pressing was designed. To reduce the material flow in the straight parts, drawbeads were used, and their positions, dimensions, and lengths were also scaled. An experimental drawing of the die is shown in Figure 2.

2.2. Material

The bathtubs used in the experiment were made from cold rolled steel sheet for enameling, which is produced by U.S. Steel Kosice. The deep drawing of the bathtub model was done using a Kosmalt 190 (i.e., DC06EK according to EN 10209) steel sheet with a thickness of 0.5 mm. This special steel had to meet two opposing requests: good formability and good enameling properties in the aspects of both the fishscale and the pinhole resistance. The steel belongs to a group of cold rolled low carbon aluminum-killed and annealed steel. This type of steel has a good hydrogen storage ability due to the numerous micro-voids generated after cold-rolling, which still exist after annealing [29]. The chemical composition of the steel is shown in Table 2; its microstructure and texture are shown in Figure 3.

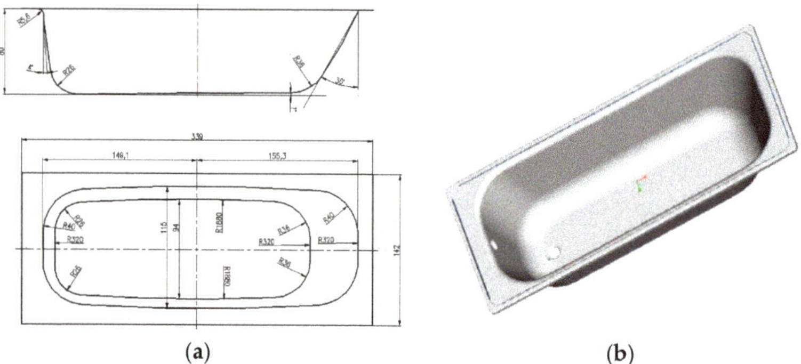

Figure 1. The box-shaped pressing—bathtub model: (**a**) 2D sketch; (**b**) 3D view.

Table 1. Similarity criteria for the box-shaped pressing—bathtub model.

Parameter	Real Bathtub	Bathtub Model
Geometry similarity (scale 1:5)		
Length [mm]	1695	339
Width [mm]	710	142
Height [mm]	400	80
Wall to bottom radius [mm] (i.e., Punch radius [mm])	130	26
Wall to flange radius [mm] (i.e., Die radius [mm])	28	5.6
Mechanical similarity		
Press	Hydraulic Fritz Muller BZE 2000	Hydraulic Fritz Muller BZE 100
Ram working velocity [mm·s^{-1}]	25	15
Die and punch material	Cast steel	Cast steel
Physical similarity		
Material	Enameling steel Kosmalt	Enameling steel Kosmalt
Lubricant	Vantol S	Vantol S

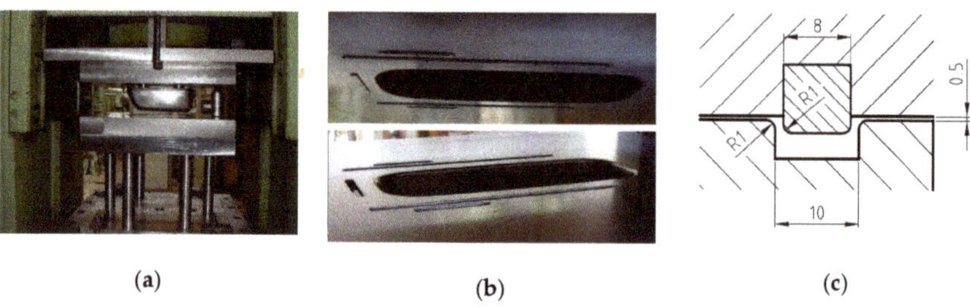

Figure 2. Experimental drawing of the die: (**a**) overview; (**b**) details of drawbeads and grooves; (**c**) dimensions of the drawbeads [mm].

Table 2. Chemical composition of the Kosmalt 190 material [wt %].

C	Mn	P	S	Al	N	Cu	Ni	Cr
0.030	0.140	0.009	0.008	0.042	0.003	0.014	0.015	0.013

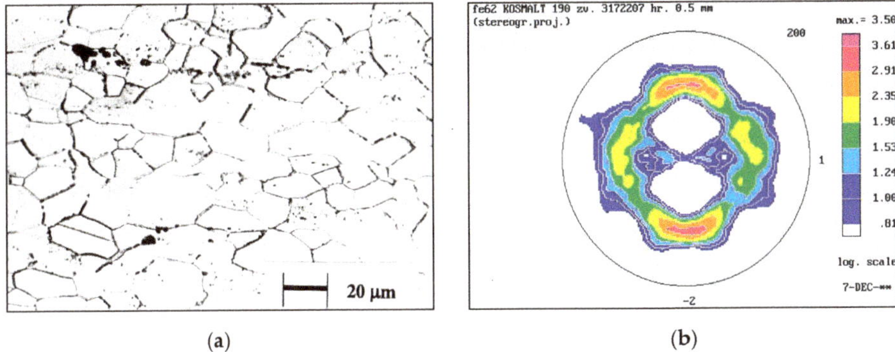

Figure 3. Microstructure and texture of the Kosmalt 190 material in the rolling direction: (**a**) microstructure; (**b**) texture.

The experimental material, Kosmalt 190, has a typical recrystallization structure. The density of the orientation components (111)<110> and (111)<112> is 10. The texture of the material is suitable for deep drawing. The structure is ferrite-pearlite-cementitic with a polyedric grain grade of 7 or occasionally grade 5. The cementite segregation is type 1A2, mainly on the grain boundaries in a few isolated cases in short rows.

For the numerical and physical experiments, Kosmalt 190 material with a nominal thickness of $a_0 = 0.5$ mm was used. From the point of view of formability, the material properties are shown in Table 3. These were measured according to the following standards: mechanical properties by STN EN ISO 6892-1, the normal anisotropy ratio by STN EN ISO 10113, and the strain hardening exponent by STN EN ISO 10275 using the TIRAtest 2300 testing machine (TIRA Maschinenbau GmbH, Rauenstein, Germany) controlled by a PC.

Table 3. Formability parameters of Kosmalt 190 material.

Dir. [°]	$R_{p0.2}$ [MPa]	Rm [MPa]	A_{80} [%]	r [–]	rm [–]	Δr [–]	n [–]	nm [–]	Δn [–]
0	158 ±0.9	280 ±1.3	45.5 ±0.3	1.58 ±0.036			0.226 ±0.002		
45	159 ±1.1	286 ±0.9	42.4 ±0.5	1.33 ±0.032	1.57	0.47	0.227 ±0.001	0.226	−0.001
90	155 ±1,0	279 ±0.5	45.4 ±0.5	2.02 ±0.052			0.225 ±0.001		

Note: $R_{p0.2}$—yield strength; Rm—ultimate tensile strength; A_{80}—elongation (80 mm initial gage length); r—plastic strain ratio (r-value); r_m—average r-value; Δr—planar anisotropy of r-value. n—strain hardening exponent (n-value); n_m—average n-value; Δn—planar anisotropy of n-value.

Five specimens were measured in each rolling direction for each test. The elongation was measured by the length extensometer and the width reduction was measured by the width extensometer, both with a precision level of ±0.001 mm. The plastic strain ratio was calculated at the engineering strain level of 20% using automatic determination. The strain hardening exponent was evaluated within an engineering strain level of 5% to 20%. The average values and planar anisotropy were calculated as follows:

$$r_m = \frac{1}{4}\left(r_{0°} + 2{\cdot}r_{45°} + r_{90°}\right), \tag{1}$$

$$\Delta r = \frac{1}{2}\left(r_{0°} - 2{\cdot}r_{45°} + r_{90°}\right), \tag{2}$$

$$n_m = \frac{1}{4}\left(n_{0°} + 2{\cdot}n_{45°} + n_{90°}\right), \tag{3}$$

$$\Delta n = \frac{1}{2}\left(n_{0°} - 2{\cdot}n_{45°} + n_{90°}\right). \tag{4}$$

2.3. Numerical Simulation Model

Numerical simulations were done using Pam-Stamp 2G software from the ESI Group (Paris, France). The simulation model was created using the 3D CAD/CAM software Creo, and its components were exported in neutral igs format. The die, punch, and blankholder were meshed using the Pam-Stamp 2G meshing module when importing CAD data. The meshed die components are shown in Figure 4a. To shorten the computing time, the symmetry of the blank along the longitudinal axis of the model was used. The blank shape (Figure 4b), dimensions, and its positioning against the die were also the same as for the real process and were scaled 1:5 to reach geometry similarity. The steel sheet rolling direction (0°) was positioned in the longitudinal axis of the bathtub model pressings.

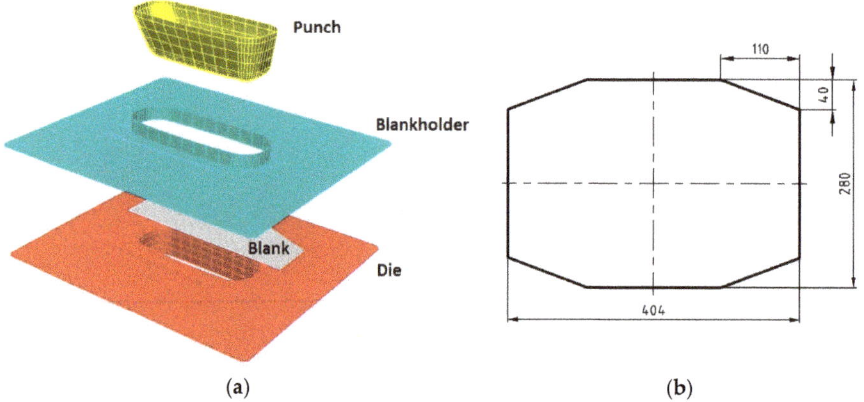

Figure 4. Simulation model and blank: (**a**) meshed components of the drawing die; (**b**) shape and dimensions of the blank.

There are two ways to represent drawing beads in a simulation model: (a) using a physical model—drawing beads need to be physically modelled on the CAD model of the blankholder and grooves on the die. This concept is a complicated way to simulate the blankholding stage individually before the drawing stage. Changing the restrictions on the bead geometry is complicated due to importation of a new blankholder model and meshing procedure; (b) using a numerical model—there is a macro within simulation software that can be used for drawbead action through restriction and the blankholing forces [27,30]. This property enables the actions of the drawbead on the blank and the blankholder to be represented without creating the geometry of the drawbead on the mesh. Thus, it is much easier to represent and modify a drawbead's shape and dimensions. Drawbeads are represented by nodes and lines that are positioned against the die and the blankholder. Restriction and opening forces are calculated through the drawbead macro calculator with respect to its geometry. A simulation concept with a physical model of drawing beads was used, and its geometry (drawbead length, width, height, number, and position) was the same as the drawbead geometry of the experimental drawing die.

The validation meshing strategy was applied with a minimum element size of 0.1 and a maximum of 30, a chordal error of 0.15, and a maximum angle of 15° for radii and curved surfaces. Belytschko–Tsay shell 3-node and 4-node elements (triangles, squares, and rectangles) with the Gauss thickness integration rule were applied to the objects (punch, die, blankholder) and blank as well. The small radii on the drawbead and groove were meshed to six rectangular elements with a height of 0.25 mm; the die radius was meshed to six rectangular elements with a height of 1.45 mm. Both fulfilled the maximum angle of 15° between shells.

Because the size of blank mesh influences the results, it should be optimized to offer accurate results and the lowest time of calculation possible. Hence, an initial blank mesh size of 10 mm was applied along with the adaptive meshing strategy. This allowed the mesh size of the object in contact to be refined when necessary, especially in small radii on the drawbeads and the die. Thus, the refinement level was set to 5, and the final blank mesh size was 0.625 mm.

2.3.1. Material Hardening Model

Nowadays, different mathematical models are used to describe the plastic behavior of steel sheets: Hollomon (or Ludwik), Krupkowski (or Swift), Hocket-Sherby, Gosh, Voce, Johnson-Cook, Cowper-Symonds etc. Some of these are strain rate and/or temperature dependent [30,31]. These were reached by fitting experimentally acquired data from tensile tests or other non-standardized tests, and they differ in terms of the effort required to calibrate the model and to reach the model's constants. Two isotropic material hardening models were tested during the numerical simulation:

- Hollomon

$$\sigma = K \cdot \varphi^n \tag{5}$$

- Krupkowski

$$\sigma = K \cdot \left(\varphi_0 + \varphi_{pl}\right)^n \tag{6}$$

where σ is the true stress, φ is the true strain, K is the strength coefficient, n is the strain hardening exponent, φ_0 is the pre-strain, and φ_{pl} is the plastic strain.

The Hollomon hardening model was measured within the strain levels of 5% to 20%. The Krupkowski model was determined on the basis of Hollomon's model parameters using the numerical iteration method. The models' constants are shown in Table 4.

Table 4. Hollomon and Krupkowski model constants.

Model	K [MPa]	n [–]	φ_0 [–]
Hollomon	496	0.226	-
Krupkowski	505	0.248	0.00899

2.3.2. Material Yield Locus

The most important criterion in the numerical simulation is the yield locus, which describes the transition from an elastic state to a plastic one. The yield locus also expresses the relationships between stress components at the moment of yielding due to the multiaxial stress state during metal forming. Thus, the yield point measured during uniaxial tension in the tensile test is not enough to describe the yield locus even if it is easily measured. [32]

The most widely used yield criteria for isotropic materials are Tresca (the "maximum shear stress criterion") and von-Mises (the "strain energy criterion"). However, the sheet metal exhibits a significant anisotropic property due to its crystallographic structure and the characteristics of the rolling process. Hence, in 1948, Hill [33] proposed an anisotropic yield criterion involving three orthogonal symmetry planes, which is expressed by the following quadratic function:

$$2f(\sigma_{ij}) = F(\sigma_y - \sigma_z)^2 + G(\sigma_z - \sigma_x)^2 + H(\sigma_x - \sigma_y)^2 + 2L\tau_{yz}^2 + 2M\tau_{zx}^2 + 2N\tau_{xy}^2 = 1, \tag{7}$$

where F, G, H, L, M, and N are constants specific to the anisotropy state of the material, and x, y, and z are the principal anisotropy axes [30,33]. Because the plane stress is assumed in numerical simulations of the sheet metal forming processes, the stress in the thickness direction is ignored due to its insignificance compared with that in the other two orthogonal directions. Furthermore, assuming that the principal directions of the stress tensor are coincident with the anisotropic axes, this criterion can be written as follows:

$$\sigma_1^2 - \frac{2r_0}{1 + r_0}\sigma_1\sigma_2 + \frac{r_0(1 + r_{90})}{r_{90}(1 + r_0)}\sigma_2^2 = \sigma_0^2. \tag{8}$$

In 1990, Hill stated that the range of validity of Hill 48 had been explored through numerous experiments, and that it is well suited to specific metals and textures. For more recently developed steels of higher grades, Hill 90 was developed [34]. This is a yield criterion for metal sheets with planar anisotropy, and it is based on a non-quadratic yield function. This criterion takes into account different behaviors during the bending/unbending phase. The model is a generalization of Hill 48 with non-integer powers of the principal values of the deviatoric stresses. Its constitutive relation for plane stress conditions in terms of principal stress components can be written as follows: [30,34]

$$|\sigma_1 + \sigma_2|^m + \alpha \cdot |\sigma_1 - \sigma_2|^m + \left|\sigma_1^2 - \sigma_2^2\right|^{\frac{m}{2}-1} \cdot cos(2\Phi)\left[\beta \cdot \left(\sigma_1^2 - \sigma_2^2\right) + \gamma \cdot (\sigma_1 - \sigma_2)^2 \cdot cos(2\Phi)\right] = 2 \cdot \sigma_Y^m, \quad (9)$$

where α, β, γ, and m are constants derived from the measured material data, Φ is the angle between the principal axes of the in-plane stress and the principal axes of anisotropy, and σ_Y is the equi-biaxial yield stress.

The Hill 48 yield locus was defined by Lankford's coefficients r_0, r_{45}, r_{90} (i.e., plastic strain ratios), and these were measured using tensile tests performed on the specimens taken at 0°, 45°, and 90° to the rolling direction (see Table 3).

To calculate the coefficients α, β, γ, and m for the Hill 90 yield locus, the Pam-Stamp 2G wizard was used. This calculation is based on an iterative method that minimizes a function whose variables are the yield stresses and the anisotropy coefficients (least-squares method). The user must define the uniaxial yield values $R_{p0.2}$ for each Lankford's coefficient r_α and either the Hill 90 coefficient m or the equi-biaxial yield stress σ_y. An equi-biaxial yield stress of σ_y = 220 MPa was used for the calculations, which was obtained from a hydraulic bulge test using the HYDROTEST device. Uniaxial yield values $R_{p0.2}$ and Lankford's coefficients for the rolling directions of 15°, 30°, 60°, and 75° were additionally tested according to standards shown previously (Table 5). The values of the Hill 90 yield locus constants α, β, γ, and m are shown in Table 6.

Table 5. Additional values of yield strength and the plastic strain ratio.

Direction	$R_{p0.2}$ [MPa]	r [–]
15	166 ±1.1	1.56 ±0.037
30	166 ±1.3	1.46 ±0.041
60	168 ±0.9	2.03 ±0.029
75	165 ±1.2	2.14 ±0.040

Table 6. Calculated values of the Hill 90 yield locus.

α	β	γ	m
1.56158	1.19317	20.2109	3.02902

2.3.3. Failure Criteria

Keeler-Brazier's model of forming limit curve was used to determine the material fracture in a numerical simulation. The model was implemented in Pam-Stamp 2G software, and the value of φ_1 when $\varphi_2 = 0$ was calculated as follows: [30,35]

$$\varphi_{1(0)} = \ln\left[1 + (23.3 + 14.13a_0)\frac{n}{0.21}\right] \quad (10)$$

$$\text{when } \varphi_2 < 0 \; \varphi_1 = \varphi_{1(0)} - \varphi_2 \quad (11)$$

$$\text{when } \varphi_2 > 0 \; \varphi_1 = \varphi_{1(0)} + 0.6[\exp(\varphi_2 - 1)] \quad (12)$$

where a_0 is the material thickness, and n is the strain hardening exponent. The left side of the FLC curve was calculated from Equation (11) and the right side from Equation (12).

2.3.4. Boundary Conditions

To perform numerical simulations, other boundary conditions or process parameters such as the blankholder force, friction, punch speed, and blankholder speed needed to be defined. Based on recommendations in the software manual, the blankholder speed was set to 2 m·s^{-1}, and the punch speed was set to 5 m·s^{-1}. In the z direction, both increased linearly from zero to the final value to prevent dynamic effects at contact. This was defined as the imposed velocity by the curve [30]. Contact conditions were defined by the friction coefficient and the Coulomb laws defined the friction. Water-based lubricant is used when real bathtubs are produced, so the friction coefficient was set to a constant value of 0.09 due to the friction between the steel sheet and tool steel [28,36]. The blankholder force was set to 340 kN when the bathtub model pressing free of wrinkles and fracture was drawn.

3. Results

Within the numerical simulations, each hardening model was combined with each yield criterion. Overall, four simulation concepts were done by changing the material hardening law and the yield locus. These combinations are shown in Table 7.

Table 7. Minimal thicknesses evaluated from numerical simulations and measured in the experiment.

| Simulation Number | Yield Locus/Hardening Law | Minimal Thickness [mm] | | Section B-B |
| | | Section A-A | | |
		C-E	G-H	D-E
S1	Hill 48/Hollomon	0.421	0.330	0.380
S2	**Hill 48/Krupkowski**	**0.417**	**0.363**	**0.375**
S3	Hill 90/Hollomon	0.405	0.398	0.416
S4	Hill 90/Krupkowski	0.412	0.396	0.417
	Experiment	**0.413 ± 0.006**	**0.368 ± 0.008**	**0.363 ± 0.006**

Qualitative evaluation of the numerical simulations was done using the FLD diagram in order to evaluate the deep drawing process from the point of view of both the wrinkles appearance and fracture. In all simulation concepts, fracture did not occur, while a wrinkle tendency was identified in the same areas. However, on the physical model, wrinkles were not identified. The result for simulation 2 is shown in Figure 5. The material is appeared to have very good formability, as shown by the value of the strain hardening exponent and the plastic strain ratios presented in Table 3.

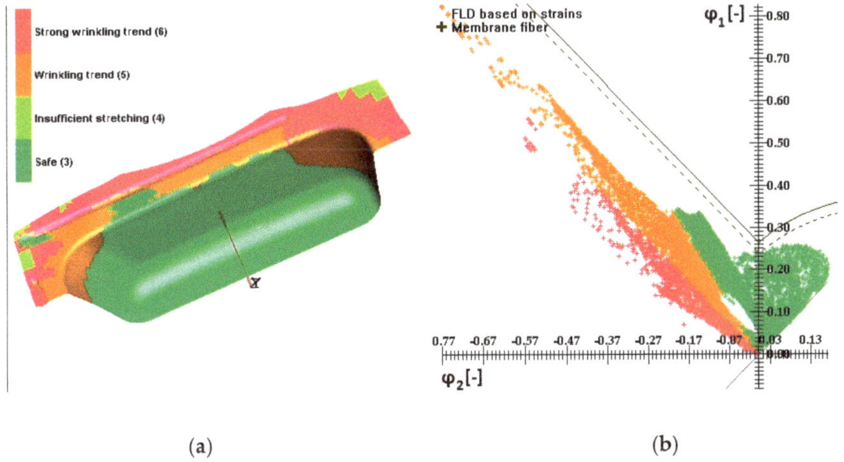

(a)   (b)

Figure 5. Results of simulation 2: (**a**) qualitative evaluation; (**b**) forming limits diagram.

To validate the material hardening law/yield locus combination, the wall thickness change was evaluated in selected sections. The sections used to measure the wall thickness change are shown in Figure 6. On the bathtub model pressing reached by the physical experiment, the thickness was measured using a micrometer (conical tips with flat spot of Ø 1 mm) three times at a single point, and the average thickness value was calculated. There was a distance of 5 mm between measurement points along the bathtub wall's length.

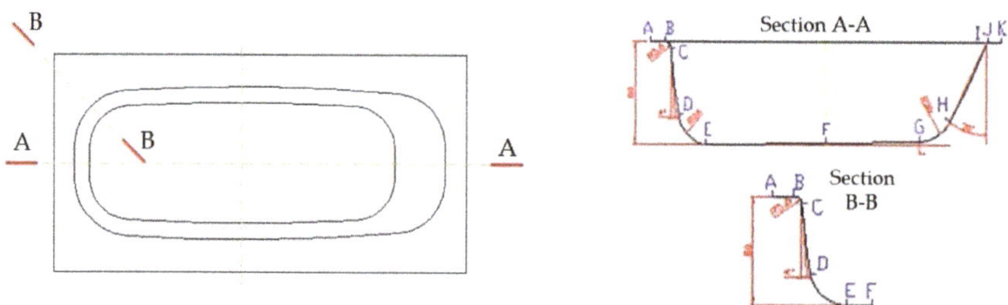

Figure 6. Section of the bathtub model used to measure the thickness (the rolling direction 0° was on the longitudinal axis).

The minimal thicknesses for the individual simulations and measured on the physical model are shown in Table 7. The minimal thickness was identified in the G-H area in the longitudinal section A-A for each simulation, i.e., at the radius of the inclined wall to the bottom. The second minimal thickness was identified in the D-E area, i.e., at the radius of the wall to the bottom for Hill 48 yield locus, but in the C-D area, i.e., at the wall, for Hill 90 yield locus. The minimal thickness in section B-B was found in the D-E area too. There was good agreement between the numerical simulation and the real measurement of thickness for each simulation representing the material hardening law/yield locus combination when considering the minimum thickness position. A comparison of the relative thickness change evaluated from the numerical simulations and measured on the bathtub model pressing is shown in Figure 7.

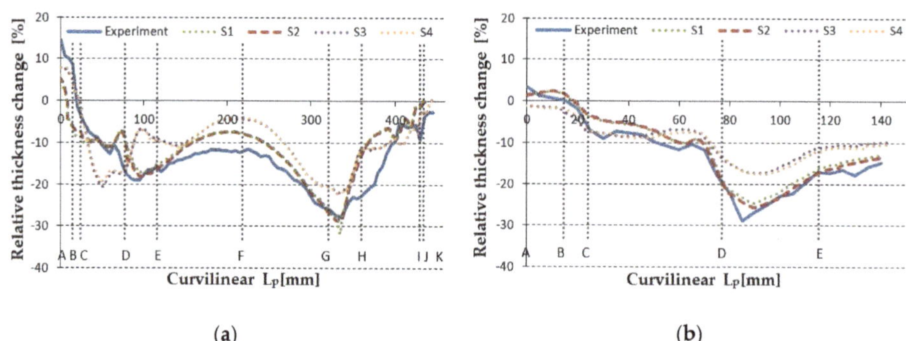

(a)

(b)

Figure 7. Relative thickness change of the bathtub model: (**a**) Sec. A-A; (**b**) Sec. B-B.

To assess the hardening law/yield locus combination, the deviation of the thickness evaluated from the numerical simulation and measured on the physical model was calculated as follows:

$$\text{Dev} = \frac{(\text{Th}_{\text{simul}} - \text{Th}_{\text{real}})}{\text{Th}_{\text{nom}}} \cdot 100\,[\%] \tag{13}$$

where Th_{simul} is the thickness evaluated from the numerical simulation, Th_{real} is the thickness measured on the physical model, and Th_{nom} is the nominal steel sheet thickness. The results are shown in Table 8 and in Figure 8. Based on the evaluation using the minimal thickness criterion, the best combination seems to be the Hill 48 yield locus and the Krupkowski material hardening law. The values of thickness at both critical regions were the closest to the thicknesses measured on the bathtub model pressing. The graph on Figure 8 shows the deviation of thickness evaluated from the numerical simulation and the thickness measured on the bathtub model pressing for each combination material hardening law/yield locus.

Table 8. Deviation of the local minimal thicknesses for the simulation and physical experiment.

Simulation Number	Section A-A		Section B-B
	C-E	G-H	D-E
S1	1.7%	−7.6%	3.4%
S2	**0.7%**	**−1.0%**	**2.4%**
S3	−1.7%	5.9%	10.6%
S4	−0.3%	5.7%	10.8%

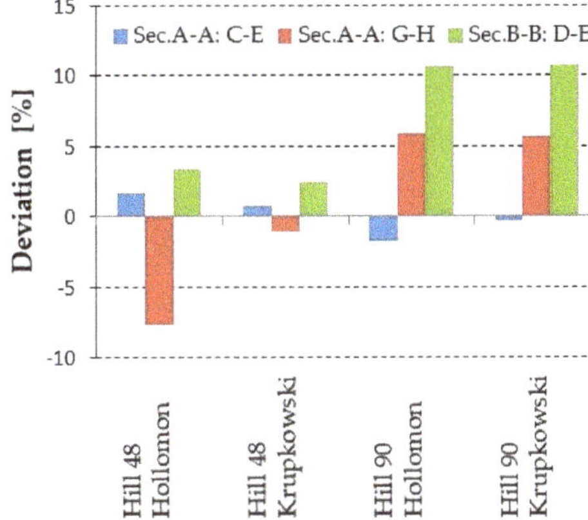

Figure 8. Thickness deviation calculated for the local minima in selected sections.

As mentioned previously, enameling steel must meet two opposing requests: good drawability and good enameling properties. Steel used in the physical experiment and numerical simulation belongs to the group of cold rolled low carbon aluminum-killed and annealed steel. From the point of view of drawability, this steel shows anisotropy due to its production process [37]. Thus, the effect of anisotropy was researched by numerical simulation to determine the minimal values when pressing free of fracture was reached in the model.

Based on the material properties shown in Table 3, other two materials were modelled from the point of view of its anisotropy. Directional values of plastic strain ratios were lowered, as shown in Table 9: about 0.1 in the rolling direction, about 0.05 at 45°, and about 0.2 at 90°. The differences in each direction were chosen on the basis of the formability evaluation of different grades of Kosmalt steel sheet, as presented in [37]. Then, the deep drawing process was numerically simulated, and the results are shown in Figures 9 and 10.

Table 9. Anisotropy parameters of the models used in the numerical simulation.

Material	r_0	r_{45}	r_{90}	r_m	Result
Kosmalt 190	1.58	1.33	2.02	1.57	Ok
Virtual B	1.48	1.28	1.82	1.47	Necking
Virtual C	1.38	1.23	1.62	1.37	Fracture

Note: $n_m = 0.226$.

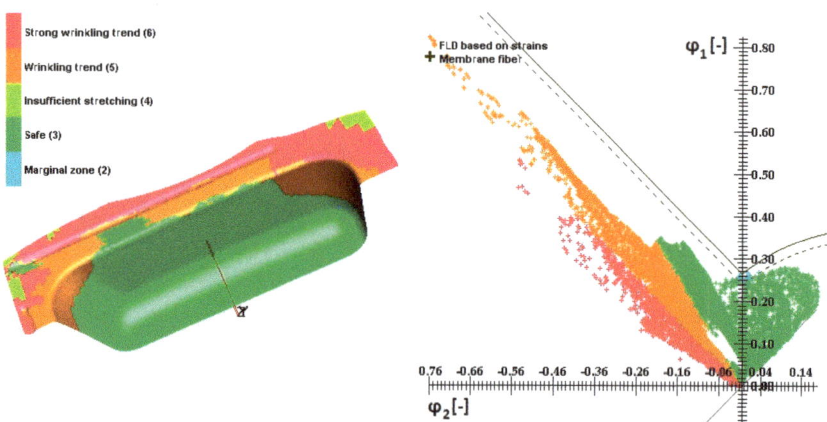

Figure 9. Results of the numerical simulation for virtual B material with $r_m = 1.47$.

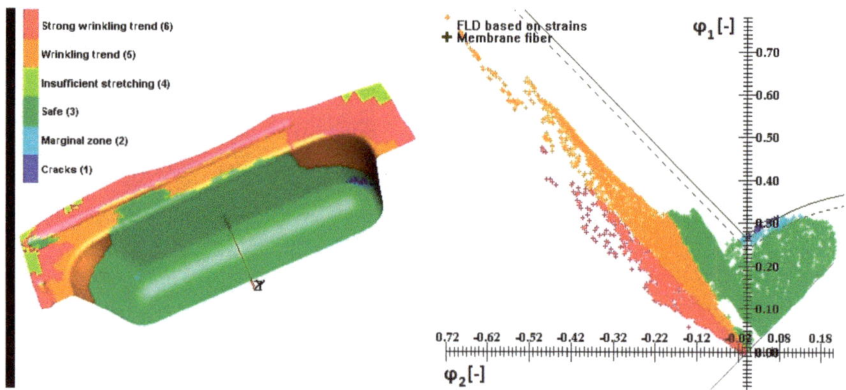

Figure 10. Results of the numerical simulation for virtual C material with $r_m = 1.37$.

When analyzing the results, the critical area of the bathtub model pressings was found to be the bottom radii at the corner. Thus, localized necking or fracture occurred when the plastic strain ratios were lowered in the 0°, 45°, and 90° directions. Considering the blank positioning (rolling direction of 0° in the longitudinal axis) and the directional dependence of the plastic strain ratio, i.e., its anisotropy, it is the location of r_{45} where the lowest value was measured. The results comply with those of [37], because the plastic strain ratio expresses the steel sheet's resistance to thinning. Thus, greater thinning of the thickness appears in the direction of the lowest plastic strain ratio.

4. Discussion

Previously published research on the bathtub deep drawing process by numerical simulation was focused on determining an optimum drawbead distribution on the die face [27] and the effects of the blankholder pressure and friction on the occurrence of fracture and wrinkling [28], but the material model definition was not clear. The bathtub model deep drawing process was researched in the presented study from the point of view of, firstly, mathematic modelling of the yield locus and the hardening law and, secondly, modelling the material anisotropy limit.

The bathtub model deep drawing process was designed to reach geometric and physical similarity by using scaled modelling die and the same material, deformation process, and friction characteristics [7,9–13]. Thus, the Buckingham π theorem was applied and a physical model of the drawing die was designed at a scale of 1:5. Consequently, the simulation model was defined in PamStamp 2G software when the friction was set to 0.09 [30,36] and the blankholder force was 340 kN (the specific blankholder pressure was 6 MPa) to reach a model pressing free of wrinkles and fractures.

Because special steel for enameling is used for the production of bathtubs, four simulations were done to verify the combination yield locus/hardening law. It is desirable to note that the hardening curves described in Equations (5) and (6) are consistent with experimental ones only at a certain strain stage, and with large strains, the flow curve reaches saturation [38]. Thus, necking and fracture were judged by a forming limit curve [39]. Concerning the yield locus, both anisotropic criteria came from directionally measured plastic strain ratios, while Hill 48 is well adapted for strongly anisotropic materials ($r > 1$) and Hill 90 is used for materials with planar anisotropy ($r < 1$) [40]. The validity of the combination yield locus/hardening law was judged by comparing the thickness change in selected sections [16,18,20]. As a result, the Hill 48 yield locus combined with the Krupkowski hardening law offered the lowest thickness deviation (2.4%) when values calculated from numerical simulations and the physical model were compared. The results comply with [24] and were also used for mild steel and drawing quality steel when the deep drawing process was numerically simulated [16,17,21,25]. The selection of a proper mathematical model for the yield locus and hardening law is even more important for metal sheets and processes, especially when new types of steel or springback phenomena are simulated [17,25,26]. To define these models, more tests than just measuring uniaxial tension with a tensile test or biaxial tension with the Bulge test are necessary [15,32].

Keeler-Brazier's model determined the fracture limit and necking (marginal zone) in numerical simulations. Because it outcomes from a large number of experiments, it is well adopted in the numerical simulation of low carbon steels. It is easy to define using the strain hardening exponent and the thickness of material, while both influence the formability of the material [41–43]. However, constant strain paths are required during deep drawing, and their linearity was confirmed using software. This is also supported by the fact that the strain pathways in the first drawing of axisymmetric and uniform sectioned parts are satisfactorily linear, and the FLD for the as-received condition can be used comfortably when making formability assessments [39].

As an outcome of numerical simulations done for virtual materials, cold rolled low carbon aluminum-killed steel used for enameling must meet the conditions of $r_m \geq 1.47$ and $n_m \geq 0.23$ to obtain bathtub model pressings free of fracture. To improve the deep drawing process of the bathtub model pressing or the real bathtub in production, it is recommended that these values are higher. This is because the plastic strain ratio improves the steel sheet's resistance to thinning of the thickness, while the strain hardening exponent unifies the strain distribution during deformation and prevents necking [37].

To improve the drawability, new steels have been developed such as ultra-low carbon Ti-IF steel, decarbonized (De-C) steel [29], and boron-microalloyed steel [44,45]. These possess excellent deep drawability with a lower yield stress and a higher elongation, plastic strain ratio, and strain hardening exponent compared with those of low carbon Al-killed steels. For Ti-IF and De-C steel, this is because of the very low carbon component when Ti is added, which stabilizes all of the carbon and nitrogen atoms, eliminates ageing, and improves elongation in the rolling direction [29,44–46]. Modern ideas to improve the ductility and preserve the high strength of sheet metal for deep drawing lies in the use of materials with ultrafine grained structures when complemented with coarse grained elements [47].

5. Conclusions

The bathtub model deep drawing process was researched in the presented study from the point of view of, firstly, mathematic modelling of the yield locus and hardening law and, secondly, modelling of the material anisotropy limit. Based on the numerical simulations and physical experiment performed, the following outputs can be concluded:

- The Hill 48 and Hill 90 yield locus mathematical models and the Hollomon and Krupkowski hardening law mathematical models for cold rolled low carbon aluminum-killed steel for enameling were determined from tensile tests at angles of $0°$, $45°$, and $90°$ to the rolling direction and bulge tests. Experimental Kosmalt 190 steel with a thickness of $a_0 = 0.5$ mm showed extra deep drawing quality with $r_m = 1.57$ and $n_m = 0.226$.
- In all numerical simulations and physical experiments, the bathtub model pressing was drawn free of fracture and wrinkles when simulated at the same blankholder force (340 kN) and friction (0.09) values. Keeler-Brazier's mathematic model was used to define the forming limit curve and to determine material fracture in numerical simulations.
- The best yield locus/hardening law combination appeared to be Hill 48/Krupkowski. This was determined by comparing the wall thicknesses of model pressing in selected sections after simulations and physical experiments. The deviations at the local minima were 0.7% and -1.0% in section A-A (longitudinal) and 2.4% in section B-B (corner). The course of relative thickness change evaluated from numerical simulations and experimental measurements showed good conformity.
- The material's anisotropy limits were found to be $r_m = 1.47$ and $n_m = 0.23$ when the model pressing free of fracture was drawn in a numerical simulation. Virtual materials were defined from experimentally measured values of the plastic strain ratio.

Author Contributions: M.T. and E.E. designed the experiment; M.T. and J.K. performed and evaluated the numerical simulations; M.T. and J.H. designed the model die and performed and evaluated the physical experiment; M.T., E.E., J.K., and J.H. analyzed the results; M.T. wrote the article.

Funding: This research received no external funding.

Acknowledgments: The work was accomplished under the grant project VEGA 2-0080-19 "Prediction of weldability and formability for laser welded tailored blanks made of combined high strength steels with CAE support" and project APVV-0273-12 "Supporting innovations of autobody components from the steel sheet blanks oriented to the safety, the ecology and the car weight reduction".

Conflicts of Interest: The authors declare no conflict of interest.

References

1.   Menezes, L.F.; Teodosiu, C. Three-dimensional numerical simulation of the deep-drawing process using solid finite elements. *J. Mater. Process. Technol.* **2000**, *97*, 100–106.
2.   Thomas, W.; Oenoki, T.; Altan, T. Process simulation in stamping—Recent applications for product and process design. *J. Mater. Process. Technol.* **2000**, *98*, 232–243. [CrossRef]
3.   Spišák, E. *Mathematical Modelling and Simulation of Technological Processes—Deep Drawing*, 1st ed.; ELFA, Ltd.: Košice, Slovakia, 2000; p. 156. ISBN 80-7099-530-0.
4.   Jha, A.; Sedaghati, R.; Bhat, R. Dynamic Testing of Structures Using Scale Models. In Proceedings of the 46th AIAA/ASME/ASCE/AHS/ASC Structures, Structural Dynamics & Materials Conference, Austin, TX, USA, 18–21 April 2005; AIAA 2005-2259. American Institute of Aeronautics and Astronautics: Reston, VA, USA, 2005.
5.   Langhaar, H.L. *Dimensional Analysis and Theory of Models*, 1st ed.; John Wiley & Sons: New York, NY, USA, 1951; ISBN 9780471516781.
6.   Szucz, E. *Similitude and Modeling: Fundamental Studies in Engineering*, 1st ed.; Elsevier: Amsterdam, The Netherlands, 1980; Volume 2.

7. Coutinho, C.P.; Baptista, A.J.; Rodrigues, J.D. Reduced scale models based on similitude theory: A review up to 2015. *Eng. Struct.* **2016**, *119*, 81–94. [CrossRef]

8. Sedov, L.I. *Similarity and Dimensional Methods in Mechanics*, 10th ed.; CRC Press: Boca Raton, FL, USA, 1993; p. 496. ISBN 9780849393082.

9. Gronostajski, Z.; Hawryluk, M. Analysis of metal forming processes by using physical modeling and new plastic similarity condition. In Proceedings of the CP907, 10th ESAFORM Conference on Material Forming, Zaragoza, Spain, 18–20 April 2007.

10. Davey, K.; Darvizeh, R.; Al-Tamimi, A. Finite Similitude in Metal Forming. In Proceedings of the NUMIFORM 2016: The 12th International Conference on Numerical Methods in Industrial Forming, Troyes, France, 4–7 July 2016.

11. Al-Tamimi, A.; Darvizeh, R.; Davey, K. Experimental investigation into finite similitude for metal forming processes. *J. Mater. Process. Technol.* **2018**, *262*, 622–637. [CrossRef]

12. Krishnamurthy, B.; Bylya, O.; Davey, K. Physical modelling for metal forming processes. *Procedia Eng.* **2017**, *207*, 1075–1080. [CrossRef]

13. Keran, Z.; Piljek, P.; Ciglar, D. Metal forming similarity theory in numerical simulations. *J. Technol. Plast.* **2016**, *41*, 38–45.

14. Ajiboye, J.S.; Jung, K.H.; Im, Y.T. Sensitivity study of frictional behavior by dimensional analysis in cold forging. *J. Mech. Sci. Technol.* **2010**, *24*, 115–118. [CrossRef]

15. Roll, K. Simulation of sheet metal forming—Necessary developments in the future. In Proceedings of the NUMISHEET, 7th International Conference on Numerical Simulation of 3D Sheet Metal Forming Processes, Interlaken, Switzerland, 1–5 September 2008; ETH Zürich: Zürich, Switzerland, 2008; pp. 3–11.

16. Silva, M.B.; Baptista, R.M.S.O.; Martins, P.A.F. Stamping of automotive components: A numerical and experimental investigation. *J. Mater. Process. Technol.* **2004**, *155*, 1489–1496. [CrossRef]

17. Choudhury, I.A.; Lai, O.H.; Wong, L.T. PAM-STAMP in the simulation of stamping process of an automotive component. *Simul. Model. Pract. Theory* **2006**, *14*, 71–81. [CrossRef]

18. Padmanabhan, R.; Oliveira, M.C.; Alves, J.L.; Menezes, L.F. Numerical simulation and analysis on the deep drawing of LPG bottles. *J. Mater. Process. Technol.* **2008**, *200*, 416–423. [CrossRef]

19. Vafaeesefat, A. Finite Element Simulation for Blank Shape Optimization in Sheet Metal Forming. *Mater. Manuf. Process.* **2011**, *26*, 93–98. [CrossRef]

20. Frącz, W.; Stachowicz, F.; Pieja, T. Aspects of verification and optimization of sheet metal numerical simulations process using the photogrammetric system. *Acta Metall. Slovaca* **2013**, *19*, 51–59. [CrossRef]

21. Čada, R.; Tiller, P. Evaluation of Draw Beads Influence on Intricate Shape Stamping Drawing Process. *Technol. Eng.* **2014**, *11*, 5–10. [CrossRef]

22. Labergere, C.; Badreddine, H.; Msolli, S.; Saanouni, K.; Martiny, M.; Choquart, F. Modeling and numerical simulation of AA1050-O embossed sheet metal stamping. *Procedia Eng.* **2017**, *207*, 72–77. [CrossRef]

23. Schrek, A.; Švec, P.; Brusilová, A.; Gábrišová, Z. Simulation process deep drawing of tailor welded blanks DP600 and BH220 materials in tool with elastic blankholder. *J. Mech. Eng. Stroj. Caopis* **2018**, *68*, 95–102. [CrossRef]

24. Neto, D.M.; Oliveira, M.C.; Alves, J.L.; Menezes, L.F. Influence of the plastic anisotropy modelling in the reverse deep drawing process simulation. *Mater. Des.* **2014**, *60*, 368–379. [CrossRef]

25. Neto, D.M.; Oliveira, M.C.; Santos, A.D.; Alves, J.L.; Menezes, L.F. Influence of boundary conditions on the prediction of springback and wrinkling in sheet metal forming. *Int. J. Mech. Sci.* **2017**, *122*, 244–254. [CrossRef]

26. Mulidran, P.; Šiser, M.; Slota, J.; Spišák, E.; Sleziak, T. Numerical Prediction of Forming Car Body Parts with Emphasis on Springback. *Metals* **2018**, *8*, 435. [CrossRef]

27. Chen, F.K.; Chiang, B.H. Analysis of die design for the stamping of a bathtub. *J. Mater. Process. Technol.* **1997**, *72*, 421–428. [CrossRef]

28. Hojny, J.; Wozniak, D.; Glowacki, M.; Zaba, K.; Nowosielski, M.; Kwiatkowski, M. Analysis of die design for the stamping of a bathtub. *Arch. Metall. Mater.* **2015**, *60*, 661–666. [CrossRef]

29. Sun, Q.; Xu, C.; Jiang, W. Study on the enameling properties of cold-rolled sheet steels containing different alloyed elements. *Glass Enamel* **2015**, *6*, 1–5.

30. ESI Group. *Pam-Stamp 2015 User's Guide*; ESI Group: Paris, France, 2015.

31. Gerlach, J.; Kessler, L.; Linnepe, M. A study regarding material aspects for numerical modelling of mild steels. In Proceedings of the IDDRG 2009 Conference—Material Property Data for More Effective Numerical Analysis, Golden, CO, USA, 1–3 June 2009; pp. 351–360.

32. Wang, L.; Lee, T.C. The effect of yield criteria on the forming limit curve prediction and the deep drawing process simulation. *Int. J. Mach. Tools Manuf.* **2006**, *46*, 988–995. [CrossRef]

33. Hill, R.A. Theory of the yielding and plastic flow of anisotropic metals. *Proc. R. Soc. Lond. Ser. A Math. Phys. Sci.* **1948**, *193*, 281–297.

34. Hill, R. Constitutive modelling of orthotropic plasticity in sheet metals. *J. Mech. Phys. Solids* **1990**, *38*, 405–417. [CrossRef]

35. Keeler, S.P.; Brazier, W.G. Relationship between Laboratory Material Characterization and Press-Shop Formability. *Microalloying* **1975**, *75*, 517–530.

36. Hrivňák, A.; Evin, E. *Formability of Steels*, 1st ed.; Elfa: Košice, Slovakia, 2004; ISBN 80-89066-93-3.

37. Pollák, L.; Tomáš, M.; Kuzmišin, J. Material formability and anisotropy of steel sheets for enamelling. In Proceedings of the 2005 International Scientific Conference on Structural Materials, Trnava, Slovakia, 22 June 2005; STU Bratislava: Bratislava, Slovakia, 2005; pp. 1–9.

38. Beygelzimer, Y.; Kulagin, R.; Toth, L.S.; Ivanisenko, Y. The self-similarity theory of high pressure torsion. *Beilstein, J. Nanotechnol.* **2016**, *7*, 1267–1277. [CrossRef]

39. Paul, S.K. Path independent limiting criteria in sheet metal forming. *J. Manuf. Process.* **2015**, *20*, 291–303. [CrossRef]

40. Koc, M. *Hydroforming for Advanced Manufacturing*, 1st ed.; Woodhead Publishing: Amsterdam, The Netherlands, 2008; ISBN 9781845693282.

41. Banabic, D. *Sheet Metal. Forming Processes—Constitutive Modelling and Numerical Simulation*, 1st ed.; Springer: Berlin/Heidelberg, Germany, 2010.

42. Shaw, J.; Chen, M.; Watanabe, K. Metal Forming Characterization and Simulation of Advanced High Strength Steels. *SAE Trans.* **2001**, *110*, 926–935.

43. Tisza, M.; Kovácx, Z.P. New methods for predicting the formability of sheet metals. *Prod. Process. Syst.* **2012**, *5*, 45–54.

44. Denes, E.; Gergely, J.; Szabados, O.; Vero, B. A Comparative Study of the Properties of Low Carbon Aluminium-killed and Boron-microalloyed Steels for Enamelling. In Proceedings of the XXI International Enamellers Congress, Shanghai, China, 18–22 May 2008; pp. 274–281.

45. Zhao, Y.; Huang, X.; Yu, B.; Chen, L.; Liu, X. Influence of boron addition on microstructure and properties of a low-carbon cold rolled enamel steel. *Procedia Eng.* **2017**, *207*, 1833–1838.

46. Yi, Z.; Hongyan, W.; Linxiu, D. Effect of Continuous Annealing Process on Microstructure and Properties of Ultra-Low Carbon Cold Rolled Enamel Steel. In Proceedings of the 24th International Enamellers Congress, Chicago, IL, USA, 28 May–1 June 2018; pp. 115–122.

47. Estrin, Y.; Beygelzimer, Y.; Kulagin, R. Design of Architectured Materials Based on Mechanically Driven Structural and Compositional Patterning. *Adv. Eng. Mater.* **2019**. [CrossRef]

© 2019 by the authors. Licensee MDPI, Basel, Switzerland. This article is an open access article distributed under the terms and conditions of the Creative Commons Attribution (CC BY) license (http://creativecommons.org/licenses/by/4.0/).

MDPI

St. Alban-Anlage 66

4052 Basel

Switzerland

Tel. +41 61 683 77 34

Fax +41 61 302 89 18

www.mdpi.com

Metals Editorial Office

E-mail: metals@mdpi.com

www.mdpi.com/journal/metals

www.ingramcontent.com/pod-product-compliance
Lightning Source LLC
LaVergne TN
LVHW070119170726
843515LV00007B/1719